D0138481

**ESTIMATING
CONSTRUCTION
COSTS**

**McGRAW-HILL
BOOK COMPANY**
New York
St. Louis
San Francisco
Auckland
Düsseldorf
Johannesburg
Kuala Lumpur
London
Mexico
Montreal
New Delhi
Panama
Paris
São Paulo
Singapore
Sydney
Tokyo
Toronto

R. L. PEURIFOY

*Formerly Professor of Construction Engineering
Texas A. & M. University
and
Oklahoma State University*

Estimating Construction Costs

THIRD EDITION

This book was set in Times New Roman.
The editors were B. J. Clark and J. W. Maisel;
the cover was designed by Joseph Gillians;
the production supervisor was Leroy A. Young.
Kingsport Press, Inc., was printer and binder.

Library of Congress Cataloging in Publication Data

Peurifoy, Robert Leroy, date
    Estimating construction costs.

    Includes index.
    1. Building—Estimates—United States.   I.   Title.
TH435.P47   1975   692′.5   74-31373
ISBN 0-07-049738-9

ESTIMATING CONSTRUCTION COSTS

Copyright © 1958, 1975 by McGraw-Hill, Inc. All rights reserved.
Copyright 1953 by McGraw-Hill, Inc. All rights reserved.
Printed in the United States of America. No part of this publication may be reproduced,
stored in a retrieval system, or transmitted, in any form or by any means,
electronic, mechanical, photocopying, recording, or otherwise,
without the prior written permission of the publisher.

1 2 3 4 5 6 7 8 9 0 KPKP 7 9 8 7 6 5

Contents

Purpose of this book. Purpose of estimating. Types of estimates. Approximate estimates. Detailed estimates. Form for preparing estimates. Check list of operations. Lump-sum estimates. Unit-cost estimates. Material take-off. Equipment costs. Cost of labor. Wage rates. Production rates for labor. Overhead. Social security tax. Unemployment compensation tax. Workmen's compensation and employer's liability insurance. Fire insurance. Extended coverage insurance. Automatic builder's risk insurance. Complete value builder's risk insurance. Comprehensive general liability insurance. Contractor's protective liability insurance. Installation floater policy. Contractor's equipment floater. Bid bond. Contractor's performance bond.

crete pavement. Equipment for hot-mix asphaltic-concrete pavement. Factors affecting the cost of asphalt pavement. Flexible-base and double-surface-treatment asphalt pavement.

General information. Shoring trenches. Sheeting trenches. Pile-driving equipment. Pile-driving hammers. Steel-sheet piling. Wood piles. Driving wood piles. Concrete piles. Precast concrete piles. Cast-in-place concrete piles. Monotube piles. Raymond step-tapered concrete piles. Cost of cast-in-place concrete piles. Steel piles. Jetting piles into position. Drilled shafts and underreamed footings. Cost of concrete piles in drilled holes.

Costs of concrete structures. Forms. Forms for concrete structures. The cost of materials for forms. Nails required for forms. Form oil. Labor required to build forms. Forms for footings. Forms for foundation walls. Design of forms for foundation walls. Design of forms for concrete walls using tables. Materials required for forms for foundation walls. Form ties. Labor building wood forms for foundation walls. Plywood forms. Prefabricated form panels. Commercial prefabricated forms. Forms for concrete columns. Materials required for forms for concrete columns. Cost of lumber for forms. Quantity of nails required for forms. Cost of adjustable steel column clamps. Labor making and erecting forms for concrete columns. Economy of reducing the size of concrete columns. Shores. Wood shores. Adjustable shores. Quantity of lumber required for forms for concrete beams using wood shores. Quantity of lumber required for forms for concrete beams using adjustable shores. Labor required to build forms for concrete beams. Forms for flat-slab-type concrete floors. Design of forms for flat-slab-type concrete floors. Quantity of lumber required for forms for flat-slab-type concrete floors. Labor required to build forms for flat-slab-type concrete floors. Forms for slabs for beam-and-slab-type concrete floors. Quantity of lumber required for forms for beam-and-slab-type concrete floors. Labor required to build forms for beam-and-slab-type concrete floors. Forms for metal-pan and concrete-joist-type concrete floors. Lumber required for metal-pan and concrete-joist construction. Labor required to build forms and install metal pans for concrete floors. Concrete stairs. Lumber required for concrete stairs. Labor required to build forms for concrete stairs. Types and sources of reinforcing steel. Properties of reinforcing bars. Estimating the quantity of reinforcing steel. Cost of reinforcing steel. Size extras for reinforcing steel. Quantity extras for reinforcing steel. Cost for detailing and listing reinforcing steel. Cost for fabricating reinforcing steel. Cost for reinforcing steel delivered to a project. Labor placing reinforcing steel bars. Welded-wire fabric. Labor placing welded wire-fabric. Concrete. Cost of concrete. Quantities of materials for concrete. Output of concrete mixers. Labor mixing and placing concrete. Ready-mixed concrete. Labor placing ready-mixed concrete. Lightweight concrete. Perlite concrete aggregate. The cost of Perlite concrete. Tilt-up concrete walls. General description. Concrete bridge piers. Forms for piers. Steel forms. Wood forms. Forms. The cost of forms.

General. Concrete-floor finishes. Monolithic topping. Materials required for monolithic topping. Labor finishing concrete floors using monolithic topping. Separate concrete topping. Materials required for a separate topping. Labor mixing, placing, and finishing separate topping. Labor finishing concrete floors with a power machine. Terrazzo floors. General. Terrazzo topping bonded to concrete floors. Terrazzo placed on wood floors. Labor required to place terrazzo floors. Asphalt tile. General. Laying asphalt tile on concrete floors. Laying asphalt tile on a wood floor. Labor laying asphalt tile.

General. Steel-joist system. General description. Floor and ceiling. Bridging. Metal lath. Ceiling extensions. Joist wall anchors. Joist beam anchors. Bolted end connections. Welded end connections. Sizes and dimensions of steel joists. Cost of steel joists. Labor erecting steel joists. Labor installing metal lath on top of steel joists. Labor placing welded-wire fabric. Concrete for slabs. Combined corrugated-steel forms and reinforcement for floor system. Description. Installing corrugated sheets. Labor installing corrugated sheets.

Masonry units. Estimating the cost of masonry. Mortar. Lime for mortar. Brick masonry. Sizes of bricks. Thickness of brick walls. Estimating the quantity of bricks. Quantity of mortar. Types of joints for brick masonry. Bonds. Labor laying bricks. Face brick with common-brick backing. Concrete blocks.

Quantities of materials required for concrete blocks. Labor laying concrete blocks. Load-bearing clay tile. Quantities of materials required for clay-tile walls. Labor laying clay tile. Stone masonry. Bonds for stone masonry. Mortar for stone masonry. Weights of stone. Cost of stone. Labor setting stone masonry.

Chapter 11   **Carpentry**   **238**

General information. Lumber sizes. Grades of lumber. Cost of lumber. Nails and spikes. Bolts. Timber connectors. Installing toothed rings and spike grids. Fabricating lumber. Rough carpentry houses. House framing. Sills. Floor girders. Floor and ceiling joists. Studs. Rafters. Subfloors. Roof decking. Wall sheathing. Board and batten siding. Drop siding. Bevel siding. Ceiling and partition. Wood shingles. Fascia, frieze, and corner boards. Wood furring and grounds. Door bucks. Finished wood floors. Timber structures. Fabricating timber for structures. Mill buildings. General descriptions. Columns. Beams and girders. Floors. Labor fabricating and erecting columns. Labor fabricating and erecting girders and floor beams. Roof trusses. General information. Types of roof trusses. Estimating the cost of wood roof trusses. Labor fabricating and assembling wood roof trusses. Hoisting wood roof trusses.

Chapter 12   **Interior Finish, Millwork, and Wallboard**   **284**

Cost of interior finish and millwork. Labor required to install interior finish. Wood window frames. Fitting and hanging wood sash. Wood door frames and jambs. Fitting and hanging wood doors. Wallboards. Gypsum wallboards. Installing gypsum wallboards. Labor installing gypsum wallboards.

Chapter 13   **Lathing and Plastering**   **291**

Estimating the cost of lathing and plastering. Grades of workmanship. Lathing. Types of lath. Wood lath. Labor applying wood lath. Metal lath. Applying metal lath on wood studs, joists, and furring strips. Applying metal lath on steel studs for partition walls. Applying metal lath for suspended ceilings. Labor applying metal lath. Plastering. Number of coats applied. Total thickness of plaster. Materials used for plaster. Covering capacity of plaster. Labor applying plaster.

Chapter 14   **Painting**   **404**

General information. Materials. Covering capacity of paints. Preparing a surface for painting. Labor applying paint. Equipment required for painting. The cost of painting.

charges for size and section. Extra charges for quantity.
Estimating the weight of structural steel. The cost of preparing
shop drawings. The cost of shop handling and fabricating structural steel. The cost of applying a coat of shop paint to structural steel. The cost of shop overhead and profit. The cost of
transporting structural steel. The cost of fabricated structural
steel delivered to a project. Erecting structural steel. Equipment for erecting structural steel. Labor erecting structural
steel. Labor bolting structural steel. Welded structures. General information. Advantages of welded connections. Erection
equipment required for welded-steel structures. Erecting steel
structures with welded connections. Types of welds. Arc-
welding terminology. Methods of producing the most economical welds. Electrodes. The cost of welding. Quantity of details
and welds required for a structural-steel frame building. Analysis of the cost of welded connections compared with riveted
connections. Painting structural steel.

Cost of a water-distribution system. Pipe lines. Bell-and-spigot
cast-iron pipe. Push-on joint cast-iron pipe. Fittings for bell-
and-spigot cast-iron pipe. Mechanical-joint cast-iron pipe.
Valves. Service lines. Fire hydrants. Tests of water pipes.
Sterilization of water pipes. Cost of cutting cast-iron pipe.
Labor required to lay cast-iron pipe. Labor required to lay
cast-iron pipe with mechanical joints. Labor required to lay
push-on joint cast-iron pipe. The cost of a cast-iron pipe water-
distribution system. Reinforced-concrete pressure pipe.

Items included in sewerage systems. Sewer pipes. Fittings for
sewer pipe. Constructing a sewer system. Installing sewer
pipe. Service wye branches. Manholes. Cleanout boots. Digging trenches for sewer pipe. Labor shaping the bottoms of
trenches. Labor laying sewer pipe. Backfilling and tamping
earth around sewer pipes. Excavating for sewer manholes.
Building sewer manholes. Quantity of concrete for manhole
bottoms. Quantity of bricks for manholes. Quantity of mortar for manholes. Labor building manholes.

Cost of land, right of way, and easements. Legal expenses.
Bond expense. Cost of construction. Engineering expense.
Interest during construction. Contingencies.

Preface

In preparing the third edition of this book, the author has rewritten and revised the material appearing in the second edition to improve the presentation of this material and to permit the use of costs that are applicable to 1974.

Even though the costs used in the book are representative of the costs for materials, equipment, and labor existing in certain locations in the United States at the time the book was written, the reader should realize that costs will vary with locations and the passage of time.

However, it is hoped that the methods and procedures used in the book for estimating construction costs will continue to apply regardless of changes in the costs of materials, equipment, and labor.

Commercial estimators who have applied the methods and procedures presented in earlier edition of this book have advised the author that use of the book has assisted them in preparing more accurate estimates. It is hoped that this beneficial service will be continued with the third edition.

Comments from readers will be welcomed.

R. L. PEURIFOY

List of Symbols and Abbreviations

$$
\begin{aligned}
\text{ac} &= \text{alternating current} \\
\text{ASA} &= \text{American Standards Association} \\
\text{AWWA} &= \text{American Water Works Association} \\
\text{bhp} &= \text{brake horsepower} \\
\text{bm} &= \text{bank measure, volume of earth before loosening} \\
\text{cfm} &= \text{cubic feet per minute} \\
\text{C. I.} &= \text{cast iron} \\
\text{cu ft} &= \text{cubic feet} \\
\text{cu yd} &= \text{cubic yard} \\
\text{cwt} &= \text{100 pounds} \\
\text{d} &= \text{pennyweight for nails} \\
D \text{ and } M &= \text{dressed and matched lumber} \\
\text{dbhp} &= \text{drawbar horsepower} \\
\text{dc} &= \text{direct current} \\
\text{deg} &= \text{degree} \\
\text{diam} &= \text{diameter} \\
\text{ed} &= \text{edition} \\
F &= \text{Fahrenheit}
\end{aligned}
$$

fbm = feet board measure of lumber
fob = free on board
fpm = feet per minute
ft = feet
ft-lb = foot-pound
fwhp = flywheel horsepower
gal = gallon
gpm = gallons per minute
hp = horsepower
hp-hr = horsepower-hour
hr = hour
in. = inch
kw = kilowatt
kwhr = kilowatthour
lb = pound
lin ft = linear feet
M = 1,000
M cu ft = 1,000 cubic feet
M fbm = 1,000 feet board measure
min = minute
mph = miles per hour
No. = number
pc = piece
psf = pounds per square foot
psi = pounds per square inch
pt = pint
S4S = surfaced on 4 sides
sk = sack
sq = square, 100 square feet
sq in. = square inch
tph = tons per hour
wt = weight

**ESTIMATING
CONSTRUCTION
COSTS**

1
Introduction

Purpose of this book The primary purpose of this book is to enable the user to gain fundamental knowledge of estimating the cost of projects to be constructed. It is not intended that use of or reference to the information contained in the book will produce cost estimates that may be applied as given to a project containing similar items. There are so many variations in the costs of materials, equipment, and labor from one locality to another and with time that no book can dependably give costs that may be applied directly to estimates. However, if the estimator learns to determine the quantities of materials, equipment, and labor for a given project and applies proper unit costs to these items, he should be able to estimate the direct costs accurately.

Purpose of estimating Construction estimates are prepared before a project is constructed in order to determine the probable cost of the project. Thus an estimate is, at best, a close approximation of the actual cost, the true value of which will not be known until the project is completed and all costs are recorded. An estimator does not establish the cost of a project. If a contract for the construction of a project is based on an estimate, this simply establishes the amount which the contractor will receive for constructing the project.

It is the responsibility of an estimator to apply costs which are already established to the various materials, equipment, operations, and services required to construct a project.

Types of estimates Construction estimates may be divided into at least two different types, depending on the purposes for which they are prepared. They are approximate estimates and detailed estimates. Each of these may be subdivided.

Approximate estimates For certain purposes the use of short-cut or approximate methods of estimating is justified. The prospective owner of a project may wish to know the approximate cost of a project before making a final decision to construct it. A governmental agency will need to know the approximate cost before holding a bond election. Sometimes an estimate of the cost of replacing a project with similar construction is desirable for tax purposes. Usually an approximate estimate is sufficiently accurate for these purposes.

An architect will reduce a building to square feet of area, or cubic feet of volume, then multiply the number of units by the estimated cost per unit; or an engineer will multiply the number of cubic yards of concrete in a structure by the estimated cost per cubic yard to determine the probable cost of the project. Considerable experience and judgment are required to obtain a dependable approximate estimate of the cost, as the estimator must adjust the unit costs to allow for variations in costs resulting from qualities of materials, workmanship, location, and construction difficulties. Approximate estimates are not sufficiently accurate for bid purposes.

Detailed estimates A detailed estimate of the cost of a project is prepared by determining the costs of materials, construction equipment, labor, overhead, and profit. Such estimates are almost universally prepared by contractors prior to the submitting of bids or the entering into contracts for important projects.

When preparing a detailed estimate for a given project, the estimator should divide the project into as many operations as are required. Table 1-1 illustrates the meaning of an operation. In so far as possible the operations should appear in the estimate in the order that they will be performed in constructing the project. Thus the first direct cost may be for clearing the site on which the project is to be constructed. This item should be followed with the cost of temporary construction, such as an office, sheds, etc., excavation, and foundation, continuing in proper order through the last operation performed, such as the cost of general cleanup and moving away from the project. If this order is followed, the danger of omitting the costs of one or more operations is reduced.

Form for preparing estimates Experienced estimators will readily agree that it is very important to use a good form when preparing an estimate. As previously stated, the form should provide for the treatment of each operation to be performed in constructing a project. For each operation

TABLE 1-1   Building-estimate Check List and Summary

Item	Operation	Amount
100	Wrecking, moving, and clearing site	$ 464.80
200	Temporary construction: office, sheds, etc.	1,683.90
300	Excavation, grading, backfill, and special fill	3,943.76
400	Foundation support: piles, caissons, cribs	X*
500	Shoring, sheeting: temporary and permanent	X
600	Underpinning: temporary and permanent	X
700	Water lines, gas, drains, sewers, conduits	1,954.66
800	Paving, curbs, sidewalks, and drives	2,756.80
900	Concrete, except items 800 and 2700	58,324.60
1000	Concrete surfacing and cement work	747.58
1100	Concrete forms: wood, metal; metal arches	28,964.40
1200	Reinforcing rods and mesh; metal inserts	11,284.95
1300	Concrete block	26,852.40
1400	Common brick	15,862.78
1500	Face brick	29,430.50
1600	Firebrick; flue lining and special brick	X
1700	Hollow tile: load bearing and partition	6,890.47
1800	Terra-cotta fireproofing: arch, furring, etc.	7,416.96
1900	Gypsum block: lintels, furring, book tile	X
2000	Water and dampproofing	1,916.40
2100	Architectural terra cotta	X
2200	Artificial stone	3,658.00
2300	Cut stone: lime, sand, blue, etc.	X
2400	Exterior marble and granite	X
2500	Structural steel and iron	16,586.00
2600	Metal lumber	X
2700	Stacks: metal, brick, concrete	X
2800	Ornamental and miscellaneous iron	3,816.25
2900	Iron doors, grates, and shutters	942.30
3000	Vault lights	246.40
3100	Metal store fronts	X
3200	Bronze and other art metal	367.10
3300	Lumber, wallboard, and rough carpentry	1,387.30
3400	Insulation and soundproofing	4,921.70
3500	Millwork and finished carpentry	9,627.54
3600	Special cabinet, panel, stair, and door work	4,342.46
3700	Artwood floors: parquetry	1,696.83
3800	Rough hardware	2,283.28
3900	Finished hardware	3,715.60
4000	Roofing: nonmetal, slate, tile, composition	4,218.30
4100	Sheet-metal work, plain; metal roofing	2,940.80
4200	Pressed sheet metal: ceilings, panels, etc.	X
4300	Metal furring, lathing, corner beads, grounds	1,827.64

TABLE 1-1   Building-estimate Check List and Summary (*Continued*)

Item	Operation	Amount
4400	Plastering: plain, ornamental; plasterboard	14,943.70
4500	Stucco	X
4600	Imitation marble and stone, and scagliola	X
4700	Interior marble, slate, stone	X
4800	Tile, mosaic, and terrazzo	3,286.93
4900	Composition, cork, and patented floors	4,819.57
5000	Kalamein and other fire doors, windows	X
5100	Hollow metal doors, windows, and trim	$ 9,654.80
5200	Steel sash and partitions	14,564.78
5300	Calking	967.90
5400	Weather strips	326.30
5500	Glass and glazing	5,795.45
5600	Priming and painting	6,346.80
5700	Decorating	2,856.20
5800	Plumbing and gas fitting	46,374.94
5900	Electrical wiring	12,230.00
6000	Electrical fixtures	3,750.50
6100	Heating and ventilating; boiler setting	32,375.80
6200	Elevators, dumb-waiters, hoists; controls	6,856.00
6300	Elevator doors and enclosures	940.00
6400	Mail chutes	X
6500	Pneumatic system	X
6600	Vacuum system	X
6700	Refrigeration system; refrigerators	X
6800	Sprinkler system	X
6900	Tanks	X
7000	Vaults and vault doors	1,687.40
7100	Revolving doors	X
7200	Awning, shades, and blinds	3,126.80
7300	Fly screens	1,788.70
7400	Floor coverings	X
7500	Special equipment not otherwise listed	X
7600	Special interior fixtures	3,274.90
7700	....	X
7800	....	X
7900	....	X
8000	....	X
8100	....	X
8200	....	X
8300	....	X
8400	....	X
8500	....	X
8600	....	X

TABLE 1-1   Building-estimate Check List and Summary (*Continued*)

Item	Operation	Amount
8700	Landscape: leveling, sodding, planting	243.00
8800	Job overhead: equipment, fees, permits, etc. (see Table 1-4)	69,285.00
8900	Compensation, liability, contingent insurance	10,464.96
9000	Taxes, sales and use tax, tax on payrolls, etc.	8,578.40
9100	Total cost: labor, material, subbids	$525,611.29
9200	Profit	30,413.16
9300	Subtotal	$556,024.45
9400	Bond	3,064.16
9500	Amount of bid	$559,088.61

* The symbol X indicates that there is no cost for this operation.

there should be a systematic listing of materials, equipment, labor, and any other items, with space for all calculations, number of units, unit costs, and total costs.

Each operation should be assigned a code number, which number should be reserved exclusively for that operation on this as well as on estimates for other projects within a given construction organization. For example, in Table 1-1, item 900 refers to concrete, whereas item 1400 refers to common brick. The accounting department should use the same item numbers when preparing cost records.

Example   This example is intended to illustrate a form which might be used in preparing an estimate.

1400 Common brick	
Brick, 8,000 sq ft × 6 brick per sq ft	= 48.00 M
Add for waste, 1% × 48 M	= 0.48 M
Total quantity	= 48.48 M
Cost of bricks, 48.48 M × $45.00 per M	= $2,181.60
Bricklayers, 48 M × 10 hr per M = 480 hr @ $8.23	= $3,950.40
Helpers, 48 M × 6 hr per M = 288 hr @ $5.79	= $1,667.52

Check list of operations   When preparing an estimate, an estimator should use a check list which includes all the operations necessary to construct the project. Before completing an estimate, one should check with this list to be sure that no operations have been omitted. Table 1-1 illustrates a list which may be used when preparing an estimate for a building. It is desirable for the operations to appear as nearly as possible in the same order that they will be performed when constructing the project. Other check lists, serving the same purpose, should be prepared for projects such as highways, water systems, sewerage systems, etc.

The check list may be used to summarize the costs of a project by providing a space for entering the cost of the operation, as illustrated. A suitable symbol should be used to show no cost for those operations which are not required. The figure shown under amount should be the total cost of materials, equipment, and labor for the particular operation, as determined in the detailed estimate.

Table 1-1 illustrates a form which may be used as a check list and as a building-estimate summary.

Lump-sum estimates　The estimates for some projects, such as buildings, are prepared for the purpose of submitting lump-sum bids on the projects. When estimating the cost of or bidding a project on this basis, only one final cost figure is quoted, namely, the amount given for item 9500 in Table 1-1. Unless there are revisions in the plans and specifications, or in the quantities of work required for a project, this figure represents the amount which the owner will pay to the contractor for the completed project.

A lump-sum estimate must include the cost of all materials, construction equipment, labor, overhead, insurance, taxes, profit, and bonds required. It is desirable to estimate the costs of materials, equipment, and labor separately for each operation, obtain a subtotal of these costs for the entire project, then estimate the cost of overhead, insurance, taxes, profit, and bonds.

Unit-cost estimates　Many projects are bid on a unit-cost basis. Such projects include pavements, curbs and gutters, earthwork, various kinds of pipe lines, clearing and grubbing land, etc. The cost per unit, submitted in a bid, includes the furnishing of materials, equipment, labor, supervision, insurance, taxes, profit, and bonds, as required, for completely installing a unit. The units designated include square yards, cubic yards, linear feet, tons, acres, etc. A separate estimate should be prepared for each type or size unit.

The costs of materials, equipment, and labor are determined for each unit. These are designated direct costs. To these costs there must be added a proportionate part of each indirect cost, such as moving in, temporary construction, overhead, insurance, taxes, profit, and bonds since indirect costs are not bid separately.

A unit-cost bid might have the following form:

　42 acres of clearing and grubbing @ $360.00 per acre
　6,240 lin ft of 6-in. Class B cast-iron pipe in place at $4.68 per lin ft
　8,564 lin ft of 8-in. Class B cast-iron pipe in place at $6.24 per lin ft

Material takeoff　It is the duty of an estimator to prepare cost estimates from plans and specifications which are usually produced by other persons. The first step is a quantity takeoff. This will involve all materials placed in the project, plus earth excavation and fill.

Materials for each operation should be separately listed, in the correct quantities, according to their classification and unit costs. The unit costs of the different materials should be obtained from reliable sources and used as the basis of estimating the costs of materials for the project. If the prices quoted for materials do not include delivery costs, the estimator must include appropriate costs for transporting them to the project.

Equipment costs   Equipment costs are discussed in Chap. 2.

Cost of labor   The cost of labor for a project should be estimated by dividing the project into operations, such as earthwork, concrete forms, concrete, common brick, face brick, etc., then estimating the cost of labor for each operation. The laborers should be classified according to the work which they perform and the wages which they receive, and for each labor classification the total amount of time required should be estimated. Usually the time is expressed in man-hours. A man-hour is one man working one hour. In order to estimate the cost of labor, it is necessary for the estimator

TABLE 1-2   Approximate Average Base Wage Rates in the United States in Dollars per Hour

Trade	Wage rate
Bricklayers	$8.23
Carpenters	7.67
Electricians	8.47
Painters	7.24
Plasterers	7.69
Plumbers	8.39
Building laborers	5.79
Ironworkers, structural	7.64
Ironworkers, reinforcing steel	7.64
Stonemasons	8.23
Mason's helpers	7.43
Cement finishers	7.35
Glaziers	7.24
Sheetmetal workers	7.33
Welders	6.75
Power shovel operators	6.81
Hoisting engineers	6.93
Air compressor operators	5.73
Air tool operators	5.25
Tractor operators	6.81
Truck drivers	4.51

to know the effective wage rates and the time required to complete each operation.

Wage rates Wage rates vary greatly with the locations of projects, as indicated by such publications as the *Engineering-News Record* and *Construction Review* [1]. Table 1-2 gives current average or approximately average wage rates by trade or classification in the United States.

The rates listed in Table 1-2 are base rates. In addition to paying the base rates, an employer must pay or contribute additional amounts for such items as Social Security taxes, Workmen's Compensation Insurance, Unemployment Compensation Taxes, and possibly others, the costs of which may vary considerably, as indicated in Table 1-3 for the Pacific Coast Division of the United States, for bricklayers and carpenters only. Similar costs and variations may apply to all labor trades and locations. An estimator must determine the effective total wage cost for each trade or classification that will be employed on any given project.

The wage rates for work done in excess of 8 hr per day, or 40 hours per week, or on Saturdays, Sundays, and holidays will be increased, generally to 1½ or 2 times the basic scale.

The wage rates listed in Tables 1-2 and 1-3 apply for construction performed in the designated areas. Sometimes the rates will be lower for work performed outside urban areas, or for work performed on heavy construction.

In general, the national average basic wage rates will be used to determine the costs in the examples appearing in this book. The effect of fringe benefits, taxes, and insurance will not be included in the rates.

Production rates for labor A production rate is defined as the number of units of work produced by a man in a specified time, usually an hour

TABLE 1-3 Hourly Wage Rates and Employer's Contribution to Fund for Building Trades*

Location	Bricklayers			Carpenters		
	Basic scale	Employer contribution	Total rate	Basic scale	Employer contribution	Total rate
Fresno, California	$6.98	$1.82	$8.80	$8.10	$1.92	$10.02
Los Angeles, California	7.79	1.80	9.59	6.75	2.14	8.89
San Diego, California	7.60	1.88	9.48	7.06	2.01	9.07
Portland, Oregon	7.60	1.09	8.69	6.88	1.23	8.11
Seattle, Washington	7.53	1.03	8.56	6.80	1.22	8.02

* Rates effective as of January 2, 1973, *Construction Review*, page 52, April, 1973.

or a day. Production rates may also specify the time in man-hours or man-days required to produce a specified number of units of work, such as 12 man-hours to lay 1,000 bricks. This book will use an hour as the unit of time. Production rates should be realistic to the extent of including an allowance for the fact that a man usually will not work 60 min during an hour.

The time which a laborer will consume in performing a unit of work will vary between laborers and between projects and with climatic conditions, job supervision, complexities of the operation, and other factors. It requires more time to fabricate and erect lumber forms for concrete stairs than for concrete foundation walls. An estimator must analyze each operation in order to determine the probable time required for the operation.

Information on the rates at which work has been performed on similar projects is very helpful. Such information may be obtained by keeping accurate records of the production of labor on projects as construction progresses. For the information to be most valuable to an estimator, there should be submitted with each production report an accurate record showing the number of units of work completed, the number of laborers employed, by classification, the time required to complete the work, and a description of job conditions, climatic conditions, and any other conditions or factors which might affect the production of labor. The reports should be for relatively short periods of time, such as a day or a week, in order that the conditions described will accurately represent the true conditions for the given period. Reports covering a complete project, lasting for several months, will give average production rates but will fail to indicate varying rates resulting from changes in working conditions. It is not sufficiently accurate for an estimator to know that a bricklayer laid an average of 800 bricks per day on a project. The estimator should know the rate at which each type of brick was laid under different working conditions, considering the climatic and any other factors which might have affected production rates. All experienced construction men know that the production of labor is usually low during the early stages of construction. As the organization becomes more efficient, the production rates will improve; then as the construction enters the final stages, there will usually be a reduction in the production rates. This is important to an estimator. For a small job it is possible that labor will never reach its most efficient rate of production because there will not be sufficient time. If a job is of such a type that laborers must frequently be transferred from one operation to another or if there are frequent interruptions, the production rates will be lower than when the laborers remain on one operation for long periods of time without interruptions.

In this book numerous tables are included giving the rates at which laborers should perform various operations. These rates include an adjustment for nonproductive time, by assuming that a man will actually work about 45 to 50 min per hr. Conditions on some projects may justify a further adjustment in the rates. The frequent use of a range in rates instead of a single rate will permit the estimator to select the rate which he believes is most appropriate for his project.

Illustrative Examples The examples which follow will illustrate methods of estimating the labor required to perform units of work for different types of projects under varying conditions.

Example This example illustrates the probable rates of excavating earth by hand under different conditions.

For the first project, the soil is a sandy loam which requires light loosening with a pick before shoveling. The maximum depth of the trench or pit will be 4 ft. Climatic conditions are good, with a temperature of about 70 deg. A laborer should easily loosen ½ cu yd of earth per hour. Using a long-handled round-pointed shovel, it should require about 150 loads to remove a cubic yard of earth, bank measure. If a laborer can handle 2½ shovel loads per minute, he will remove 1 cu yd per hr. These rates of production will require 2 man-hours to loosen and 1 man-hour to remove a cubic yard of earth from the trench, or a total of 3 man-hours per cu yd, bank measure.

For a second project, the soil is a tough clay which is difficult to dig and which lumps badly. The maximum depth of the trench or pit will be 5 ft. The temperature will be about 100 deg, which will reduce the operating efficiency of the laborers. Under these conditions a laborer may not loosen more than ¼ cu yd per hr. Because of the physical condition of the loosened earth, it will require about 180 shovel loads to remove a cubic yard of earth. If a laborer can handle 2 shovel loads per minute, he can remove 120 shovel loads per hour. This is equivalent to ⅔ cu yd per man-hour or 1½ man-hours per cu yd, bank measure. For these rates of production it will require 4 man-hours to loosen and 1½ man-hours to remove a cubic yard, or a total of 5½ man-hours per cu yd.

Example This example illustrates a method of determining the probable rate of placing reinforcing steel for a given project.

Steel bars are to be used to reinforce a concrete slab 57 ft wide by 70 ft long. The bars will be ½ in. in diameter, with no bends, maximum length limited to 20 ft, and spaced 12 in. apart both ways. All laps will be 18 in. Precast concrete blocks, spaced not over 6 ft apart each way, will be used to support the reinforcing. The bars will be tied at each intersection, using bar ties. The steel will be stored in orderly stock piles, according to length about 80 ft, average distance from the center of the slab. The slab will be constructed on the ground.

The length of the bars parallel to the 57-ft side will be		
Length of the side	=	57 ft
Length of laps, 2 × 18 in.	=	3 ft
		———
Total length of bars per row	=	60 ft
Use 3 bars 20 ft long	=	60 ft
Total number of bars required, 3 × 70	=	210
Total length of the bars, 210 × 20 ft	=	4,200 ft
The length of the bars parallel to the 70-ft side will be		
Length of side	=	70 ft
Length of laps, 3 × 18 in.	=	4 ft 6 in.
		———
Total length of bars per row	=	74 ft 6 in.
Use 4 bars 18 ft 7 in. long	=	74 ft 6 in.
Total number of bars required, 4 × 57	=	228
Total length of the bars, 228 × 18 ft 7 in.		4,246 ft

The weight of the reinforcing will be

20-ft bars, 4,200 ft @ 0.668 lb per ft	= 2,806 lb
18-ft 7-in. bars, 4,246 ft @ 0.668 lb per ft	= 2,836 lb

Total weight	= 5,642 lb

The number of intersections of bars will be 57 × 70 = 3,990

The time required to place the reinforcing, using 2 steel setters, should be about as follows:

Carrying the bars to the slab site:

Time to walk 160 ft @ 100 fpm	= 1.6 min
Add time to pick up and put down reinforcing	= 1.0 min

Time for round trip	= 2.6 min

Assume that the 2 men can carry 6 bars, weighing approximately 80 lb each trip

Number of trips required, 438 bars ÷ 6 bars per trip	= 73
Total time to carry reinforcing, 73 × 2.6 ÷ 60	= 3.17 hr

Placing the bars on blocks and spacing them:
Assume that 2 men working together can place 2 bars per min, or 120 bars per hr

Time to place, 438 bars ÷ 120 per hr	= 3.67 hr

Tying the bars at intersections:
Assume that a man can make 5 ties per min
2 men will make 2 × 5 × 60 = 600 ties per hr

Time to tie reinforcing, 3,990 ties ÷ 600 per hr	= 6.65 hr

The total working time will be

Carrying the reinforcing	= 3.17 hr
Placing the reinforcing	= 3.67 hr
Tying the reinforcing	= 6.65 hr

Total working time	= 13.49 hr

On a project a man will seldom work more than 45 to 50 min per hr, because of necessary delays. Based on a 45-min hour, the total clock time to handle and place the reinforcing will be

$$\frac{13.49 \times 60}{45} = 18.0 \text{ hr}$$

Total man-hours for the job, 2 × 18	= 36
No. of tons placed, 5,642 ÷ 2,000	= 2.821
Man-hours per ton placed, 36 ÷ 2,821	= 12.75

If the reinforcing steel in the previous example is bent to furnish negative reinforcing over the beams or if it has hooks on the ends and if, in addition, it must be hoisted to the second floor of a building, it will require extra time to hoist and place it. An estimator should adjust the production rate accordingly instead of using flat rates for all projects.

The previous examples are included to illustrate how production rates may be determined. If the rates used in the examples are based on actual observations, they serve as an excellent guide in establishing probable production rates for work performed under similar conditions. The more information an estimator has on actual production rates from projects previously completed, the better he is prepared to estimate the probable rates for future projects.

Overhead The overhead costs chargeable to a project involve many items which cannot be classified as materials, construction equipment, or labor. Some firms divide overhead into two categories, job overhead and general overhead.

Job overhead includes costs which can be charged specifically to a project. These costs are the salaries of the project superintendent and other staff personnel and the costs of utilities, supplies, engineering, tests, drawings, rents, permits, insurance, etc., which can be charged directly to the project.

General overhead is a share of the costs incurred at the general office of the company. These costs include salaries, office rent, utilities, insurance, taxes, shops and yards, and other company expenses not chargeable to a specific project.

Table 1-4 gives a list of the items which might be included in overhead costs, both job and general.

Some estimators follow the practice of multiplying the direct costs of a project, materials, equipment, and labor by an assumed percent to determine the probable cost of overhead. While this method gives quick results, it may not be sufficiently accurate for most estimates.

While it is possible to estimate the cost of job overhead for a given project, it is usually not possible to estimate accurately the cost of general overhead chargeable to a project. Since the cost of general overhead is incurred in operating all the projects constructed by a contractor, it is reasonable to charge to each project a portion of this cost. The actual amount charged may be based on the duration of the project, the amount of the contract, or a combination of the two.

Example This example illustrates a method of determining the amount of general overhead chargeable to a given project.

Average annual value of construction   = $6,000,000
Average annual cost of general overhead   =   240,000
Amount of general overhead chargeable to a project,

$$\frac{\$240,000 \times 100}{\$6,000,000} = 4\% \text{ of the total cost of the project}$$

Social security tax The Federal government and some states require an employer to pay a tax for the purpose of providing old-age benefits to

TABLE 1-4   Check List and Summary of Overhead Costs

Item	Class of expense	Amount
8801	Equipment, tools, accessories	
.1	Rental	$ 360.00
.2	Freight	175.00
.3	Hauling	85.00
.4	Loading, unloading, erecting, dismantling	240.00
.5	Other	
8802	Job organization	
.1	Superintendent	15,780.00
.2	Timekeeper, material clerk	7,360.00
.3	Bookkeeper, accounting	3,580.00
.4	Clerical, stenographer, other	3,220.00
.5	Other; watchman, safety, list titles	4,200.00
.6	Other	
8803	Light, power, water, connections	
.1	Electricity, light and power	320.00
.2	Carbide gas, cutting, welding	80.00
.3	Gasoline, heating oils, etc.	180.00
.4	Water, sprinkling, mixing, etc.	110.00
.5	Other	
8804	Supplies	
.1	Office, stationery, forms, books	80.00
.2	Job shop	X*
.3	Other	130.00
8805	Traveling and hotel expenses	
.1	Officials and members of organization	660.00
.2	Other	175.00
8806	Express, freight, etc.	580.00
8807	Demurrage allowance	90.00
8808	Hauling, hired for odd jobs	360.00
8809	Advertising, labor, materials	180.00
8810	Signs, company, warnings, notices, etc.	80.00
8811	Engineering, surveys, inspections	
.1	Layout, lines, levels, batter boards, etc.	225.00
.2	Public inspectors, wiring, plumbing, heating	95.00
.3	Inspection of subcontract work	80.00
.4	Lot survey	120.00
.5	Other, describe	

TABLE 1-4   Check List and Summary of Overhead Costs (*Continued*)

Item	Class of expense	Amount
8812	Tests	
.1	Soil, tests, pits, borings, etc.	320.00
.2	Materials, cement, aggregate, steel, etc.	160.00
.3	Structure, floor loading	X
.4	Other, describe	
8813	Drawings	
.1	Shop and settings	80.00
.2	Drafting	160.00
.3	Extra prints	80.00
.4	Other, describe	
8814	Photographs	40.00
8815	Patents and royalties	X
8816	Legal, attorney and notary fees	290.00
8817	Medical and hospital expenses	420.00
8818	Telephone and telegraph	
.1	Telephone installation	30.00
.2	Telephone service	60.00
.3	Long distance calls	160.00
.4	Telegrams	30.00
.5	Other, describe	
8819	Rents	
.1	Job office, warehouse, etc.	X
.2	Land, unloading and storage facilities	120.00
.3	Other, describe	
8820	Permits	
.1	Building	160.00
.2	Water and sewer	50.00
.3	Street	40.00
.4	Other, describe	
8821	Insurance	
.1	Fire	240.00
.2	Tornado	80.00
.3	Earthquake	X
.4	Theft	120.00
.5	Automobile	60.00
.6	Other, describe	
8822	Petty cash items	750.00

TABLE 1-4   Check List and Summary of Overhead Costs (*Continued*)

Item	Class of expense	Amount
8823	Interest	
.1	On deposits and job funds	2,400.00
.2	On delayed payments for purchases	120.00
.3	Other, describe	
8824	Cutting and patching for trades	650.00
8825	Contingencies: guarantees, strikes, wages, rains, freezing, floods, tornados, earthquakes, material shortages, price increases, transportation, subsoil conditions, ambiguous contract provisions	4,200.00
8826	Cold weather expenses	
.1	Thawing materials	180.00
.2	Weather protection, window and door closures, temporary walls, and closures	850.00
.3	Temporary heat, installing and maintaining	240.00
.4	Other, describe	
8827	Repairs: street, sidewalks, property	500.00
.1	Other, describe	
8828	Pumping: dewatering. May be included under excavation, sheeting, of main summary	450.00
8829	Final cleanup	
.1	Windows	140.00
.2	Walls	160.00
.3	Floors	120.00
.4	Premises	260.00
.5	Other, describe	
8830	Association dues: job share	120.00
8831	Taxes: occupation, property, etc., not included elsewhere	250.00
8832	Share of general company overhead	16,650.00
8833	Other items, describe	
8834		
8835		
8836		
8837	TOTAL COST	69,285.00

* X   This symbol indicates that there is no cost for this item.

persons who become eligible. The present Federal rate requires the employer to pay 5.85 percent of the gross earnings of an employee up to $13,200 per year. The employee contributes an equal amount through the employer. This rate is subject to change by Congress.

Unemployment compensation tax This tax, which is collected by the states, is for the purpose of providing funds with which to compensate workers during periods of unemployment. The base cost of this tax is usually 3 percent of the wages paid to the employees, all of which is paid by the contractor. This rate may be reduced by establishing a high degree of employment stability, with few layoffs, during a specified period of time.

Workmen's compensation and employers liability insurance Most states require contractors to carry workmen's compensation and employers liability insurance as a protection to the workers on a project. In the event of an injury or death of an employee working on the project, the insurance carrier will provide financial assistance to the injured person or to his family. Although the extent of financial benefits varies within the several states, in general they cover reasonable medical expenses plus the payment of reduced wages during the period of injury. Each state which requires this coverage has jurisdiction, through a designated agency, over the insurance to the extent of specifying the minimum amounts to be carried, the extent of the benefits, and the premium rates paid by the employer.

The base or manual rates for workmen's compensation insurance vary considerably among states, and within a state they vary according to the classification of work performed by an employee. A higher premium rate is charged for work that subjects workers to a greater risk of injury. A contractor who establishes a low record of accidents on his jobs for a specified period of time will be granted a credit, which will reduce the cost of his insurance. A contractor who establishes a high record of accidents over a period of time will be required to pay a rate higher than the base rate, thus increasing the cost of the insurance.

The premium rate for this insurance is specified to be a designated amount for each $100.00 of wages paid under each classification of work. The base rates for workmen's compensation insurance, effective on the indicated date, are given in Appendix A. In order to determine the cost of insurance for a given project, it is necessary to estimate the amount of wages that will be paid under each classification of work, then apply to each wage classification the appropriate rate. As the base rates are subject to changes, an estimator should verify them before preparing an estimate.

Fire insurance This insurance affords a contractor protection against loss resulting from fire damage during the period of construction. As the premium rates vary with the location and the type of construction, it is necessary to obtain the rate for a given project before estimating the cost of this insurance.

Extended coverage insurance Instead of purchasing several types of insurance it may be desirable to group all the protection under extended coverage. This insurance can be obtained to provide protection against loss resulting from fire, tornado, explosion, hail, riot, smoke damage, damage from aircraft, and vehicle damage. The premium rates for this insurance should be obtained from the contractor's insurance agent.

Automatic builder's risk insurance This insurance, which provides protection against loss to temporary and permanent buildings, tool houses, tools, supplies, machinery, and materials located on the premises and used in connection with the construction of the project, becomes effective as soon as property is placed at the site. The contractor is required to report monthly the insurable value of the project on the last day of each month. The premium cost is based on the rate and the reported values. In the event of a loss, the recovery is limited to the actual value at the time of the loss.

Completed value builder's risk insurance Under the terms of this coverage the insurance is based on the estimated completed value of the project. However, since the actual value varies from zero at the beginning of construction to the full value when the project is completed, the premium rate usually is set at 55 percent of the rate applicable to builder's risk insurance. In the event of a loss, the recovery is limited to the actual value at the time of the loss.

Although the cost of protection will vary with the type of structure and its location, the following rates for each $100.00 of ultimate value of a concrete building will indicate representative costs:

For fire insurance	= $0.13
For vandalism insurance	= 0.027
For explosion insurance	= 0.11
Total	= $0.267

Comprehensive general liability insurance As the result of construction operations, it is possible that persons not employed by the contractor may be injured or killed. Also, it is possible that property not belonging to the contractor may be damaged. Comprehensive general liability insurance should be carried as a protection against loss resulting from such injuries or damage. This insurance provides the protection ordinarily obtainable through public liability and property damage insurance. The coverage should be large enough to provide the necessary protection for the given project. The premium rate varies with the limits of liability specified in the policy.

Contractor's protective liability insurance This is a contingent insurance which protects a contractor against claims resulting from accidents caused by subcontractors or their employees, for which the contractor may be held liable.

Installation floater policy This insurance provides protection to the contractor against loss resulting from the collapse of a structure during erection.

Contractor's equipment floater This insurance provides protection to the contractor against loss or damage to his equipment because of fire, lightning, tornado, flood, collapse of bridges, perils of transportation, collision, theft, landslide, overturning, riot, strike, and civil commotion.

The cost of this insurance, which will vary with the location, should be about $0.55 per $100.00 of equipment value per year.

Bid bond It is common practice to require each bidder on a project to furnish with his bid a bid bond, a cashier's check, or a certified check, in an amount equal to 5 to 20 percent of the amount of the bid. In the event that the contract to construct the project is tendered to a bidder and he refuses or fails to sign the contract, the owner may retain the bond or check as liquidated damages. Bid bonds covering construction usually cost $5.00.

For some projects, cashier's checks are specified instead of bid bonds. These checks, which are issued to the owner of the project by a bank, are purchased by the bidder. They can be cashed easily, whereas it is necessary for the owner to secure payment on a bid bond through the surety, and the surety may challenge the payment. The use of cashier's checks requires bidders to tie up considerable sums of money for periods which may vary from a few days to several weeks in some instances.

There is no uniform charge for cashier's checks. Some banks charge $0.25 per $100.00 for small checks, with reduced rates for large checks, while some make no charge for checks furnished for regular customers. However, the interest cost for a cashier's check for $100,000 for a period of 2 weeks at 6 percent interest will amount to approximately $240.00.

Contractor's performance bond All government agencies and many private owners require a contractor to furnish a performance bond to endure during the period of construction of a project. The bond is furnished by an acceptable surety to assure the owner that the contract will be completed at the specified cost and that all wages and bills for materials will be paid. In the event a contractor fails to complete a project, it is the responsibility of the surety to secure completion. Although the penalty under a performance bond is specified as 25, 50, or 100 percent of the amount of the contract, the cost of the bond usually is based on the amount of the contract.

Representative costs of performance bonds are as follows:

For buildings and similar projects
    For the first $100,000   = $10.00 per $1,000.00
    For the next $2,400,000 = $ 6.50 per $1,000.00
    For the next $2,500,000 = $ 5.25 per $1,000.00
    For the next $2,500,000 = $ 5.00 per $1,000.00
    For all over $7,500,000 = $ 4.70 per $1,000.00

For highways and engineering construction
    For the first $100,000   = $7.50 per $1,000.00
    For the next $2,400,000 = $5.00 per $1,000.00
    For the next $2,500,000 = $4.00 per $1,000.00
    For the next $2,500,000 = $3.90 per $1,000.00
    For all over $7,500,000 = $3.60 per $1,000.00

Profit   Profit is defined as the amount of money, if any, which a contractor retains after he has completed a project and has paid all costs for materials, equipment, labor, overhead, taxes, insurance, etc. The amount included in a bid for profit is subject to considerable variation, depending on the size of the project, the extent of risk involved, the desire of the contractor to get the job, the extent of competition, and other factors. A contractor might include 2 to 5 percent profit on a $1,000,000 highway paving project, when the risk is low and competition is high, whereas he might include 15 to 20 percent profit, or more, on a foundation or river project, when the risk is high and there is little competition.

Representative estimates   Numerous examples of estimates are presented in this book to illustrate the steps which should be followed in determining the probable cost of the project. Nominal amounts are included for overhead and profit in some instances to give examples of complete estimates for bid purposes. In other instances only the costs of materials, construction equipment, and labor are included. The latter three costs are referred to as direct costs. They represent the most difficult costs to estimate, and it is with them that we are primarily concerned.

In preparing the sample estimates, unit prices for materials, equipment, and labor are used primarily to show how an estimate is prepared. The reader should realize that these unit costs will vary with the time and the location of a project. An estimator must obtain and use unit prices which are correct for the particular project. Estimators do not establish prices: they simply use them.

It should be remembered that estimating is not an exact science. Experience, judgment, and care should enable an estimator to prepare an estimate which will reasonably approximate the ultimate cost of the project.

Figure 1-1 illustrates a form which might be used in preparing a detailed estimate. When a project includes several operations, the direct costs for material, equipment, and labor should be estimated separately for each operation, then the indirect costs, for insurance, taxes, overhead, profit, performance bond, etc., should be estimated for the entire project.

Instructions to the reader   In the illustrative examples which are presented in this book, a uniform method of calculating and expressing the time units for equipment and labor and the total cost is used. For equipment the time is expressed in equipment-hours, and for labor it is expressed in

SAMPLE ESTIMATE

Item No.		Description	Calculations

400	0	Furnish and drive 200 creosote-treated piles
		    Drive piles to full penetration into normal
		    soil. Length of piles, 50 ft. Size of piles, 14-
		    in.-butt and 6-in.-tip diameters	

	10	Materials	
		Piles, add 5 for possible breakage	205 piles $\times$ 50 ft
	20	Equipment	
		Moving to and away from the job	
		Crane, 12-ton crawler type	200 piles $\div$ $2\frac{1}{2}$ per hr
		Hammer, single-acting, 15,000 ft-lb	
		Boiler, water, fuel, etc.	
		Leads and sundry equipment	
	30	Labor, add 16 hr to set up and take down equipment	
		Foreman	80 + 16
		Fireman	
		Crane operator	
		Crane oiler	
		Men on hammer	2 $\times$ 96
		Helpers	2 $\times$ 96
		Subtotal, direct cost	
	50	Overhead	10% $\times$ \$27,325.00
	60	Social security tax	5.85% $\times$ \$4,223.00
	62	Workmen's compensation insurance	8.68% $\times$ \$4,223.00
	64	Unemployment tax	3% $\times$ \$4,223.00
	70	Subtotal cost	
	80	Profit	10% $\times$ \$30,797.80
	85	Subtotal cost	
	90	Performance bond	$\frac{3}{4}$% $\times$ \$33,877.58
	94	Total cost, amount of bid	
	95	Cost per lin ft.	\$34,131.68 $\div$ 10,000 lin ft

Fig. 1-1 Form used to estimate construction costs.

No. units	Unit	Unit cost		Material cost		Equipment cost		Labor cost		Total cost	
10,250	lin ft	2	00	20,500	00					20,500	00
		Lump sum				750	00			750	00
80	hr	12	60			1,008	00			1,008	00
80	hr	3	50			280	00			280	00
80	hr	4	25			340	00			340	00
80	hr	2	80			224	00			224	00
96	hr	8	25					792	00	792	00
96	hr	5	40					518	40	518	40
96	hr	6	25					600	00	600	00
96	hr	4	80					460	80	460	80
192	hr	4	90					940	80	940	80
192	hr	4	75					911	00	911	00
				20,500	00	2,602	00	4,223	00	27,325	00
										2,732	50
										247	05
										366	56
										126	69
										30,797	80
										3,079	78
										33,877	58
										254	10
										34,131	68
										3	41

man-hours. A man-hour is one man working one hour or two men working one-half hour.

If a job which requires the use of four trucks will last 16 hr, the time units for the truck will be the product of the number of trucks times the length of the job, expressed in hours. The unit of cost will be for 1 truck-hour. The calculations are as follows:

Trucks, 4 × 16 hr = 64 tk-hr @ $6.00 = $384.00

In a similar manner the time and cost for the truck drivers will be shown as follows:

Truck drivers, 4 × 16 hr = 64 m-hr @ $4.00 = $256.00

The terms 64 tk-hr and 64 m-hr may be shortened to read 64 hr without producing ambiguity.

Production rates   In order to determine the time required to perform a given quantity of work, it is necessary to estimate the probable rates of production of the equipment or labor. These rates are subject to considerable variation, depending on the difficulty of the work, job and management conditions, and the condition of the equipment.

A production rate is the number of units of work produced by a unit of equipment or a man in a specified unit of time. The time is usually 1 hr. The rate may be determined during an interval of time when production is progressing at the maximum possible speed. It is obvious that such a rate cannot be maintained for a long period of time. There will always be interruptions and delays which will reduce the average production rates to less than the ideal rates. If a machine works at full speed only 45 min per hr, the average production rate will be 0.75 of the ideal rate. The figure 0.75 is defined as an operating factor.

A power shovel with a 1-cu-yd dipper may be capable of handling three dippers per minute under ideal conditions. However, on a given job the average volume per dipper may be only 0.8 cu yd, with the shovel actually operating only 45 min per hr.

The ideal output will be 180 × 1 = 180 cu yd per hr
The dipper factor will be 0.8
The time factor will be 0.75
The combined operating factor will be 0.8 × 0.75 = 0.6
The average output will be 0.6 × 180 = 108 cu yd per hr

The average output should be used in computing the time required to complete a job.

Tables of production rates In this book numerous tables give production rates for equipment and men. In all tables the rates are adjusted to include an operating factor, usually based on a 45- to 50-min working hour. If this factor is too high for a given job, the rates should be reduced to more appropriate values.

If an estimator has access to production rates that were obtained on projects constructed under conditions similar to the conditions that will exist on a project for which he is preparing an estimate, he should use them instead of using rates appearing in tables prepared by someone else.

PROBLEMS

1-1 What will be the total cost of workmen's compensation insurance for a project constructed in your state, using the base rates given in Appendix A, for which the labor costs by classification will be as follows?

Excavation, earth	$1,845.00
Carpentry, general	$4,692.00
Concrete work, general	$27,365.00
Electrical wiring	$1,690.00
Glazing	$545.00
Painting	$1,460.00
Plastering	$3,940.00
Plumbing	$2,815.00
Roofing	$885.00
Sheetmetal work	$825.00

What is the average cost per $100.00 of wages paid?

1-2 A contractor who constructs projects having an average value of $4,680,500.00 per year determines that the cost of labor averages 23 percent of the total cost of all of his work. The average cost of workmen's compensation insurance has been equal to 7.56 percent of the cost of labor.

What is the maximum amount that the contractor can afford to spend each year to install a safety program that will reduce the cost of insurance by 25 percent? The cost of the safety program should not exceed the reduction in the total cost of insurance.

REFERENCES

1 "Construction Review," U.S. Department of Labor and U.S. Department of Commerce.

2
Cost of Construction Equipment

General information   Most projects involve the use of construction equipment. The purchase of equipment represents a capital investment by the owner for the purpose of accomplishing the work which it will do, and at the same time make a profit on the investment. If a profit is to be realized from the use of equipment, it is first necessary for the owner to recover from the use of the equipment during its useful life sufficient money to pay the entire cost of the equipment, plus the cost of maintenance and repairs, interest, insurance, taxes, storage, fuel, lubrication, etc., plus an additional amount for profit. Any estimate must provide for the cost of equipment used on the project.

Sources of equipment   The use of equipment may be secured through purchase or rental. For each method there are several plans.

When equipment is purchased, either one of the following plans may be used:

1 Cash purchase
2 Purchase on a deferred-payment plan

Equipment may be rented under one of the following plans:

1 The lessee will pay a specified price per month, week, day, or hour for the use of each unit,

a. With the lessee to pay for the operator, fuel, lubrication, and all repairs

b. With the lessor to pay for the operator, fuel, lubrication, and all repairs

c. With some other combination of *a* and *b*

2 The lessee will pay a specified price for each unit of work performed by the equipment.

3 The lessee will pay a specified rental rate for the use of the equipment, with an option to purchase the equipment at a later date, with the provision that all or a part of the money paid for rent shall apply toward the purchase of the equipment.

Equipment costs When equipment is to be rented, the estimator should include the cost in his estimate. Appendix C gives representative rental rates based on a national survey made by the Associated Equipment Distributors in 1973 [1].

When equipment is purchased, it is necessary to determine the cost of owning and operating each unit, which will include several or all of the following items:

1 Depreciation

2 Maintenance and repairs

3 Investment

4 Fuel and lubrication or another type of energy, such as electricity

In this book the general practice will be to charge for equipment that is owned on an hourly basis. Discussions and examples which follow will illustrate methods of estimating the hourly cost of owning and operating equipment. These examples are intended to show the estimator how one may determine the probable hourly cost for any type of construction equipment. The costs which are determined in these examples apply for the given conditions only, but by following the same procedure and using appropriate prices for the particular equipment, the estimator may determine hourly costs which are suitable for use on any project.

Appendix B gives approximate costs per hour for owning and operating equipment. An estimator should exercise judgment by increasing or decreasing the rates when operating conditions on a given project will be materially different from those for which the costs were determined.

Depreciation costs Depreciation is the loss in value of equipment resulting from use and age. The owner of equipment must recover the original cost of the equipment during its useful life or sustain an equipment loss on those projects where the equipment is used. The cost of a unit of equipment should include the purchase price and the cost of transporting it to the purchaser, plus the cost of unloading and assembling it at its destination.

While any reasonable method may be used for determining the cost of depreciation, the following three are most commonly used:

1 Straight-line method
2 Declining-balance method
3 Sum-of-the-years-digits method

Each of these methods is approved by the U.S. Bureau of Internal Revenue for income tax purposes. However, an owner may select a life different from that given in the table, provided that it is reasonable.

If a unit of equipment is continued in use after it is completely depreciated, no additional depreciation charge may be made against it when determining profit or loss from its use for income tax purposes.

Straight-line depreciation When the cost of depreciation is determined by this method, it is assumed that a unit of equipment will decrease in value from its original total cost at a uniform rate. The depreciation rate may be expressed as a cost per unit of time, or it may be expressed as a cost per unit of work produced. The depreciation cost per unit of time is obtained by dividing the original cost, less the estimated salvage value to be realized at the time it will be disposed of, by the estimated useful life, expressed in the desired units of time, which may be years, months, weeks, days, or hours. For example, a given unit of equipment, whose original cost is $12,000, may have a useful life of 2,000 hr per yr for 5 yr and a salvage value of $2,000. The cost of depreciation is determined as follows:

Total depreciation, $12,000–$2,000   = $10,000
Annual cost of depreciation, $10,000 ÷ 5   = $ 2,000
Hourly cost of depreciation, $2,000 ÷ 2,000 = $1.00

Another method of estimating the straight-line cost of depreciation is to divide the original cost, less estimated salvage value, by the probable number of units of work which it will produce during its useful life. This method is satisfactory for equipment whose life is determined by the rate at which it is used instead of by time. Examples of such equipment include the pump and discharge pipe on a hydraulic dredge, rock crushers, rock-drilling equipment, rubber tires, and conveyor belts.

Declining-balance method Under this method of determining the cost of depreciation, the estimated life of the equipment in years will give the average percent of depreciation per year. This percent is doubled for the 200 percent declining-balance method. The value of the depreciation during any given year is determined by multiplying the resulting percent by the value of the equipment at the beginning of that year. While the estimated salvage value is not considered when determining depreciation, the depreciated value is not permitted to drop below a reasonable salvage value.

When the cumulative sum of all costs of depreciation is deducted from the original total cost, the remaining value is designated as the book value. Thus, if a unit of equipment whose original cost was $10,000 has been depreciated a total of $6,000, the book value will be $4,000.

Example This example illustrates how the declining-balance method of determining the cost of depreciation may be applied to a unit of equipment.

Total cost, $10,000
Estimated salvage value, $1,000
Estimated life, 5 yr
Average rate of depreciation, 20% per yr
Double this rate of depreciation, 2 × 20   = 40%
Cost of depreciation, first year, 0.40 × $10,000 = $4,000.00
Book value at the start of the second year   = $6,000.00
Cost of depreciation, second year, 0.40 × $6,000 = $2,400.00

Table 2-1 gives the schedule of depreciation costs for this equipment.

This method may be used for any reasonable useful life. The depreciation per year may be continued until the book value of the equipment is reduced to a reasonable salvage value.

Sum-of-the-years-digits method Under this method of determining the cost of depreciation, all the digits representing each year of the estimated life of the equipment are totaled. For an estimated life of 5 years, the sum of the digits will be $1 + 2 + 3 + 4 + 5 = 15$. Deduct the estimated salvage value from the total cost of the equipment. During the first year, the cost of depreciation will be $5/15$ of the cost less salvage value. During the

TABLE 2-1 Annual Cost of Depreciation Using the Declining-balance Method

End of year	Percent depreciation	Depreciation for the year	Book value
0	0	$ 0	$10,000.00
1	40	4,000.00	6,000.00
2	40	2,400.00	3,600.00
3	40	1,440.00	2,160.00
4	40	864.00	1,296.00
5	40	518.40	777.60
5*	. . .	296.00	1,000.00

* The value of the equipment may not be depreciated below a reasonable minimum salvage value. If this value is $1,000, the lower figures will apply; otherwise the upper figures will apply.

TABLE 2-2 Annual Cost of Depreciation Using the Sum-of-the-years-digits Method

End of year	Depreciation ratio	Total depreciation	Depreciation for the year	Book value
0	0	$9,000	$ 0	$10,000
1	$5/15$	9,000	3,000	7,000
2	$4/15$	9,000	2,400	4,600
3	$3/15$	9,000	1,800	2,800
4	$2/15$	9,000	1,200	1,600
5	$1/15$	9,000	600	1,000

second year, the cost of depreciation will be $4/15$ of the cost less salvage value. Continue this process for each year through the fifth year. Table 2-2 gives the schedule of depreciation costs for a unit of equipment under the stated conditions.

Total cost   = $10,000
Estimated salvage value =   1,000
                          ──────────
Total cost of depreciation = $ 9,000
Estimated useful life, 5 yr
Sum of the years digits, 15

Costs of maintenance and repairs The costs for maintenance and repairs include the expenditures for replacement parts and the labor required to keep the equipment in good working condition. These costs will vary considerably with the type of equipment and the service for which it is used. If a power shovel is used to excavate soft earth, the replacement of parts will be considerably less than when the same shovel is used to excavate hard clay or rock. The extent of variation in the costs for repairs is illustrated by the following examples, which are based on the cost records kept by the owner of the equipment.

Example This example covers information for two draglines of identical make and size, each operated for a period of 6 yr.

Unit	Time operated, hr	Cost of repairs	
		Total	Per hr
1	9,768	$39,472	$4.03
2	12,448	16,316	1.31

Example This example covers information for two motor graders of identical make and size, each operated for a period of 5 yr.

| | Time operated, | Cost of repairs | |
Unit	hr	Total	Per hr
1	11,680	$8,239	$0.71
2	12,168	3,696	0.31

The manufacturers of tractors have released information showing that the average costs for maintenance and repairs for crawler tractors are approximately 100 percent of the cost of depreciation during a 5-yr period of use. The costs for individual units will run substantially higher or lower than the average, depending on the service provided and the types of jobs on which the equipment is used.

The Power Crane and Shovel Association [2] suggests that the rates shown in the accompanying table be applied when determining the average costs for maintenance, repairs, and supplies for the indicated equipment.

| | Useful life | | Percent of total cost | |
	Years	Hours	Per year	Per hour
For shovels and hoes				
Size, cu yd:				
⅜–¾	5	10,000	20.00	0.0100
1–1½	6	12,000	16.67	0.0083
2–2½	8	16,000	12.50	0.00625
For draglines and clamshells				
Size, cu yd:				
⅜–¾	5	10,000	16.00	0.00800
1–1½	9	18,000	8.89	0.00445
2–2½	12	24,000	6.66	0.00333
For lifting cranes				
Capacity, tons:				
2½– 5	5	10,000	12.00	0.00600
10–15	9	18,000	6.67	0.00333
20 and over	12	24,000	5.00	0.00250

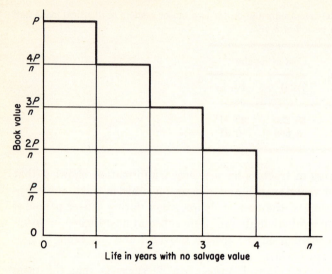

Fig. 2-1 Value of equipment by year.

The costs for maintenance and repairs given in this article should be used as a guide only. Estimators should increase or decrease the suggested rates when they believe that operating conditions will be more or less severe, respectively, than average conditions.

The cost of rubber tires Many types of construction equipment use rubber tires, whose life usually will not be the same as for the equipment on which they are used. The cost of depreciation and repairs for tires should be estimated separately from that for the equipment.

A set of tires for a unit of equipment, whose cost is $1,200.00, may have an estimated life of 5,000 hr, with the repairs during the life of the tires costing 15 percent of the initial cost of the tires. The cost is determined as follows:

Depreciation, $1,200 ÷ 5,000 hr = $0.24 per hr
Repairs, 0.15 × $0.24   = 0.04 per hr
——————————
    Total cost   = $0.28 per hr

Investment costs It costs money to own equipment, regardless of the extent to which it is used. These costs, which are frequently classified as investment costs, include interest on the money invested in the equipment and taxes of all types which are assessed against the equipment, insurance, and storage. The rates for these items will vary somewhat among different owners, with location, and for other reasons.

There are several methods of determining the cost of interest paid on the money invested in the equipment. Even though the owners pay cash for equipment, they should charge interest on the investment, because the money spent for the equipment could be invested in some other asset which would produce interest for the owner if not invested in equipment.

The average annual cost of interest should be based on the average value of the equipment during its useful life. This value may be obtained by establishing a schedule of values for the beginning of each year that the equipment will be used. The calculations given below illustrate a method of determining the average value of equipment:

Original cost of equipment, $25,000
Estimated useful life, 5 yr
Average annual cost of depreciation, $25,000 ÷ 5 = $5,000

Beginning of year	Cumulative depreciation	Value of equipment
1	0	$25,000
2	$ 5,000	20,000
3	10,000	15,000
4	15,000	10,000
5	20,000	5,000
6	25,000	0

Total of values in column 3 = $75,000
Average value, $75,000 ÷ 5 = $15,000
Average value as % of original cost,

$$= \frac{\$15,000 \times 100}{\$25,000} = 60 \text{ percent}$$

The average value of equipment may be determined from the following equation:

$$\bar{P} = \frac{P(n + 1)}{2n} \tag{2-1}$$

where P = total initial cost
$\bar{P}$ = average value
n = useful life, yr

Equation (2-1) assumes that a unit of equipment will have no salvage value at the end of its useful life. If a unit of equipment will have salvage value when it is disposed of, the average value during its life of use will be obtained from Eq. (2-2).

$$\bar{P} = \frac{P(n + 1) + S(n - 1)}{2n} \tag{2-2}$$

where P = total initial cost
$\bar{P}$ = average value
S = salvage value
n = useful life, yr

Example   Consider a unit of equipment costing $25,000, with an estimated salvage value of $5,000 after 5 yr. Using Eq. (2-2), we get

$$\bar{P} = \frac{25,000(5+1) + 5,000(5-1)}{2 \times 5}$$

$$= \frac{150,000 + 20,000}{10}$$

$$= \$17,000$$

Because insurance and taxes are usually paid on the depreciated value of equipment, it is proper to use the average value in determining the average annual cost of insurance and taxes. See Table 2-3

It is common practice to combine the cost of insurance, interest, taxes, and storage and to estimate them as a fixed percent of the average value of the equipment. The present national average rate is about 14 percent, which includes interest at 9 percent, insurance, taxes, and storage at 5 percent of the average value per year.

Operating costs   Construction equipment which is driven by internal-combustion engines requires fuel and lubricating oil, which should be considered as an operating cost. Whereas the amounts consumed and the unit cost of each will vary with the type and size of equipment, the conditions under which it is operated, and location, it is possible to estimate the cost reasonably accurately for a given condition.

An estimator should be reasonably familiar with the conditions under which a unit of equipment will be operated. Whereas a tractor may be equipped with a 200-hp engine, this tractor will not demand the full power

TABLE 2-3   Average Value of Equipment with No Salvage Value After Useful Life

Estimated life, yr	Average value as percent of original cost
2	75.00
3	66.67
4	62.50
5	60.00
6	58.33
7	57.14
8	56.25
9	55.55
10	55.00
11	54.54
12	54.17

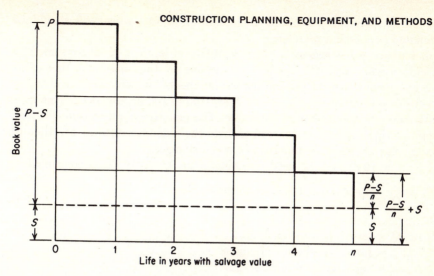

CONSTRUCTION PLANNING, EQUIPMENT, AND METHODS

Fig. 2-2 Value of equipment by year.

of the engine at all times, possibly only when it is used to load a scraper or to negotiate a steep hill. Also, equipment is seldom, if ever, used 60 min during an hour. Thus, the fuel consumed should be based on the actual operating conditions. Perhaps the average demand on an engine might be 50 percent of its maximum power for an average of 45 min per hr.

Fuel consumed When operating under standard conditions, namely at a barometric pressure of 29.9 in. of mercury and at a temperature of 60°F, a gasoline engine will consume approximately 0.06 gal of fuel for each actual horsepower-hour developed. A diesel engine will consume approximately 0.04 gal of fuel for each actual horsepower-hour developed.

Consider a power shovel with a diesel engine rated at 160 hp. During a 20-sec cycle the engine may be operated at full power while filling the dipper in tough ground, requiring 5 sec. During the balance of the cycle, the engine may be operated at not more than 50 percent of its rated power. Also, the shovel may not operate more than 45 min per hr on an average. For this condition the approximate amount of fuel consumed during an hour would be determined as follows:

Rated power, 160 hp
Engine factor:
    Filling the dipper, 5/20   = 0.250
    Rest of cycle, 15/20 × 0.5   = 0.375
                            ———
    Total factor   = 0.625
Time factor, 45/60   = 0.75
Operating factor, 0.625 × 0.75   = 0.47
Fuel consumed per hr, 0.47 × 160 × 0.04 = 3.1 gal

For other operating conditions the quantity of fuel consumed should be estimated in a similar manner.

Lubricating oil consumed The quantity of lubricating oil consumed by an engine will vary with the size of the engine, the capacity of the crankcase, the condition of the pistons and the number of hours between oil changes. It is common practice to change oil every 100 to 200 hr, unless extreme dust makes more frequent changes desirable. The quantity of oil consumed by an engine during a change cycle will include the amount added at the time of change plus the makeup oil added between changes.

Equation (2-3) may be used to estimate the quantity of oil consumed.

$$Q = \frac{\text{hp} \times 0.6 \times 0.006 \text{ lb per hp-hr}}{7.4 \text{ lb per gal}} + \frac{c}{t} \tag{2-3}$$

where Q = quantity consumed, gph
hp = rated horsepower of engine
c = capacity of crankcase, gal
t = number of hours between oil changes

Equation (2-3) is based on an operating factor of 0.60 or 60 percent. It assumes that the quantity of oil consumed between oil changes will be 0.006 gal per rated horsepower-hour. Using this equation, a 100-hp engine with a crankcase capacity of 4 gal, requiring a change every 100 hr, we find that the quantity consumed per hour will be

$$Q = \frac{100 \times 0.6 \times 0.006}{7.4} + \frac{4}{100} = 0.089 \text{ gal}$$

Lubricants other than crankcase oil will be required for motor-driven equipment. Although the costs of such lubricants will vary, an average cost equal to 50 percent of the cost of the crankcase oil will be satisfactory.

Examples illustrating the cost of owning and operating equipment

Example Determine the probable cost per hour of owning and operating a 2-cu-yd diesel-engine-powered crawler-type power shovel. The following information will apply.

Engine, 160 hp
Crankcase capacity, 6 gal
Hours between oil changes, 100
Operating factor, 0.60
Fuel consumed per hr, $0.6 \times 0.04 \times 160 = 3.9$ gal
Lubricating oil consumed per hr,

$$\frac{160 \times 0.6 \times 0.006}{7.4} + \frac{6}{100} = 0.138 \text{ gal}$$

Shipping weight, 136,000 lb
Useful life, 6 yr
Hours used per year, 2,000

Cost to owner:

Price fob factory	= $97,240
Freight, 136,000 × $1.80 per cwt	= 2,448
Unloading and assembling	= 286
Total cost	= $99,974
Average investment, 0.5833 × $99,974	= $58,315

Annual cost:

Depreciation, $99,974 ÷ 6 yr	= $16,662
Maintenance and repairs, 100% of depreciation	= 16,662
Investment, 0.14 × $58,315	= 8,164
Total annual fixed costs	= $41,488

Hourly cost:

Fixed cost, $41,488 ÷ 2,000 hr	= $ 20.75
Fuel, 3.9 gal @ $0.40	= 1.56
Lubricating oil, 0.138 gal @ $1.00	= 0.14
Other lubricants, 0.5 × $0.14	= 0.07
Total cost per hr, excluding labor	= $ 22.52

Example Determine the probable cost per hour for owning and operating a 25-cu-yd heaped-capacity bottom-dump wagon with six rubber tires. Because the tires will have a different life than the wagon, they will be treated separately. The following information will apply:

Engine, 250 hp, diesel
Crankcase capacity, 14 gal
Hours between oil changes, 80
Operating factor, 0.60
Fuel consumed per hr, 0.6 × 0.04 × 250 = 6.0 gal
Lubricating oil consumed per hr,

$$\frac{250 \times 0.60 \times 0.006}{7.4} + \frac{14}{80} = 0.30 \text{ gal}$$

Useful life, 5 yr
Hours used per year, 2,000
Maintenance and repairs, 50% of depreciation
Life of tires, 5,000 hr
Repairs to tires, 15% of depreciation of tires

Cost to owner:

Cost delivered to owner	= $82,700
Less cost of tires	= 15,620
Net cost less tires	= $67,080
Average investment, 0.6 × $82,700	= $49,620

Annual cost:

Depreciation, $67,080 ÷ 5 yr	= $13,416
Maintenance and repairs, 0.5 × $13,416	= 6,708
Investment, 0.14 × $49,620	= 6,947
Total annual fixed cost	= $27,071

Hourly cost:

Fixed cost, $27,071 ÷ 2,000	= $ 13.54
Tire depreciation, $15,620 ÷ 5,000	= 3.12
Tire repairs, 0.15 × $3.12	= 0.47
Fuel, 6.0 gal @ $0.40	= 2.40
Lubricating oil, 0.30 gal @ $1.20	= 0.36
Other lubricants, 0.5 × $0.36	= 0.18
Total cost per hr, excluding labor	= $ 20.07

The hourly cost of owning and operating construction equipment, as illustrated in the previous examples, will vary with the conditions under which the equipment is operated. The job planner should analyze each job to determine the probable conditions that will affect the cost and adjust the cost if such seems applicable.

Tables giving equipment costs   The table in Appendix B gives estimated hourly costs for owning and operating construction equipment.

PROBLEMS

2-1   The original cost of a power shovel is $48,640. The estimated useful life is assumed to be 5 yr, with no salvage value. Determine its book value at the end of 2 yr, using each of the following methods of determining depreciation:

(a)   Straight line
(b)   Declining balance
(c)   Sum of the years digits

2-2   If the power shovel of Prob. 2-1 is assumed to have a salvage value of $4,500 at the end of 5 yr, find the book value at the end of 4 yr using the straight-line, declining-balance, and sum-of-the-years-digits methods.

2-3   A 1½-cu-yd power shovel, whose total cost is $82,520, is assumed to have a useful life of 12,000 hr. Using the method suggested by the Power Crane and Shovel Association, determine the probable cost per year and per hour for maintenance, repairs, and supplies.

2-4   What is the average annual investment cost for a tractor whose original cost is $76,540, and whose useful life is estimated to be 5 yr, with no salvage value?

What is the average annual investment cost of this tractor if it is assumed to have a salvage value of $8,500 at the end of the 5 yr?

Use an investment cost equal to 14 percent of the average value of the tractor.

2-5   Determine the probable cost per hour for owning and operating a crawler tractor for the given conditions.

Engine, 325-fwhp diesel
Shipping weight, 48,640 lb
Freight rate, $1.34 per cwt
Capacity of crankcase, 9 gal
Estimated life, 6 yr
Estimated operating time, 2,000 hr per yr
Time between oil changes, 120 hr
Assume a 50% operating factor and a 45-min hour
Cost, fob factory, $53,890

Probable salvage value, $7,500
Investment cost, 14% of average value per yr
Cost of diesel fuel, $0.35 per gal
Cost of lubricating oil, $1.00 per gal
Assume that repairs cost 75 percent of depreciation

2-6 Determine the probable cost per hour for owning and operating a wheel-type tractor and scraper for the given conditions.

Engine, 280-fwhp diesel
Shipping weight, 46,560 lb
Freight rate, $1.25 per cwt
Capacity of crankcase, 7 gal
Time between oil changes, 100 hr
Estimated life, 5 yr
Operating time, 2,000 hr per yr
Assume an operating factor of 55% and a 45-min hr
Cost, fob factory, $73,650
Cost of tires, $3,780
Life of tires, 5,000 hr
Cost of repairing tires, 15% of tire cost per hr
Probable salvage value, $6,450
Investment cost, 14% of average value per yr
Cost of diesel fuel, $0.35 per gal
Cost of lubricating oil, $1.00 per gal
Assume that repairs cost 80% of depreciation

2-7 Determine the probable cost per hour for owning and operating the tractor-scraper of Prob. 2-6 when the operating time is 1,200 hr per yr, with a useful life of 6 yr. The other conditions will be the same as those given in Prob 2-6.

2-8 Determine the probable cost per hour for owning and operating the tractor of Prob 2-5 when the unit will be operated 1,200, 1,600, 2,000, and 2,400 hr per year.

All other conditions will be the same as those given in Prob 2-5.

REFERENCES

1   "1973 Rental Compilation," 24th ed., Associated Equipment Distributors, Chicago, 1974.
2   Technical Bulletin No. 2, "Operating Cost Guide," Power Crane and Shovel Association, New York.

3
Handling and Transporting Material

General In many instances, construction materials are delivered by the seller or producer directly to the project in trucks. However, in other instances the materials are delivered by railroad cars to a siding, where they are left to be unloaded and hauled by the purchaser.

Some projects require the use of aggregates, sand and gravel, or crushed stone, which are produced from natural deposits or quarries and hauled to the project in trucks or tractor-pulled wagons.

The handling and hauling may be done by a contractor, using his laborers and equipment, or it may be accomplished through a subcontractor. Regardless of the method used, it will involve a cost which must be included in the estimate for a project.

When estimating the time required by a truck for a round trip, the estimator should divide the round-trip time into four elements as follows:

1 Loading
2 Hauling, loaded
3 Unloading
4 Returning, empty

The time required for each element should be estimated. If elements 2 and 4 require the same time, they may be combined. As the time required for hauling and returning will depend on the distance and the effective speed, it is necessary to determine the probable speed at which a vehicle can travel along the given haul road for the conditions that will exist. Speeds are dependent on the vehicle, traffic congestion, condition of the road, and other factors. An appropriate operating factor should be used in determining production rates. For example, if a truck will operate only 45 min per hr, this time should be used in determining the number of round trips the truck will make in an hour.

While the discussions and examples given in this chapter do not include all types of handling and transporting, they should serve as guides to illustrate methods which may be applied to any job.

Capacities of railroad cars The minimum weight for a railroad car-load of materials varies with the materials and also with respect to whether the shipment originates within the state or outside the state. Table 3-1 gives the approximate minimum weight per carload for some of the more commonly used construction materials. Actual loads may exceed the minimum loads up to the volumetric or weight capacities of the cars.

Hauling lumber from railroad cars to the job Lumber is usually loaded by laborers directly onto flat-bed trucks, hauled to the job, and stacked according to size.

TABLE 3-1 Approximate Minimum Quantities of Construction Materials per Railroad Carload

Material	Minimum weight, lb
Asphalt	40,000
Asphalt rock	100,000
Brick	60,000
Cast-iron pipe	40,000
Cement, bulk or sacks	50,000
Clay building tile	60,000
Concrete blocks	60,000
Concrete sewer pipe	26,000
Gravel	100,000
Lime in sacks	50,000
Lumber	34,000
Reinforcing steel	40,000
Rolled-steel shapes	40,000
Sand	100,000
Steel pipe	40,000
Vitrified-clay sewer pipe	26,000
Wood piles	34,000

A laborer should be able to unload lumber from a car or a truck at a rate of 2,000 to 4,000 fbm per hr, using the lower rate for small pieces and the higher rate for large pieces. A fair average rate should be about 3,000 fbm per hr. Two or three men may work in a car.

Trucks of the type generally used will haul 2 to 6 tons, corresponding to 1,000 to 3,000 fbm per load. The average speed will vary with the distance, type of road, traffic congestion, and weather.

Example Estimate the cost of unloading 40,000 fbm of lumber from railroad cars onto trucks, which will carry 2,000 fbm per load, and hauling it 2 miles to a job.

An examination of the haul road indicates an average speed of 20 mph for the trucks, including necessary delays to check oil, gasoline, water, etc. Assume that a man will handle 3,000 fbm of lumber per hour.

Based on using two men, a truck driver and a laborer, and one truck, the length of the job would be determined as follows:

Rate of loading a truck, 2 × 3,000	= 6,000 fbm per hr
Time to load a truck, 2,000 ÷ 6,000	= 0.33 hr
Time to unload a truck, 2,000 ÷ 6,000	= 0.33 hr
Travel time, round trip, 4 miles ÷ 20 mph	= 0.20 hr
Total time per load	= 0.86 hr
No. trips per hr, 1 ÷ 0.86	= 1.16
Quantity hauled per hr, 1.16 × 2,000	= 2,320 fbm
Total time for the job, 40,000 ÷ 2,320	= 17.2 hr

An alternate method of determining the length of the job is as follows:

No. truckloads required, 40,000 ÷ 2,000 = 20
Round-trip time per load, 0.86 hr
Total time for the job, 20 × 0.86 = 17.2 hr
The cost will be

Truck, 17.2 hr @ $4.25	= $ 73.10
Truck driver, 17.2 hr @ $3.75	= 64.50
Laborer, 17.2 hr @ $3.25	= 55.90
Total cost	= $193.50
Cost per M fbm, $193.30 ÷ 40	= 4.84

Based on the use of two laborers who remain at the cars, two laborers who unload the trucks at the job, and a driver for each truck, with the drivers to assist in loading and unloading the trucks, the costs will be determined as follows:

Rate of loading truck, 3 × 3,000 = 9,000 fbm per hr	
Time to load a truck, 2,000 ÷ 9,000	= 0.222 hr
Time to unload a truck, 2,000 ÷ 9,000	= 0.222 hr
Travel time, round trip, 4 miles ÷ 20 mph	= 0.200 h
Total time per load	= 0.644 hr

No. trips per hr per truck, 1 ÷ 0.644 = 1.55
No. trucks required,

$$\frac{\text{Round-trip time}}{\text{Time to load}} = \frac{0.644}{0.222} \qquad = 2.9$$

An alternate method of determining the number of trucks required is as follows:

No. trips per hr per truck   = 1.55
Quantity of lumber hauled per hr per truck,
   1.55 trips × 2,000 fbm   = 3,100 fbm
No. trucks required, 9,000 ÷ 3,100   = 2.9
Determine the cost using three trucks.

Total time for the job, 40,000 fbm ÷ 9,000 fbm per hr = 4.44 hr

The cost will be
   Trucks, 3 × 4.44 = 13.32 hr @ $4.25   = $ 56.51
   Truck drivers, 13.32 hr @ $3.75   =   49.95
   Laborers, 4 × 4.44 = 17.76 hr @ $3.25   =   57.72
                                                    ————
     Total cost   = $164.18

Determine the cost using two trucks.
As there are not enough trucks to keep the laborers busy, the rate of hauling will determine the length of the job.

Quantity hauled per hr per truck, 1.55 × 2,000   = 3,100 fbm
Quantity hauled per hr by two trucks   = 6,200 fbm
Total time required for the job, 40,000 ÷ 6,200   = 6.45 hr
The cost will be
   Trucks, 2 × 6.45 = 12.9 hr @ $4.25   = $ 54.83
   Truck drivers, 12.9 hr @ $3.75   =   48.38
   Laborers, 4 × 6.45 = 25.8 hr @ $3.25   =   83.85
                                                    ————
     Total cost   = $187.06

It should be noted that the total costs are based on labor and truck costs for the actual time at work, with no allowance for laborers and trucks while they are waiting their turn to start working. If there will be costs during such nonproductive times, the total cost of the job will be higher than the values calculated.

Unloading sand and gravel from railroad cars, and hauling to the job with trucks   When sand and gravel are delivered by railroad, the material is usually shipped in gondola cars, which are spotted on a siding in the vicinity of the job. If a clamshell is used to unload the material into trucks, it may be desirable to use one or two laborers to assist the clamshell in removing material from the corners of the cars.

The following example illustrates a method of determining the cost of unloading and hauling the material to a job.

Example   Estimate the probable cost of unloading 160 cu yd of gravel from railroad cars into trucks and hauling it 3 miles to a project. Use a ½-cu-yd truck-mounted clamshell to load 6-cu-yd dump trucks. The trucks can maintain an average speed of 30 mph hauling and returning to the cars. The truck time at the dump will average 8 min, including the time required to check, service, and refuel a truck.

The clamshell should handle an average of 32 cu yd per hr,* allowing time for delays in removing gravel from the corners of each car, for moving the clamshell from one car to another, and for other minor delays. The estimated cost will be determined as follows:

Time to load a truck, 6 cu yd ÷ 32 cu yd per hr	= 0.188 hr
Travel time, round trip, 6 miles ÷ 30 mph	= 0.200 hr
Time at the dump, 8 min ÷ 60 min per hr	= 0.133 hr
Total round-trip time	= 0.521 hr

No. loads per hr per truck, 1.00 ÷ 0.521 = 1.93
Volume hauled per truck per hr, 6 × 1.93 = 11.58 cu yd
No. trucks required, 32 cu yd ÷ 11.58 = 2.77

If three trucks are used, the time required to complete the job will be 160 cu yd ÷ 32 cu yd per hr = 5 hr.

The cost will be

Moving the clamshell to and from the job	=	$ 30.00
Cost of clamshell, 5 hr @ $9.95	=	49.75
Trucks, 3 × 5 = 15 hr @ $7.12	=	106.80
Clamshell operator, 5 hr @ $6.20	=	31.00
Clamshell oiler, 5 hr @ $4.30	=	21.50
Truck drivers, 3 × 5 = 15 hr @ $3.90	=	58.50
Laborers, 2 × 5 = 10 hr @ $3.60	=	36.00
Total cost	=	$333.55

If two trucks are used, the length of the job will be determined by the rate at which the trucks will haul the gravel. As previously determined, a truck should haul 11.58 cu yd per hr. Two trucks should haul 2 × 11.58 = 23.16 cu yd per hr.

The length of the job will be 160 ÷ 23.16 = 6.9 hr.
The cost will be

Moving the clamshell to and from the job	=	$ 30.00
Cost of clamshell, 6.9 hr @ $9.95	=	68.65
Trucks, 2 × 6.9 = 13.8 hr @ $7.12	=	98.26
Clamshell operator, 6.9 hr @ $6.20	=	42.78
Clamshell oiler, 6.9 hr @ $4.30	=	29.67
Truck drivers, 2 × 6.9 = 13.8 hr @ $3.90	=	53.82
Laborers, 2 × 6.9 = 13.8 hr @ $3.60	=	49.68
Total cost	=	$372.86

Thus it is cheaper to use three trucks.

If a foreman is used to supervise the job, his salary should be added to the cost of the project.

The costs determined in the previous examples are the minimum amounts, based on the assumption that the cost durations can be limited to 5 hr or 6.9 hr. It may be impossible

---

* See Table 4-8 for the output of a clamshell.

or impractical to limit the costs to these respective durations. If such is the case, the appropriate durations should be used.

Unloading and hauling bricks If bricks are shipped by rail to a siding near a project, they are hauled to the project with flat-bed trucks, which carry about 2,000 to 3,000 bricks per load. Popular sizes of building bricks weigh about 4 lb each.

Workers, using brick tongs, will carry 6 to 10 bricks per load. A worker should be able to pick up his load, walk to the truck, which is spotted at the car door, deposit the bricks, and return for another load in ½ to 1 min per trip. If it is assumed that the average time for a trip is ¾ min and the worker carries 8 bricks per trip, in 1 hr he will handle 640 bricks. For other conditions, the rates will be given in Table 3-2.

If several cars of bricks are to be unloaded and hauled to a job, it may be more economical to keep a gang of three or four men in the cars unloading bricks and a similar gang at the job to assist in unloading the trucks. Two or more trucks should be used for hauling the bricks, depending on the distance between the cars and the job. An ideal condition is to balance the number of trucks with the rate at which the laborers handle the bricks. It is not always possible to do this exactly, so there will be some time lost by the trucks or by the laborers. In some instances the truck driver will assist in loading and unloading his truck, while in other instances he will not handle bricks.

Example Estimate the cost of hauling 60,000 bricks from railroad cars to a job 4 miles distant. Each truck will haul 3,000 bricks per load and can travel at an average speed of 20 mph loaded and at 30 mph empty, allowing for lost time. Three laborers will be stationed at the cars to load the trucks, and another three will be at the job to unload the trucks. The truck drivers will assist in loading and unloading the trucks. Each laborer will carry

TABLE 3-2 Rates of Handling Bricks from Cars to Trucks

Bricks carried per trip	Trip time, min	Bricks handled per hr	Hours per M brick
6	0.5	720	1.39
8	0.5	960	1.04
10	0.5	1,200	0.83
6	0.75	480	2.08
8	0.75	640	1.56
10	0.75	800	1.25
6	1.00	360	2.78
8	1.00	480	2.08
10	1.00	600	1.67

eight bricks at the car and at the job. Assume that a laborer will average ¾ min per trip at the car and at the job.

No. bricks handled per labor-hour will be 8 × 60 ÷ ¾ = 640
Rate of loading a truck, 4 × 640 = 2,560 per hr

Time to load a truck, 3,000 ÷ 2,560	= 1.170 hr
Time to unload a truck, same	= 1.170 hr
Time to drive from car to job, 4 ÷ 20	= 0.200 hr
Time to drive from job to car, 4 ÷ 30	= 0.133 hr
	= 2.673 hr
Time for round trip	= 2.673 hr

No. bricks hauled per hr per truck, 3,000 bricks ÷ 2.673 hr = 1,125
No. trucks required to haul 2,560 bricks per hr, 2,560 ÷ 1,125 = 2.28

It will be necessary to use either two or three trucks. If two trucks are used, the cost will be determined as follows.

Rate of hauling bricks, 2 × 1,125 = 2,250 per hr
Time required to finish the job, 60,000 ÷ 2,250 = 26.7 hr
The cost will be

Trucks, 2 × 26.7 = 53.4 hr @ $5.60	= $ 299.04
Truck drivers, 53.4 hr @ $3.50	= 186.90
Laborers, 6 × 26.7 = 160.2 hr @ $3.25	= 520.65
Total cost	= $1,006.59

If three trucks are used, the rate of loading the trucks, as previously determined, will be 2,560 bricks per hr.

Time required to finish the job, 60,000 ÷ 2,560 = 23.5 hr
The cost will be

Trucks, 3 × 23.5 = 70.5 hr @ $5.60	= $ 394.80
Truck drivers, 70.5 hr @ $3.50	= 246.75
Laborers, 6 × 23.5 = 141.0 hr @ $3.25	= 458.25
Total cost	= $1,099.80

It will be more economical to use two trucks.

In some instances it is desirable to determine whether it is more economical to transport materials by trucks directly from the source to the job instead of by combination railroad and trucks. If trucks only are used, intermediate handling costs will be eliminated. Estimators should determine the cost by each method before deciding which method will be used. Also, they should consider that when less than a carload of material is required, the railroad freight rates are higher per unit than for a full carload.

Unloading and hauling cast-iron pipe

This project involves unloading cast-iron pipe from railroad cars onto trucks and hauling it to a project, where it will be laid on the ground along city streets. The average haul distance will be 5 miles.

The pipe will be 12 in. diameter, 18 ft long, and weigh 1,140 lb per joint or 18 ft. A 4-ton truck-mounted crane will be used to unload the pipe onto 8-ton trucks, which will haul 14 joints per load. The pipe will be unloaded from the trucks using an 80-hp crawler tractor, with a side boom, which will move along with the trucks.

The trucks can average 25 mph loaded and 30 mph empty.

The total labor crew will consist of the following persons:

At the railroad car
  One crane operator
  One crane oiler
  Two laborers in the car
  Two laborers on the truck
At the unloading site
  One tractor operator
  Two laborers helping unload pipe

Determine the number of trucks required and the direct cost per linear foot for unloading and hauling the pipe.

The time required by a truck for a round trip will be
  Loading truck, 14 joints × 3 min per joint = 42 min ÷ 60   = 0.70 hr
  Hauling to job, 5 miles ÷ 25 mph   = 0.20 hr
  Unloading truck, 14 joints × 3 min per joint = 42 min ÷ 60   = 0.70 hr
  Returning to car, 5 miles ÷ 30 mph   = 0.17 hr
                                                                     ———
  Total time   = 1.77 hr
No. trips per hr, 1 ÷ 1.77 = 0.565
No. joints hauled per hr per truck, 0.565 × 14 = 7.9
No. joints unloaded per hr, 60 ÷ 3 = 20
No. trucks required, 20 ÷ 7.9 = 2.54
Use three trucks.

Assume that the crane and truck will operate 50 min per hr. The average number of joints hauled per hour will be 20 × 50 ÷ 60 = 16.67.

The cost per hr will be
  Crane, 1 hr @ $10.80   = $10.80
  Trucks, 3 hr @ $6.75   =   20.25
  Tractor with boom, 1 hr @ $8.20   =   8.20
  Crane operator, 1 hr @ $6.20   =   6.20
  Crane oiler, 1 hr @ $4.30   =   4.30
  Tractor operator, 1 hr @ $5.80   =   5.80
  Truck drivers, 3 hr @ $3.80   =   11.40
  Laborers, 6 hr @ $3.25   =   19.50
  Foreman, 1 hr @ $7.60   =   7.60
                                                       ———
  Total cost   = $94.05
  Cost per lin ft, $94.05/(18 × 16.67)   =   0.314

Any cost of moving equipment to the job and back to the storage yard should be prorated to the total length of pipe handled, and added to the unit cost determined above, to obtain the total cost per unit length.

PROBLEMS

3-1 Estimate the total direct cost and the cost per M fbm of lumber for unloading 50 M fbm of lumber from railroad cars and hauling it to a job which is 3 miles from the cars. A 5-ton stake-body truck will haul 3 M per load. The trucks can average 25 mph loaded and 40 mph empty.

One laborer plus the truck driver will load lumber onto the truck from the car and another laborer plus the truck driver will unload the truck and stack the lumber at the job. Each man will handle 1,200 fbm per hr, when working.

Assume that the laborers and the trucks operate 50 min per hr.

Determine the cost of trucks from Appendix B.

Determine the most economical number of trucks to use.

Determine if placing an additional laborer at the car and at the job will reduce the cost of handling and hauling the lumber

Labor costs per hour will be

Truck drivers, $4.25
Laborers, $3.75

3-2 Using the information given in Prob 3-1, determine if it is less expensive to keep two laborers at the cars and two other laborers at the job site or to use only two laborers, who will ride with the truck driver to load and unload a truck.

3-3 The owner of a sand and gravel pit is considering the purchase of a fleet of trucks to deliver aggregate to his customers. Two sizes are being considered, namely 10-cu-yd and 15-cu-yd diesel-engine-powered dump trucks. The haul distances will vary from 6 to 20 miles, with an average distance of about 12 miles.

It is estimated that the 10-cu-yd trucks can travel at an average speed of 40 mph loaded and at 50 mph empty, while the 15-cu-yd trucks can travel at an average speed of 35 mph loaded and at 45 mph empty.

The trucks will be loaded from stockpiles of aggregate, using a 2-cu-yd clamshell with an angle of swing averaging 90 deg. See Table 4-8 for the output for the clamshell.

The average truck time at the dump will be 4 min for the 10-cu-yd truck and 5 min for the 15-cu-yd truck. Assume that the trucks and the clamshell will operate 45 min per hr.

Determine which size truck is more economical, using the truck costs from Appendix B. A truck driver will be paid $4.50 per hr.

Because the cost of the clamshell will be the same for each size truck, its cost need not be considered.

3-4 Determine if it is more economical to use a 1- or a 2-cu-yd clamshell to load sand and gravel from stockpiles into 10-cu-yd diesel-engine-powered dump trucks. Assume an average angle of swing of 90 deg for the clamshell.

Use Table 4-8 to determine the production rate for each clamshell.

In addition to the time required by a clamshell to load a truck, there will be an average delay of 2 min waiting for another truck to move into position for loading.

Assume a 50-min hour for both the clamshell and the trucks.

The costs per hr will be

> 1-cu-yd clamshell, $12.95
> 2-cu-yd clamshell, $24.05
> Truck, $12.50
> Truck driver, $4.75
> Clamshell operator, $7.25
> Clamshell oiler, $5.25

3-5   A total of 140,000 bricks are to be unloaded from railroad cars onto 5-ton gasoline-engine-powered stake-body trucks and hauled to a project 4 miles from the cars. A truck, which can haul 3,000 bricks per load, will average 30 mph loaded and 40 mph empty. Assume that the trucks operate 45 min per hr.

Determine the total cost and the cost per M bricks for each of the stated conditions.

(*a*)   Three laborers plus a truck driver will each load 450 bricks per hour onto a truck and three other laborers plus a truck driver will unload the bricks from a truck at the job at the same rate.

(*b*)   Three laborers plus a truck driver will each load 450 bricks per hr onto a truck, then the truck driver and the three laborers will ride on the truck to the job, where they and the truck driver will unload the bricks at the same rate as at the car.

The costs per hour will be

> Truck, each, $5.75
> Truck driver, $4.75
> Laborers, each, $3.75

3-6   The operator of a gravel pit is invited to bid on furnishing 2,500 cu yd of bank-run gravel for a job. The gravel is to be delivered to a job, which is 12 miles from the pit.

The gravel will be loaded into trucks using a power shovel which will load at a rate of 80 cu yd per hr. The trucks will haul 10 cu yd per load and average 35 mph loaded and 45 mph empty.

Assume that the power shovel and the trucks will operate 45 min per hr.

The following costs will apply:

> Royalty paid for gravel, $0.20 per cu yd
> General overhead, $0.15 per cu yd
> Profit, $0.25 per cu yd
> Power shovel, $12.60 per hr
> Trucks, each, $12.10 per hr
> Shovel operator, $6.25 per hr
> Shovel oiler, $4.75 per hr
> Truck drivers, $4.75 per hr
> Foreman, $7.50 per hr
> Labor taxes, 15 percent of all wages paid, including the foreman
> What price per cubic yard should be bid?

4

Earthwork and Excavation

General Most projects constructed today involve excavation to some extent. The extent of excavation varies from a few cubic yards for footings and trenches for pipes to millions of cubic yards for large earth-filled dams. It is usually a relatively simple operation to determine the quantity of material to be excavated. It is much more difficult to estimate the rate at which it will be handled by men or by earth-excavating equipment. There are a great many factors which affect the rates of production. These factors may be divided into two groups, job and management.

Job factors Job factors involve the type of material, extent of water present, weather, freedom of men and equipment to operate on the job, size of the job, length of haul for disposal, condition of the haul road, etc. It is difficult for the constructor to change job conditions.

Management factors Management factors involve organization for the job, maintaining good morale among the workers, selecting and using suitable equipment and methods, care in servicing equipment, maintaining production records, and others. These factors are under the control of the constructor.

Methods of excavating Methods of excavating vary from hand digging and shoveling for small jobs to that done by trenching machines, power shovels, draglines, clamshells, tractor-pulled scrapers, bulldozers, elevating loaders, drilling machines, and dredges. Some material, such as rock, is so hard that it is necessary to drill holes and loosen it with explosives prior to excavating it.

Hauling material When it is necessary to move the excavated material some distance for disposal or to build an earth structure, transporting equipment such as wheelbarrows, trucks, and tractor-pulled wagons should be used. Many types and sizes are available to the constructor. The size of the hauling unit should be balanced with the output of the excavating equipment if possible.

Physical properties of earth In order to estimate the cost of excavating and hauling earth intelligently, it is necessary to have a knowledge of the physical properties of earth.

In its undisturbed condition, prior to excavating, earth will weigh about 100 lb per cu ft, or about 2,700 lb per cu yd. The weight of solid rock is approximately 150 lb per cu ft, or about 4,000 lb per cu yd. However, the weight varies with the type of earth or rock, and if a more exact weight is required, it must be determined for the particular project.

When earth and rock are loosened during excavation, they assume a larger volume and a corresponding reduction in weight per unit of volume. This increase in volume is described as swell and is usually expressed as a percent gain compared with the original volume. If earth is placed in a fill and compacted with modern equipment, it usually occupies a smaller volume than in its natural state in the cut or borrow pit. This decrease in volume is described as shrinkage and is expressed as a percent of the original volume. Table 4-1 indicates the percent of swell for various soils.

Excavating by hand The rate at which a man can excavate earth varies with the type of earth, the extent of digging required, the height to which it must be lifted, and climatic conditions. If loosening is necessary, a pick is most commonly used. Lifting is generally done with a round-pointed long-handled shovel. It requires 150 to 200 shovels of earth to excavate a cubic yard in its natural state.

TABLE 4-1 Percent of Swell for Different Classes of Earth

Material	Percent swell
Sand or gravel	14–16
Loam	16–25
Ordinary earth	20–30
Dense clay	25–40
Solid rock	50–75

TABLE 4-2   Rates of Handling Earth by Hand

Operation	Cu yd per hr	Hr per cu yd
Excavating sandy loam	1–2	0.5–1.0
Shoveling loose earth into a truck	½–1	1.0–2.0
Loosening with a pick	¼–½	2.0–4.0
Shoveling from trenches to 6 ft 0 in. deep	½–1	1.0–2.0
Shoveling from pits to 6 ft 0 in. deep	½–1	1.0–2.0
Backfilling	1½–2½	0.4–0.7
Spreading loose earth	4–7	0.15–0.25

NOTE: For hours per cubic yard the lower values should be used for sandy loam and the higher values for heavy soils and clays.

Representative rates of excavating and hauling earth are given in Table 4-2.

Hauling earth with wheelbarrows   If the excavated earth is to be hauled distances up to 100 ft, wheelbarrows are frequently used. A wheelbarrow will hold about 3 cu ft loose volume. A man should be able to haul the earth 100 ft and return in 2½ min if the haul path is reasonably firm and smooth. Filling the wheelbarrow with loose earth will require about 2½ min. Thus it will require about 5 min to load and haul 3 cu ft of earth up to 100 ft. This corresponds to about 1 hr per cu yd, bank measure.

Example   Estimate the cost of excavating and backfilling a trench 3 ft 0 in. wide, 4 ft 0 in. deep, and 150 ft long in sandy loam. Twelve laborers will work at the project with a parttime foreman, who will spend one-half of his time on this project.

Volume of earth, 3 × 4 × 150 ÷ 27 = 66.7 cu yd
The cost will be
Loosening earth, 66.7 cu yd × 2 hr per cu yd = 133.4 hr @ $3.25	= $435.55
Shoveling earth from trench, 66.7 cu yd × 1 hr per cu yd = 66.7 hr @ $3.25	= 216.78
Shoveling earth back from trench, 66.7 cu yd × 0.5 hr per cu yd = 33.4 hr @ $3.25	= 108.55
Backfilling trench, 66.7 cu yd × 0.5 hr per cu yd = 33.4 hr @ $3.25	= 108.55
Foreman, 266.9/12 × 2 = 11.2 hr $5.25	= 58.80
Total cost	= $928.23
Cost per cu yd, $928.23 ÷ 66.7 cu yd	= 13.92
Cost per lin ft, $928.23 ÷ 150 lin ft	= 6.19

Excavating with trenching machines   Even though it may be economical to excavate short sections of trenches with hand labor, the use of a trenching machine is more economical for larger jobs. Once the machine is transported to the job and put into operation, the cost of excavating is considerably less than the cost by hand. For a given job the savings in excavating costs resulting from the use of the machine as compared with hand excavating must be sufficient to offset the cost of transporting the machine to the job and back to storage after the job is completed. Otherwise hand labor is more economical.

Trenching machines may be purchased or rented. Several types are available.

Wheel-type trenching machines For shallow trenches such as those required for water mains and gas and oil pipe lines, a wheel-type machine is frequently used. The wheel rotates at the rear of the machine, which is mounted on crawler tracks. A combination of teeth and buckets attached to the wheel loosens and removes the earth from the trench as the machine advances. The earth is cast into a windrow along the trench. This machine may be used to excavate trenches whose depths do not exceed approximately 8 ft.

Ladder-type trenching machines For deep trenches, such as those required for sewer pipes and other utilities, the ladder-type machine should

TABLE 4-3   Data on Trenching Machines

Depth of trench, ft	Width of trench, in.	Digging speed, ft per hr
Wheel type		
2–4	16, 18, 20	150–600
	24, 24, 26	90–300
	28, 30	60–180
4–6	16, 18, 20	40–120
	22, 24, 26	25–90
	28, 30	15–40
Ladder type		
4–6	16, 20, 24	100–300
	22, 26, 30	75–200
	28, 32, 36	40–125
6–8	16, 20, 24	40–125
	22, 26, 30	30–60
	28, 32, 36	25–50
8–12	18, 24, 30	30–75
	30, 33, 36	15–40

Fig. 4-1 Wheel-type trenching machine. (*The Parsons Company*).

be used. Machines with inclined or vertical booms are available. The boom is mounted at the rear of the machine. Cutter teeth and buckets are attached to endless chains which travel along the boom. As the machine advances, the earth is excavated and cast along the trench. The depth of cut is adjusted by raising or lowering the boom. By adding side cutters, the width of the trench may be increased considerably. This type of machine is available in a number of sizes, making it possible to excavate trenches 20 ft or more in depth and up to 6 ft or more in width.

Table 4-3 gives information on various types of trenching machines. Use the higher speeds for soft earth such as sandy loam and the lower speeds for heavy tight earth and clay.

Example Estimate the total cost and the cost per linear foot for excavating a trench 3 ft 0 in. wide, 6 ft 6 in. average depth, and 2,940 ft long in ordinary earth. A ladder-type trenching machine will be used.

There will be no obstructions to retard the progress of the machine.

Table 4-3 indicates a speed between 30 and 50 ft per hr.

Assume an average speed of 40 ft per hr.

Time required to complete the job, 2,940 ÷ 40 = 73.5 hr.

The cost will be

Transporting the machine to and from the job	=	$ 190.00
Trenching machine, 73.5 hr @ $10.60	=	779.10
Utility truck, 73.5 hr @ $4.58	=	336.63

Machine operator, 73.5 hr @ $6.75	=	496.13
Truck driver, 73.5 hr @ $3.75	=	275.63
Laborers, 2 men, 147 hr @ $3.25	=	477.75
Foreman, 73.5 hr @ $7.80	=	573.30
Total cost	=	$3,128.54
Cost per lin ft, $3,128.54 ÷ 2,940 lin ft	=	1.065

Excavating with power shovels   Power shovels are excavating machines. They will handle all classes of earth without prior loosening; but in excavating solid rock it is necessary to loosen the rock first, usually by drilling holes and discharging explosives in them. The excavated material is loaded into trucks, tractor-pulled wagons, or cars, which haul it to its final destination. For the shovel to maintain its maximum output, sufficient hauling units must be provided.

Output of power shovels   In estimating the output of a power shovel, it is necessary to know the class of earth to be excavated, the depth of dig, the ease with which hauling equipment can approach the shovel, the angle of swing from digging to emptying the dipper, and the size of the dipper. The size of a power shovel is designated by the struck capacity of the dipper, expressed in cubic yards, loose measure.

When digging hard earth the output will be less than when digging soft earth. If the face against which the shovel is digging is too shallow, it will

Fig. 4-2   Ladder-type trenching machine. (*The Parsons Company.*)

TABLE 4-4 Ideal Outputs of Power Shovels, in Cubic Yards per 60-min Hour, Bank Measure*

Class of material	Size shovel, cu yd								
	3/8	1/2	3/4	1	1¼	1½	1¾	2	2½
Moist loam or light	3.8	4.6	5.3	6.0	6.5	7.0	7.4	7.8	8.4
sandy clay	85	115	165	205	250	285	320	355	405
Sand and gravel	3.8	4.6	5.3	6.0	6.5	7.0	7.4	7.8	8.4
	80	110	155	200	230	270	300	330	390
Good common earth	4.5	5.7	6.8	7.8	8.5	9.2	9.7	10.2	11.2
	70	95	135	175	210	240	270	300	350
Hard, tough clay	6.0	7.0	8.0	9.0	9.8	10.7	11.5	12.2	13.3
	50	75	110	145	180	210	235	265	310
Well-blasted rock	40	60	95	125	155	180	205	230	275
Wet, sticky clay	6.0	7.0	8.0	9.0	9.8	10.7	11.5	12.2	13.3
	25	40	70	95	120	145	165	185	230
Poorly blasted rock	15	25	50	75	95	115	140	160	195

* SOURCE: Power Crane and Shovel Association.

not be possible for the shovel to fill the dipper in a single cut, which will reduce the output. If the face is too deep, the dipper will be filled before it reaches the top of the face, which will necessitate emptying the dipper and returning to a part face operation. As the angle of swing for the dipper from digging to dumping is increased the time required for a cycle will be increased, which will reduce the output of the shovel.

TABLE 4-5 Conversion Factors for Depth of Cut and Angle of Swing for a Power Shovel*

Percent of optimum depth	Angle of swing, deg						
	45	60	75	90	120	150	180
40	0.93	0.89	0.85	0.80	0.72	0.65	0.59
60	1.10	1.03	0.96	0.91	0.81	0.73	0.66
80	1.22	1.12	1.04	0.98	0.86	0.77	0.69
100	1.26	1.16	1.07	1.00	0.88	0.79	0.71
120	1.20	1.11	1.03	0.97	0.86	0.77	0.70
140	1.12	1.04	0.97	0.91	0.81	0.73	0.66
160	1.03	0.96	0.90	0.85	0.75	0.67	0.62

* SOURCE: Power Crane and Shovel Association.

Fig. 4-3   Power shovel on a typical job. (*Lima-Hamilton Corp.*)

The optimum depth of cut for a power shovel is that depth at which the dipper comes to the surface of the ground with a full dipper without overcrowding or undercrowding the dipper. The optimum depth varies with the class of soil and the size of the dipper. Values of optimum depths for various classes of soils and sizes of dippers are given in Table 4-4.

The outputs for power shovels given in Table 4-4 are based on operating a 60-min hour, at an angle of swing of 90 deg and at optimum depth. Table 4-5 gives factors that correct outputs for other depths and angles of swing.

Example   This example illustrates a method of determining the probable output of a power shovel for the stated conditions.

   Size of shovel, 1 cu yd
   Class of soil, good common earth
   Depth of cut, 9.5 ft
   Angle of swing, 120 deg
   Operating factor, 50-min hr
   From Table 4-4 the ideal output is 175 cu yd per hr
   Optimum depth is 7.8 ft

Percent of optimum depth, $\dfrac{9.5}{7.8} \times 100 = 122$ percent

From Table 4-5 the conversion factor for depth and swing will be 0.86

Probable output, $175 \times 0.86 = 150.5$ cu yd per hr

Probable output corrected for a 50-min hr will be $150.5 \times \dfrac{50}{60} = 125.5$ cu yd, bm

Cost of owning and operating a power shovel The cost of owning and operating a power shovel, as with other construction equipment, includes such items as depreciation, interest, taxes, insurance, maintenance and repairs, fuel, lubrication, greasing, supplies, and labor. Some of these items vary with use, while others do not. Methods of determining the costs of owning and operating power shovels are discussed in Chap. 2.

To estimate the total cost of excavating a given job, it is necessary to determine the cost of transporting the shovel to and from the job, the labor cost of setting the shovel up for operation at the job, the equipment and labor costs of excavating the material, and the labor cost of removing the shovel at the end of the job. Depending on the size of the shovel and job conditions, it is generally necessary to use one or two helpers in addition to the shovel operator. Also a foreman usually supervises the excavating and hauling. A portion of his salary should be charged to excavation.

Hauling excavated materials Trucks and tractor-pulled wagons or trailers are used to haul the materials excavated by power shovels. The capacity of a hauling unit may be expressed in tons or cubic yards. The latter capacity may be expressed as struck or heaped. The struck capacity is the volume which a unit will hold when it is filled even with but not above the sides. This volume depends on the length, width, and depth of the unit. The heaped capacity is the volume which a unit will hold when the earth is piled above the sides. While the struck capacity of a given unit is fixed, the heaped capacity will depend on the depth of the earth above the sides and the area of the bed. Some manufacturers specify the heaped capacity based on a 1:1 slope of the earth above the sides. The actual capacity of a unit should be determined by measuring the volume of earth in several representative loads, then using the average of these values. Units are available with capacities varying from 2 to 30 cu yd or more.

If it is desirable to express the volume in bank measure, a shrinkage factor must be applied to the loose volume. For ordinary earth, whose swell is 25 percent, the loose volume will be 1.25 times the bank volume. The shrinkage factor will be $1 \div 1.25 = 0.8$. Thus a truck hauling 10 cu yd loose measure will have a bank-measure load of $10 \times 0.8 = 8$ cu yd, or it may be expressed as $10 \div 1.25 = 8$ cu yd.

The size of the hauling units should be balanced against the dipper capacity of the shovel. For best results, considering output and economy, the capacity of a hauling unit should be four to six times the dipper capacity.

The volume which a truck can haul in a given time depends on the volume per load and the number of trips it can make in that time. The number of trips depends on the distance, speed, time at loading, time at the dump, and time required for servicing. Higher travel speeds are possible on good open highways than on streets with heavy traffic. For example, top speeds in excess of 50 mph may be possible on some paved highways, whereas speeds on crowded city streets may be not more than 10 to 15 mph. Since delays resulting from lost time at loading, dumping, and servicing trucks may reduce the actual operating time to 45 or 50 min per hr, an appropriate operating factor should be applied in determining the number of trips per unit of time.

Example Estimate the probable total cost and the cost per cubic yard for excavating and hauling 58,640 cu yd, bank measure, of good common earth for the stated conditions.

Pit conditions, good
Shovel, 1-cu-yd diesel-engine-operated crawler type
Trucks, 6 cu yd, loose measure, gasoline engine
Truck time at dump, 4 min
Truck time waiting at shovel to move into loading position, 3 min average
Average truck speed, 30 mph
Average depth of cut, 8 ft
Average angle of swing, 120 deg
Operating factor for shovel and trucks, 45-min hr
Haul distance, one way, 4 miles
Probable ideal output of shovel, 175 cu yd per hr

Percent of optimum depth, $\dfrac{8.0}{7.8} \times 100 = 102.5$

Conversion factor for depth of cut and angle of swing, 0.88
Corrected output, $0.88 \times 175 = 154$ cu yd for 60-min hr[1]

Output for 45-min hr, $\dfrac{45}{60} \times 154 = 116$ cu yd, bm

Volume of truck, $6 \div 1.25 = 4.8$ cu yd bm
Truck cycle time:

Loading truck, $4.8 \div 154^*$	$= 0.031$ hr
Traveling, $8 \div 30$	$= 0.267$ hr
At dump and waiting, $7 \div 60$	$= 0.117$ hr
Round-trip time	$= 0.415$ hr

No. trips per hr, $1 \div 0.415 = 2.41$

No. trips per 45-min hr, $2.41 \times \dfrac{45}{60} = 1.8$

Volume hauled per truck per hr, $1.8 \times 4.8 = 8.44$ cu yd bm
No. trucks required, $116 \div 8.44 - 13.8$

* Note that the time required to load a truck is based on using the corrected output of the shovel for a 60-min hr. The reason for using this rate is that it will be the rate of production for the shovel when it is loading a truck.

Use 14 trucks, possibly another for a standby
Time to complete the job, 58,640 ÷ 116 = 505 hr
The cost will be

Transporting shovel to and from job	= $	280.00
Shovel, 505 hr @ $13.66	=	6,898.30
Trucks, 14 × 505 = 7,070 hr @ $7.12	=	50,338.40
Labor setting up and dismantling shovel:		
Shovel operator, 16 hr @ $6.75	=	108.00
Shovel oiler, 16 hr @ $4.75	=	76.00
Laborer, 16 hr @ $3.25	=	52.00
Foreman, 16 hr @ $7.75	=	124.00
Labor excavating and hauling earth:		
Shovel operator, 505 hr @ $6.75	=	3,408.75
Shovel oiler, 505 hr @ $4.75	=	2,398.75
Truck drivers, 7,070 hr @ $3.75	=	26,512.50
Laborer, 505 hr @ $3.25	=	1,641.25
Foreman, 505 hr @ $7.75	=	3,913.75
Total cost	=	$95,751.70
Cost per cu yd, $95,751.70 ÷ 58,640	=	1.64

Excavating with draglines For excavating on some projects, the dragline is more suitable than the power shovel. When digging ditches or building levees, it can excavate and transport the earth within casting limits, thus eliminating hauling equipment. It can operate on wet ground and can dig earth out of pits containing water. It cannot excavate rock as well as can a power shovel. It has a lower output than a power shovel of the

TABLE 4-6   Ideal Output of Short-boom Draglines, in Cubic Yards per 60-min Hour, Bank Measure*

Class of material	Size of bucket, cu yd								
	⅜	½	¾	1	1¼	1½	1¾	2	2½
Moist loam or light	5.0	5.5	6.0	6.6	7.0	7.4	7.7	8.0	8.5
sandy clay	70	95	130	160	195	220	245	265	305
Sand and gravel	5.0	5.5	6.0	6.6	7.0	7.4	7.7	8.0	8.5
	65	90	125	155	185	210	235	255	295
Good common earth	6.0	6.7	7.4	8.0	8.5	9.0	9.5	9.9	10.5
	55	75	105	135	165	190	210	230	265
Hard, tough clay	7.3	8.0	8.7	9.3	10.0	10.7	11.3	11.8	12.3
	35	55	90	110	135	160	180	195	230
Wet, sticky clay	7.3	8.0	8.7	9.3	10.0	10.7	11.3	11.8	12.3
	20	30	55	75	95	110	130	145	175

* SOURCE: Power Crane and Shovel Association.

Fig. 4-4   Dragline on a typical job. (*Lima-Hamilton Corp.*)

same size. Many units can be converted from power shovels to draglines by changing the boom and substituting a bucket for the shovel dipper. The cost of a dragline is somewhat less than that for a shovel of the same size. The hourly operating costs are 5 to 10 percent less than for a shovel. The size of a dragline is indicated by the size of the bucket, expressed in cubic yards.

Table 4-6 gives ideal outputs for short-boom draglines when excavating at optimum depth with an angle of swing of 90 deg, based on a 60-min hour. Table 4-7 gives factors that may be used to determine outputs for other depths and angles of swing. The production determined from Tables 4-6 and 4-7 must be corrected for an operation less than 60 per hr.

Example   Estimate the total cost and the cost per cubic yard for excavating a drainage ditch 12 ft wide at the bottom, 36 ft wide at the top, 12 ft deep, and 12,280 ft long in good common earth. The excavated earth can be cast on one or both sides of the ditch. Use a 1½-cu-yd diesel-engine-powered crawler-mounted dragline. The average angle of swing will be 120 deg.

The volume of earth excavated will be $\dfrac{12 + 36}{2} \times 12 \times 12{,}280 \div 27 = 129{,}186$

cu yd bm

Reference to Table 4-6 indicates a percent of optimum depth of $\dfrac{12.0}{9.0} \times 100 = 1.33$

From Table 4-6 the ideal output will be 190 cu yd per hr bm

TABLE 4-7   The Effect of the Depth of Cut and Angle of Swing on the Output of Draglines*

Percent of optimum depth	Angle of swing, deg							
	30	45	60	75	90	120	150	180
20	1.06	0.99	0.94	0.90	0.87	0.81	0.75	0.70
40	1.17	1.08	1.02	0.97	0.93	0.85	0.78	0.72
60	1.24	1.13	1.06	1.01	0.97	0.88	0.80	0.74
80	1.29	1.17	1.09	1.04	0.99	0.90	0.82	0.76
100	1.32	1.19	1.11	1.05	1.00	0.91	0.83	0.77
120	1.29	1.17	1.09	1.03	0.98	0.90	0.82	0.76
140	1.25	1.14	1.06	1.00	0.96	0.88	0.81	0.75
160	1.20	1.10	1.02	0.97	0.93	0.85	0.79	0.73
180	1.15	1.05	0.98	0.94	0.90	0.82	0.76	0.71
200	1.10	1.00	0.94	0.90	0.87	0.79	0.73	0.69

* SOURCE: Power Crane and Shovel Association.

From Table 4-7 the depth-swing factor will be 0.89

Applying these quantities gives an output of 0.89 × 190 = 169.1 cu yd per hr for a 60-min hr

If the dragline operates an average of 50 min per hr, the corrected output will be 50/60 × 169.1 = 141 cu yd per hr bm

The time required to excavate the earth will be 129,186 ÷ 141 = 916 hr

The cost will be

Transporting dragline to and from job	= $	265.00
Dragline, 916 hr @ $24.05	=	22,029.80
Labor setting up and removing dragline:		
Operator, 16 hr @ $6.75	=	108.00
Oiler, 16 hr @ $4.75	=	76.00
Helper, 16 hr @ $3.25	=	52.00
Foreman, 16 hr @ $7.75	=	124.00
Labor excavating:		
Operator, 916 hr @ $6.75	=	6,183.00
Oiler, 916 hr @ $4.75	=	4,351.00
Helper, 916 hr @ $3.25	=	2,977.00
Foreman, 916 hr @ $7.75	=	7,099.00
Total direct cost	=	$43,264.80
Cost per cu yd, $43,264.80 ÷ 129,186	=	0.334

Handling materials with a clamshell   Most draglines in the popular sizes can be converted into clamshells by replacing the dragline bucket with a clamshell bucket. The size of a clamshell is indicated by the size of the bucket, expressed in cubic yards.

The clamshell is a satisfactory machine for handling sand, gravel, crushed stone, sandy loam, and other loose materials. Because of the difficulty which it encounters in loosening solid earth, it is not very satisfactory for handling compacted earth, clay, and other solid materials.

The bucket is lowered into the material to be handled with the jaws open. Its weight will cause the bucket to sink into the material as the jaws are closed. Then it is lifted vertically, swung to the emptying position, over a truck, to a spoil pile or elsewhere, and the jaws are opened to permit the load to flow out. All these operations are controlled by the clamshell operator. The output of the clamshell is affected by the looseness of the materials being handled, the type of material, the height of lift, the angle of swing, the method of disposing of the materials, and the skill of the operator. Table 4-8 gives the probable output of clamshells of different sizes for various angles of swing and materials. These rates should apply for jobs which do not require frequent interruptions.

The hourly cost of owning and operating a clamshell is about the same as for a dragline of the same size.

Excavating and hauling earth with tractors and scrapers Tractor-pulled scrapers are used to excavate and haul earth for such projects as dams, levees, highways, airports, and canals. Since these units perform both excavating and hauling operations, they are independent of the operations of other equipment. Thus, if one of several units breaks down, the rest of the units can continue to operate, whereas if a power shovel breaks down, the entire project must stop until the shovel is repaired.

TABLE 4-8 Representative Hourly Output of Clamshells, in Cubic Yards

Size bucket, cu yd	Angle of swing, deg	Material		
		Light loam	Sand gravel	Crushed stone
½	45	48	43	38
	90	40	36	31
	180	31	28	24
¾	45	63	56	49
	90	53	48	42
	180	41	37	32
1	45	81	73	63
	90	68	61	53
	180	54	48	42
2	45	134	120	104
	90	113	102	88
	180	87	78	68

These units are rugged and can operate under adverse conditions. They are available in a wide range of sizes and capacities. The capacity of a scraper is designated by the volume of earth that it will carry, either struck or heaped, expressed in cubic yards, loose measure. For example, the capacity might be designated as 12 cu yd struck, 15 cu yd heaped.

Crawler-tractor-pulled scrapers For short-haul distances, the crawler-type tractor, pulling a rubber-tired self-loading scraper, is sometimes used. The crawler tractor has a high drawbar pull for loading the scraper, has good traction with the ground, and can operate over muddy haul roads, but it has a low travel speed, which is a disadvantage for long hauls. The top speed is approximately 6 to 7 mph depending on the unit.

Although it is usually desirable to provide an auxiliary tractor to help during the loading operation, a crawler tractor can load a scraper without additional help, but at a reduced rate.

The size of a crawler tractor usually is expressed in terms of the drawbar horsepower, which is the power available at the drawbar, when it is operated at sea level on a level haul road having a rolling resistance of 110 lb per ton of gross load.

Wheel-type tractor-pulled scrapers For longer haul distances, in excess of approximately 600 ft, the wheel-type tractor, pulling a rubber-tired self-loading scraper, is more economical than the crawler-tractor-pulled unit. While the wheel-type tractor cannot deliver as great a tractive effort in

Fig. 4-5 Crawler tractor and scraper. (*Caterpillar Tractor Company.*)

Fig. 4-6 Two-wheeled tractor-pulled scraper. (*Euclid Division of General Motors Corp.*).

loading a scraper, the higher travel speed, which may exceed 30 mph for some units, will offset the disadvantage in loading when the haul distance is sufficiently long. Both two- and four-wheel tractors are available. A helper tractor, such as a bulldozer, should be used to help load the scrapers.

The size of a rubber-tired tractor usually is designated by the brake horsepower of the engine.

Cost of owning and operating a wheel-type tractor and scraper This example illustrates a method of determining the cost of owning and operating a wheel-type tractor and scraper. The unit is similar to the one illustrated in Fig. 4-6. The specified cost includes freight charges from the factory to the owner.

Because the life of the tires will usually be different from that of the basic unit, the cost of the tires should be estimated separately. The following conditions will apply.

Tractor, diesel engine, 275 fwhp
Scraper, 14 cu yd, struck capacity
Life of basic unit, 5 yr at 2,000 hr per yr
Salvage value after 5 yr, none
Life of tires, 5,000 hr
Cost of unit fob owner   = $71,685
Cost of tires   =   9,845
                                                                       ————————
Cost less tires   = $61,840
Average investment, 0.6 × $71,685   =   43,011
Annual cost:
   Depreciation, $61,840 ÷ 5   = $12,368

Maintenance and repairs, 75% of depreciation	=	9,276
Investment, 14% × $43,011	=	6,021
Total annual cost	=	$27,665
Hourly cost:		
Fixed cost, $27,665 ÷ 2,000	=	$13.88
Fuel, 9 gal @ $0.40 per gal	=	3.60
Lubricating oil, 0.4 gal @ $1.00 per gal	=	0.40
Other lubrication, 50% of $0.40	=	0.20
Depreciation of tires, $9,845 ÷ 5,000 hr	=	1.96
Tire repairs, 15% of $1.96	=	0.30
Total cost per hr	=	$20.34

Production rates for tractor-pulled scrapers   The production rate for a tractor-pulled scraper will equal the number of trips per hour times the net volume per trip. An appropriate operating factor, such as a 45- or 50-min hr, should be used in determining the number of trips per hour, to allow for nonproductive time.

Frequently it is desirable to express the speed of a tractor in feet per minute as well as miles per hour. A speed of 1 mph is equal to 5,280 ft in 60 min, which equals 88 fpm. Table 4-9 gives speeds in miles per hour and feet per minute.

Example   Determine the probable production rate for a crawler tractor, with 125 drawbar hp, and a wheel-type scraper, whose struck capacity is 10 cu yd. Assume that the scraper load will be heaped to give a volume of 11 cu yd. The material will be ordinary earth.

Haul distance, 500 ft at average speed of 4 mph
Return distance, 600 ft at average speed of 4.5 mph
Assume a 50-min hr
Net load, 11 ÷ 1.25 = 8.8 cu yd, bm

The fixed time per trip will be

Loading, 100 ft ÷ 88 fpm	= 1.14 min
Dumping	= 0.80 min
Turning and delays	= 0.36 min
Total fixed time	= 2.30 min

The travel time will be

Hauling, 500 ft ÷ 352 fmp	= 1.42 min
Returning, 600 ft ÷ 396 fpm	= 1.52 min
Total time per trip	= 5.24 min

Trips per hr, 50 ÷ 5.24 = 9.5
Volume per hr, 9.5 trips × 8.8 cu yd = 83.5 cu yd

TABLE 4-9   Speeds in Miles per Hour and Feet per Minute

Mph	Fpm	Mph	Fpm
1	88	7	616
2	176	8	704
3	264	9	792
4	352	10	880
5	440	15	1,320
6	528	20	1,760

Example   Determine the probable production rate for a wheel-type tractor with a 225-fwhp diesel engine, and a wheel-type scraper, whose struck capacity is 11 cu yd. The scraper will be loaded to an average heaped capacity of 13 cu yd, loose measure. The material will be good common earth, with a swell of 25 percent. While the tractor has a maximum speed of 23 mph, the average speed on a construction road will be considerably less, especially for short hauls over poorly maintained earth roads. The following conditions will apply to the unit.

> Haul distance, 860 ft at average speed of 9 mph loaded
> Return distance, 940 ft at average speed of 11 mph
> Net pay load, 13 ÷ 1.25 = 10.4 cu yd bm
> Assume a 45-min operating hour

> The time per round trip will be
> Loading, 148 ft ÷ 88 fpm   = 1.69 min
> Hauling, 860 ft ÷ 792 fpm   = 1.09 min
> Dumping   = 0.75 min
> Turning and delays   = 0.50 min
> Returning, 940 ft ÷ 968 fpm   = 0.97 min
> ────────
>   Total round-trip time   = 5.00 min
> No. trips per hr, 45 ÷ 5 = 9
> Volume hauled per hr, 9 × 10.4 = 93.6 cu yd bm

Cost of excavating and hauling earth with tractors and scrapers
The cost of handling earth with tractors and scrapers may be determined by assuming that the equipment will be operated 1 hr for quantity and cost purposes.

If a bulldozer is used to assist several scraper units while they are being loaded, the cost of the bulldozer should be distributed equally among the scraper units that are assisted. The number of scrapers that a bulldozer can assist will be equal to the total cycle time for a scraper to load, haul, dump, and return to the pit, divided by the time required by the bulldozer to assist in loading a scraper and get into position to load another scraper.

Consider the scraper unit in the example in the previous article.

The round-trip time is 5.00 min
The loading time is 1.69 min

If the round-trip time, 5.00 min, is divided by the loading time, 1.69 min, it appears that a bulldozer should be able to assist 2.96 scrapers. However, when a bulldozer finishes assisting one scraper, the bulldozer must move into position to assist the next scraper, which may require the bulldozer to travel more than 100 ft for its new position. For this reason it appears that the bulldozer cannot assist more than two scrapers. Thus one-half the cost of a bulldozer will be charged to each scraper.

The cost per hour will be
Tractor-scraper, 1 hr @ $14.25	=	$14.25
Bulldozer, 0.5 hr @ $14.40	=	7.20
Scraper operator, 1 hr @ $6.20	=	6.20
Bulldozer operator, 0.5 hr @ $6.20	=	3.10
Total cost	=	$30.75

Volume hauled per hr, 93.6 cu yd
Cost per cu yd, $30.75 ÷ 93.6 cu yd = $ 0.329

Fig. 4-7 Motor grader with tandem drive. (*Galion Iron Works & Mfg. Co.*)

Fig. 4-8 Three-drum sheep's-foot roller. (*Baker Manufacturing Co.*)

Shaping and compacting earthwork When earth is placed in a fill, it is necessary to spread it in uniformly thick layers and compact it to the specified density. Unless sufficient moisture is present, water should be added to produce the optimum moisture content, which will permit more effective compaction. Spreading may be accomplished with graders or bulldozers, or both, while compaction may be accomplished with tractor-pulled sheep's-foot rollers, smooth-wheel rollers, pneumatic rollers, vibrating rollers, or other types of equipment. For some projects the best results are obtained by using more than one type of equipment. Regardless of the type of equipment selected, there should be enough units to shape, wet, and compact the earth at the rate at which it will be delivered.

Example Estimate the total cost per cubic yard for shaping, sprinkling, and compacting earth in a fill such as a dam or a highway. The earth will be placed by tractor-scrapers at the rate of 396 cu yd per hr, with the volume measured after compaction. The earth will be placed in layers not to exceed 6 in. thick when compacted. The moisture content, probably averaging 8 percent when the earth is placed in the fill, must be increased to 12 percent. The water will be placed with truck-mounted sprinklers.

For the stated rate of placing the earth the surface area placed in 1 hr will be

$$\frac{396 \text{ cu yd} \times 27}{0.5} = 21{,}384 \text{ sq ft}$$

If it requires three passes of a grader to smooth each layer of earth satisfactorily, the area which the grader must cover in an hour will be $3 \times 21{,}384 = 64{,}152$ sq ft. A motor grader with a 12-ft blade should cover a strip or lane whose effective width will be about 8 ft for each pass of the grader. The grader should be able to travel at an average speed of at least 2 mph, allowing for stops and lost time turning around.

The area covered in 1 hr by a grader should be $2 \times 5{,}280 \times 8 = 84{,}500$ sq ft. Thus one grader should be sufficient to do the job.

The earth will be compacted using sheep's foot rollers pulled by a crawler tractor. It is estimated that 10 passes will be required to attain the specified compaction. The equivalent area that the rollers must cover in 1 hr will be $10 \times 21{,}384 = 213{,}840$ sq ft.

If roller drums 4 ft long are operated at an average speed of 2-¾ mph, allowing for lost time, the area covered by one drum in 1 hr will be $2\text{-}\frac{3}{4} \times 5{,}280 \times 4 = 58{,}080$ sq ft.

The number of drums required will be $213{,}840 \div 58{,}080 = 3.68$

It will be necessary to use four drums, which will be pulled by one tractor.

The compacted earth will weigh 3,150 lb per cu yd. The weight of water required per hour will be 4 percent of the weight of the earth placed $= 0.04 \times 396 \times 3{,}150 = 49{,}896$ lb, equal to $49{,}896 \div 8.33 = 5{,}990$ gal per hr. If a 2,000-gal sprinkler truck can make one trip in 40 min, two trucks will be required.

The cost per hour will be
Motor grader	= $ 8.65
Tractor and sheep's foot roller	= 12.60
Sprinkler trucks, 2 @ $7.45	= 14.90
Grader operator	= 5.95
Tractor operator	= 5.95
Truck drivers, 2 @ $3.75	= 7.50
Foreman	= 7.75
Total cost per hr	= $63.30
Cost per cu yd of earth, $63.30 ÷ 396 =	0.16

Drilling and blasting rock Before rock can be excavated, it must be loosened and broken into pieces small enough to be handled by the excavating equipment. The most common method of loosening is to drill holes into which explosives are placed and detonated.

Holes may be drilled by one or more of several types of drills, such as jackhammers, wagon drills, drifters, churn drills, rotary drills, etc., with the selection of equipment based on the size of the job, the type of rock, the depth and size holes required, the production rate required, and the topography at the site.

Jackhammers may be used for holes up to about 2½ in. in diameter and 15 to 18 ft deep. For deeper holes, the production rates are low and the costs are high.

Wagon drills may be used for holes 2 to 4½ in. in diameter, with depths sometimes as great as 40 ft, although shallower depths are more desirable.

Drifters are used to drill approximately horizontal holes, up to about 4 in. in diameter, in mining and tunneling operations.

Jackhammers, wagon drills, and drifters are operated by compressed air, which actuates the hammer that produces the percussion which disintegrates the rock and blows it out of the holes. Replaceable bits, which are attached to the bottoms of the hollow drill steels, are commonly used.

A churn drill disintegrates the rock by repeated blows from a heavy steel bit, which is suspended from a wire rope. Holes in excess of 12 in. in diameter may be drilled several hundred feet deep with this equipment.

Water, which is placed in a hole during the drilling operation, will produce a slurry with the disintegrated rock. A bailer is used to remove the slurry.

Rotary drills may be used to drill holes 3 to 8 in. or more in diameter to depths in excess of 100 ft. Drilling is accomplished by a bit, which is attached to the lower end of a drill stem. Either water or compressed air may be used to remove the rock cuttings.

Dynamite is frequently used as the explosive, although several other types of explosives are available. Dynamite is available in sticks of varying sizes, which are placed in the holes. The strength, which is specified as 40, 60, etc., percent, indicates the concentration of the explosive agent, which is nitroglycerin. The dynamite is usually exploded by a blasting cap, which is

Fig. 4-9 Jackhammer on curtain wall cut. (*Ingersoll-Rand Co.*)

Fig. 4-10 Wagon drill. (*Gardner-Denver Company.*)

detonated by an electric current. At least one blasting cap is required for each hole. The charges in several holes may be shot at one time.

The amount of dynamite required to loosen rock will vary from about ¼ to more than 1 lb per cu yd, depending on the type of rock, the spacing of holes, and the degree of breakage desired.

Cost of operating a drill The items of cost in operating a jackhammer or a wagon drill will include equipment and labor.

The equipment cost will include the drill, drill steel, bits, air compressor, and hose. Since each of these items may have a different life, it should be priced separately. Appendix B gives costs for owning and operating drills, air compressors, and hose.

Drill steel is purchased by size, length, and quality of steel used, with the cost of steel based on its weight. While the life of a drill steel will vary with the class of rock drilled and the conditions under which the steel is used, records from drilling projects show consumptions varying from $\frac{1}{20}$ to $\frac{1}{10}$ lb per cu yd of rock to be representative, with the higher consumption applicable to the harder rocks. Thus, if a drill steel for a wagon drill, 12 ft long, weighing 4.6 lb per lin ft, whose total weight is about 55 lb, is consumed at the rate of $\frac{1}{12}$ lb per cu yd of rock, this steel should drill enough hole to produce about 660 cu yd of rock.

Detachable bits are commonly used with jackhammers, drifters, and wagon drills. The depth of hole that may be drilled with a bit before it must be resharpened or discarded will vary considerably with the class of rock and the type of bit, with values ranging from less than a foot to as much as 100 ft or more for steel bits. Bits with carbide inserts will give much greater depths.

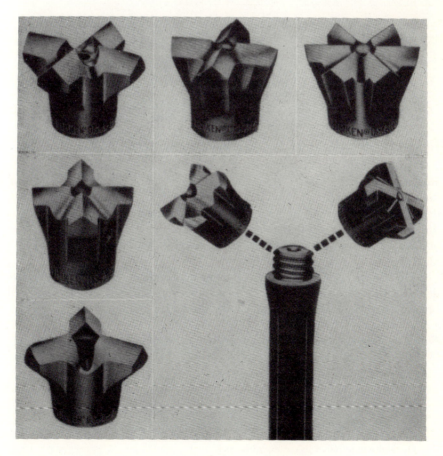

Fig. 4-11 Multiuse rock bits. (*Timken Roller Bearing Co.*)

Fig. 4-12 Blasthole drill. (*Davey Compressor Co.*)

Rates of drilling holes The rate of drilling rock will vary with several factors, including the type and hardness of the rock, type and size drill used, depth of holes, spacing of holes, topography at the site, condition of the drilling equipment, etc. Although the rates given in Table 4-10 are based on observations, they should be used as a guide only. The rates include an allowance for lost time at the job.

Holes should be drilled 1 ft or more deeper than the desired effective depth of rock loosened. This is necessary because the rock usually will not break to the full depth of the holes over the entire area blasted.

Example Estimate the cost of drilling and blasting limestone rock for the stated conditions.

Diameter of holes, 3 in.
Depth of holes, 14 ft
Effective depth of holes, 13 ft
Spacing of holes, 8 ft × 8 ft
Two medium-size wagon drills used, each with 50 ft of 1½-in. hose and connections
Dynamite required, 1 lb per cu yd of rock

TABLE 4-10 Representative Rates of Drilling Rock with Various Types of Drills

Size hole, in.	Class of rock	Rate of drilling, ft per hr				
		Jack-hammer	Wagon drill	Churn drill	Rotary drill	Diamond drill
1¾	Soft	15–20	30–45	...	.....	5–8
	Medium	10–15	25–35	...	.....	3–5
	Hard	5–10	15–30	...	.....	2–4
2⅜	Soft	10–15	30–50	...	.....	5–8
	Medium	7–10	20–35	...	.....	3–5
	Hard	4–8	15–30	...	.....	2–4
3	Soft	.....	30–50	...	.....	4–7
	Medium	.....	15–30	...	.....	3–5
	Hard	.....	8–20	...	.....	2–4
4	Soft	.....	10–25	...	.....	3–6
	Medium	.....	5–15	...	.....	2–4
	Hard	.....	2–8	...	.....	1–3
6	Soft	.....	.....	4–7	25–50	3–5
	Medium	.....	.....	2–5	10–25	2–4
	Hard	.....	.....	1–2	6–10	1–3

Cost of bits, $4.55 each
No. times bits sharpened, 3
Cost of sharpening bit, $1.10 each time
Depth of hole before resharpening bit, 36 ft
Drill steel consumed, $\frac{1}{15}$ lb per cu yd of rock
Estimated rate of drilling, 24 ft per hr
Base all quantities and costs on operating 1 hr.
Total depth of hole drilled, $2 \times 24 = 48$ ft
Effective depth of hole drilled, $48 \times {}^{13}\!/_{14} = 44.6$ ft

Volume of rock produced $\dfrac{44.6 \times 8 \times 8}{27} = 106$ cu yd

No. holes drilled, 48 ft ÷ 14 ft per hole = 3.44

The cost per hour will be
Dynamite, 106 cu yd × 1 lb = 106 lb @ $0.36	= $ 38.16
Electric caps, 3.44 @ $0.42	= 1.44
Electric wire	= 0.20
Air compressor, 1 hr @ $14.10	= 14.10
Hose, 2-in. diameter, 1 hr @ $0.56	= 0.56
Wagon drills, 2 hr @ $1.36	= 2.72
Air hoses, 2 hr @ $0.09	= 0.18
Drill steel, 106 cu yd × $\frac{1}{15}$ lb = 7.1 lb @ $0.35	= 2.54

Drill bits:

Original cost each,	=	$4.55
Sharpening, 3 × $1.10	=	3.30
Total cost per bit	=	$7.85

Total depth of hole drilled by one bit, 4 × 36 = 144 ft
Cost of bit per ft of hole, $7.85 ÷ 144 ft = $0.055

Cost of bits per hr, 48 × $0.055	=	2.64
Drill operators, 2 hr @ $6.75	=	13.50
Helpers 2 hr @ $4.25	=	8.50
Powderman, 1 hr @ $5.25	=	5.25
Helper, 1 hr @ $3.25	=	3.25
Foreman, 1 hr @ $7.25	=	7.25
Total cost	=	$100.29
Cost per cu yd, $100.29 ÷ 106 cu yd	=	$ 0.94

PROBLEMS

4-1 A truck will haul an average load of 12.5 cu yd, heaped capacity, loose measure per load. Determine the volume in bank measure when the material is sand, ordinary earth, dense clay, and well-blasted rock.

4-2 Estimate the total direct cost and the cost per linear foot for excavating a sewer trench 33 in. wide, 10 ft 6 in. deep, and 1,480 ft long in ordinary earth, using a ladder-type trenching machine. There are no obstructions to delay the machine. Assume a 50-min hr.

Use the information given in the text to estimate the digging speed of the trenching machine, and the information in Appendix B for the cost of the machine.

It will cost $180.00 to transport the machine to the job, get it ready to operate, then return it to the storage yard.

The following crew at the specified hourly wage rates, will be required:

Machine operator	$6.80
Machine oiler	4.75
Laborers, 3 each	3.25
Foreman	7.50

4-3 What is the maximum length of trench that can be excavated by hand economically rather than by a ladder-type trenching machine for the stated conditions?

Width of trench, 3 ft 0 in.
Average depth, 5 ft 6 in.
Class of soil, ordinary earth

For hand excavation it will be necessary to shovel the earth back from the edge of the trench. Assume that 10 laborers plus a foreman will be used.

It will cost $220.00 to transport a trenching machine to the job and back to the storage yard.

Use the information given in the book to estimate the rate of excavating the earth by hand, the probable speed of the trenching machine, and the cost of the machine. Assume that the machine will operate 50 min per hr. Labor costs per hour will be

Machine operator	$6.80
Machine oiler	$4.75
Laborers, each	$3.25
Foreman	$7.50

4-4 In excavating for a basement of a building, a contractor will use a ¾-cu-yd crawler-type shovel, powered by a 120-fwhp diesel engine to load earth into 6-cu-yd gasoline-engine-powered trucks, which will haul the earth to a dump. The earth is hard, tough clay.

The dimensions of the pit are 86 ft 0 in. wide, 114 ft 0 in. long, and 10 ft 0 in. deep.

The earth will be hauled to a dump 3 miles from the pit at an average hauling speed of 30 mph and a return speed of 40 mph. In addition to the time required to load a truck, haul the load to the dump, and return to the pit, a truck will consume an average of 8 min each trip at the dump and waiting at the pit to be loaded.

It will be necessary to use a 70-hp diesel-engine-operated crawler tractor to assist the trucks in getting up a ramp out of the pit.

The average angle of swing for the shovel will be 120°.

Assume that all equipment will work 50 min per hr.

It will cost $120.00 to transport the shovel to the job and back to the storage yard.

Use the information available in the text to determine the total cost and the cost per cubic yard, bank measure, for excavating and hauling the earth. The labor costs per hour will be

Shovel operator	$6.75
Shovel oiler	$4.75
Truck drivers, each	$4.25
Laborers, 3, each	$3.75
Foreman	$7.50

4-5 A project requires 486,520 cu yd, bank measure, of dense clay earth, which will be excavated by a 2-cu-yd diesel-engine-powered crawler-type power shovel, loaded into 10-cu-yd, loose measure, diesel-engine-powered dump trucks, and hauled 4 miles to the project. The trucks will average traveling 30 mph loaded and 40 mph empty.

The shovel can excavate earth at the optimum depth and at 120° average angle of swing.

Assume that the shovel and the trucks will operate 45 min per hr.

It will cost $240.00 to transport the shovel to the job, set it up to operate, and return it to the storage yard.

Use information available in the book to estimate the total cost of the job and the cost per cubic yard for excavating and hauling the earth. Labor costs per hour will be

Shovel operator	$6.75
Shovel oiler	$4.75
Truck drivers, each	$4.25
Laborers, 2, each	$3.75
Foreman	$7.50

It is estimated that the average truck time at the dump, waiting to be loaded at the shovel and for other delays will amount to 8 min per round trip.

4-6 A 1½-cu-yd diesel-engine-powered crawler-type dragline will be used to excavate a ditch in dense clay. The ditch will be trapezodial in cross section, 16 ft 0 in. wide at the bottom, 36 ft 0 in. wide at the top, 8 ft 0 in. deep and 6,480 ft long.

There will be no obstructions to affect the rate of excavating the earth. The excavated earth will be cast along the edge or edges of the ditch. The average angle of swing will be 145°.

Use of information available in the text to estimate the total cost and the cost per cubic yard, bank measure, for excavating the ditch.

Assume a 50-min hr.

It will cost $320.00 to transport the dragline to the job, set it up for operation, and return it to the storage yard. Labor costs per hour will be

Dragline operator	$6.75
Dragline oiler	$4.75
Laborer, 1 man	$4.25
Foreman	$7.50

4-7 Ordinary earth will be excavated and hauled by diesel-engine-powered wheel-type scrapers, whose capacities are 14 cu yd struck and 18 cu yd heaped, loose measure. The distance from the borrow pit to the dump site will average 2,680 ft over level earth haul road. The scrapers can average 15 mph loaded and 20 mph empty.

A 270-fwhp crawler tractor will be used to assist the scrapers in loading. Use the example on page 66 of this book to determine the number of scrapers that one crawler tractor can serve. *Note:* Generally it should require from 50 to 70 sec to load a scraper if adequate tractor power is provided. Thus a bulldozer tractor should be able to serve a scraper about every 2½ to 3 min.

Use one bulldozer and the appropriate number of scrapers to estimate the cost per cubic yard for excavating and hauling the earth. Base your costs on bank-measure volume.

The fixed time for the scrapers, including loading, dumping, turning, waiting to load or dump, etc., should average about 2½ min per trip.

Assume a 45-min hr. Each tractor operator will be paid $6.75 per hr.

4-8 In constructing a fill it is necessary to excavate ordinary earth and haul it 1,800 ft along an approximately level haul road. The earth will be excavated and hauled with units, each consisting of a two-wheel tractor and a two-wheel scraper. Each scraper will haul 15 cu yd, loose measure, per load and will be pulled by a 180-fwhp tractor. One 180-hp crawler-type tractor will be required to assist the four scraper units during loading.

The scrapers will average 12 mph loaded and 18 mph empty. The fixed time, including loading, dumping, and waiting in the pit will average 2½ min per trip. Assume a 50-min working hour.

Use the information given in Appendix B to determine the probable direct cost per cubic yard, bank measure, for excavating and hauling the earth. Labor costs per hour will be

Tractor operators, each	$6.75
Foreman	$7.50

4-9 Estimate the total direct cost and the cost per cubic yard for drilling and blasting limestone rock, classified as medium rock, for a pit 80 ft wide, 108 ft long, and 12 ft deep. Holes 2½ in. in diameter will be spaced in patterns 6 by 8 ft, over the entire area, with the outside holes being along the sides and ends of the pit. The holes will be drilled 13 ft deep.

The holes will be drilled with two wagon drills, each with 50 ft of 1½-in. hose with fittings. The air will be supplied by a 600-cfm compressor powered with a diesel engine, equipped with 100 ft of 2-in. hose.

It is estimated that drill steel will be consumed at a rate of 1 lb per 20 cu yd of rock. Each bit is expected to drill 32 ft of hole, after which it will be discarded.

It will require ¾ lb of dynamite for each cubic yard of rock blasted.

Assume a 45-min working hour.

Use the information in this book to determine the rate of drilling and the costs of equipment. Other costs and labor costs per hour will be

Transporting equipment to and from the job	$180.00
Drill steel, per lb	$0.52
Bits, each	$1.60
Dynamite, per lb	$0.35
Electric caps, each	$0.38
Generator and lead wire for the job	$12.50
Compressor operator	$5.75
Drill operators, 2, each	$6.75
Drill helpers, 2, each	$4.75
Powderman	$5.75
Powderman helper	$4.25
Foreman	$8.25

4-10 In operating a quarry for the production of crushed limestone, two heavy wagon drills will be used to drill holes 3 in. in diameter, 16 ft deep, spaced 9 ft apart each way. The effective depth of the holes will be 15 ft.

One 600-cfm portable air compressor, operated by a diesel engine, will be used to supply air for the drills. Each drill will require 50 ft of 1½-in. air hose with fittings.

The consumption of drill steel will be $\frac{1}{12}$ lb per cu yd of rock produced.

Each bit will drill 36 ft of hole, after which it will be resharpened for four additional uses.

It will require ¾ lb of dynamite for each cubic yard of rock produced. One detenator cap will be required for each hole.

Use the information in this book, including Appendix B, to determine the direct cost per cubic yard for drilling and blasting the rock.

As the equipment will be left at the quarry, there will be no cost for transporting it to and from the job.

Use the following costs for supplies and labor.

Drill steel, per lb	$0.48
Bits, each	$2.20
Sharpening each bit	$0.95
Dynamite, per lb	$0.35
Detonator caps, each	$0.38
Air compressor operator, 1,	$5.75
Drill operators, 2, each	$6.75
Helpers, 2, each	$3.75
Powderman, 1	$5.75
Powderman helper, 1	$4.25
Foreman	$8.25

5

Highways and Pavements

Operations included Operations to be discussed in this chapter include clearing and grubbing land and placing concrete and asphalt pavements. Even though the coverage is limited to only a few of the methods used, the discussions and examples presented should illustrate how estimates can be prepared for projects constructed by other methods.

CLEARING AND GRUBBING LAND

Land-clearing operations Clearing land may be divided into several operations, depending on the type of vegetation to be removed, the type and condition of the soil and topography, the amount of clearing required, and the purpose for which the clearing is done, as listed below.

    1 Complete removal of all trees and stumps, including tree roots
    2 Removal of all vegetation above the surface of the ground only, leaving the stumps and roots in the ground
    3 Disposal of the vegetation by stacking and burning it

Types of equipment used Several types of equipment are available for use in clearing land, including the following:

1 Tractor-mounted bulldozers
2 Tractor-mounted special cutting blades
3 Tractor-mounted rakes

Tractor-mounted bulldozers Whereas bulldozers were used extensively in the past to clear land, they are now being replaced by special blades mounted on tractors. There are some objections to using bulldozers for this work. Before felling large trees, the bulldozers must excavate the earth from around the trees and cut the main roots, which leaves objectionable holes in the ground and requires considerable time. Also, when stacking the felled trees and other vegetation they transport earth to the piles of trees, which makes burning more difficult.

Tractor-mounted special blades Two types of special blades are used to fell trees, both of which are mounted on the front ends of tractors.

One is a single-angle blade with a projecting stinger on the lead side, extending ahead of the blade, so that it may be forced into and through the tree to split and weaken it. If a tree is too large to be felled in one pass, the trunk may thus be split and removed in parts. Also, the tractor may

Fig. 5-1 Land-clearing blade piercing tree with stinger. (*Rome Plow Company.*)

Fig. 5-2 Tractor-mounted V blade for clearing land. (*Fleco Corp.*)

make a pass around the tree with the stinger penetrating the ground to cut the main horizontal roots of the tree. Figure 5-1 illustrates this blade in use. The unit may be used to remove stumps and to stack material for burning.

Another type of special blade is a V blade, with a protruding stinger at its lead point, as illustrated in Fig. 5-2. The sole effect of the blade permits it to slide along the surface of the ground, thereby cutting vegetation flush with the surface. However, it can be lowered below the surface of the ground to remove stumps. Also, the blade may be raised to permit the stinger to pierce the tree above the ground, as illustrated in Fig. 5-3.

Tractor-mounted rakes Figure 5-4 illustrates a tractor-mounted rake which can be used to grub and pile trees, boulders, and other materials without transporting excessive amounts of earth. This can be a very effective machine for stacking materials in piles for burning.

Disposal of brush When brush is to be disposed of by burning, it should be piled in stacks and windrows, with a minimum amount of earth included. Shaking the rake while it is moving the brush will help remove the earth.

Fig. 5-3   Tractor-mounted V blade splitting a large tree. (*Fleco Corp.*)

Fig. 5-4   Tractor-mounted land-clearing rake. (*Fleco Corp.*)

Fig. 5-5 Burning brush with forced draft and fuel oil. (*Fleco Corp.*)

Because burning is usually necessary while the brush contains considerable moisture, it may be desirable to provide a continuous external source of fuel and heat to assist in burning the material. The burner illustrated in Fig. 5-5, which consists of a gasoline-engine-driven pump and a propeller, is capable of discharging a liquid fuel onto the pile.

Rates of clearing land As previously stated, the rates of clearing land will depend on several factors, including, but not limited to, the following: (1) the density of vegetation, (2) sizes and varieties of trees, (3) type of soil, (4) topography, (5) rainfall, (6) types of equipment used, (7) skill of equipment operators, and (8) the requirements of the specifications governing the project.

Formula (5-1) may be used as a guide in estimating the required time to fell trees only, using a shear-type cutting blade, as illustrated in Fig. 5-1, mounted on a crawler tractor of the size indicated in Table 5-1 [1]. Before preparing an estimate, the estimator should visit the project to be cleared in order to obtain information needed to evaluate the variable factors in the formula. With this information, reasonably applicable values can be assigned to the factors listed in Table 5-1. Thus we have

$$T = B + M_1 N_1 + M_2 N_2 + M_3 N_3 + M_4 N_4 + DF \qquad (5\text{-}1)$$

where T = time per acre, min

B = base time for a tractor to cover an acre with no trees requiring splitting or individual treatment, min

TABLE 5-1 Representative Times Required to Cut Trees with Tractor-mounted Blades, Min

Size tractor, fwhp	Base time B	Time to cut a tree*				Time per foot for diameters above 6 ft F
		1–2 ft diameter M_1	2–3 ft diameter M_2	3–4 ft diameter M_3	4–6 ft diameter M_4	
93	40	0.8	4.0	8.0	25	
130	28	0.5	2.0	4.0	12	4.0
190	21	0.3	1.5	2.5	7	2.0
320	18	0.3	0.5	1.5	4	1.2

* The listed times are for cutting trees flush with the surface of the ground. If it is necessary to remove the stumps, the time should be increased by 50 percent.

M = time required per tree in each diameter range, min

N = number of trees per acre in each diameter range, obtained from a field survey

D = sum of the diameters in feet of all trees per acre, if any, larger than 6 ft in diameter at ground level

F = time required per foot of diameter to fell trees larger than 6 ft in diameter, min

Formula (5-1) may also be used to estimate the time required to stack felled trees into windrows spaced approximately 200 ft apart, by letting M_1, M_2, ..., represent the time required to move a tree into a windrow. Table 5-2 gives representative values for the time required to pile trees.

TABLE 5-2 Representative Times Required for Stacking Trees with Tractor-mounted Blades or Rakes, Min

Size tractor, fwhp	Base time B	Time to stack a tree				Time per foot for diameters above 6 ft F
		1–2 ft diameter M_1	2–3 ft diameter M_2	3–4 ft diameter M_3	4–6 ft diameter M_4	
93	35	0.3	0.6	2.5	...	
130	28	0.2	0.4	1.5	3.0	
190	24	0.1	0.3	1.0	2.0	0.4
320	20	0.0	0.1	0.7	1.2	0.2

TABLE 5-3 Types of Equipment Used, Species, Sizes, and Densities of Trees Removed

Plot no.	Blade used*	Percent by species		Percent by size trees, in.		No. trees per acre
		Hardwood	Pine	To 6	Above 6	
1	B	79	21	87	13	375
2	B	98	2	74	26	285
3	B	97	3	76	24	385
4	B	56	44	87	13	585
5	B	53	47	93	7	680
6	B	78	22	87	13	755
7	S	80	20	86	14	690
8	S	29	71	98	2	1,545
9	S	72	28	82	18	445
10	S	60	40	98	2	710
11	S	89	11	72	28	410
12	S	75	25	76	24	400

* B denotes a bulldozer and S denotes a shearing blade.

Cost of clearing land Very little information on this subject has been released. However, in 1958 the Agricultural Experiment Station of Auburn University, Auburn, Alabama, conducted tests to determine the cost of clearing land using three sizes of crawler tractors, equipped with bulldozer blades and with shearing blades, such as the one illustrated in Fig. 5-1. The results of the tests have been published in a booklet [2].

For test purposes an area of 24 acres was divided into 12 plots of 2 acres each, with dimensions 198 ft wide and 440 ft long. Each size tractor cleared two plots using a bulldozer blade and two plots using a shear blade. The net time required to fell, stack, and burn the material from each plot was determined. The trees consisted of pine, oak, hickory, and gum, distributed by species, size, and density as listed in Table 5-3. The diameters of the trees were measured at breast height.

The trees were felled and then pushed along the surface of the ground and stacked in windrows not more than 198 ft apart, after which they were burned. During the burning operation the timber was pushed into stacks to increase the burning effectiveness, using a tractor-mounted blade.

Table 5-4 shows the average time required by each size crawler tractor and type of blade to fell, stack, and dispose of an acre of timber. The smaller times required to dispose of trees felled with the shearing blades were the result of smaller amounts of soil in the roots of the trees felled with this type of blade.

TABLE 5-4   Average Machine Time Required, in Hours, to Clear an Acre of Land Based on Size Tractor and Blade Used

| Operation | Time per acre, hr | | | | | |
| | 93 fwhp | | 130 fwhp | | 190 fwhp | |
	B*	S*	B	S	B	S
Felling	2.19	1.58	1.71	1.14	0.92	0.71
Stacking	0.52	0.55	0.56	0.60	0.48	0.46
Disposal	1.75	0.84	1.80	0.78	1.93	0.70
Total	4.46	2.97	4.07	2.52	3.33	1.87

* B denotes a bulldozer and S denotes a shearing blade.

Example   Estimate the direct cost per acre for clearing, grubbing, and disposing of the vegetation on 46 acres of land. All trees and shrubs are to be pushed down, and all roots larger than 1 in. in diameter in the top 18 in. of soil are to be removed, stacked, and burned on the site.

The area is the right of way for a highway 150 ft wide. After the trees are felled, they will be pushed into windrows about 150 ft apart and burned.

The trees and shrubs will be felled using tractor-mounted V blades as illustrated in Fig. 5-2. Roots and any remaining stumps will be removed by rippers mounted on the rear of tractors.

All the material will be pushed into windrows using tractor-mounted rakes similar to the one illustrated in Fig. 5-4.

The area is covered with elm and oak trees, plus smaller shrubs, whose average count per acre is as follows:

Elm trees, 24 to 36 in. in diameter, 6 per acre
Elm trees, 12 to 24 in. in diameter, 20 per acre
Oak trees, 24 to 36 in. in diameter, 8 per acre
Oak trees, 12 to 24 in. in diameter, 18 per acre
Smaller trees, 124 per acre

The soil is sandy clay, reasonably well drained, with no standing water or ponds of water.

The work will be done during the summer, when rainfall should average 3 to 4 in. per month.

As the material is burned, the tractor-mounted rake will push and restack it into tighter piles to provide better burning. Fuel oil will be applied as needed to assure good burning.

Tractors having the following sizes will be used:

For felling trees, 180 fwhp
For the rake, 180 fwhp
For removing stumps and roots, 180 fwhp

Using Table 5-1 and formula (5-1), the time required to cut and fell the trees on an average acre should be

$$T = 21 + 0.3 \times 38 + 1.5 \times 14$$

$$= 21 + 11.4 + 21.0$$

$$= 53.4 \text{ min, or } 0.89 \text{ hr}$$

Using a tractor-mounted rooter with four teeth to cut a swath 6 ft wide each pass, it will require 35 passes to cover a square acre, 208 by 208 ft. The total distance traveled by the rooter will be $35 \times 208 = 7,280$ ft. At an average speed of 1 mph it will require 0.92 hr per acre to remove the stumps and roots.

Using the information in Table 5-2 and formula (5-1), the time required to stack the trees from an acre of land will be

$$T = 28 + 0.2 \times 38 + 0.4 \times 14$$

$$= 28 + 7.6 + 5.6$$

$$= 41.2 \text{ min, or } 0.68 \text{ hr}$$

Additional stacking of the material by a tractor-mounted rake during the burning will require at least one-half as much time as the initial stacking. Assume that this time will be 0.35 hr per acre.

A summary of times for the several operations, based on a 60-min hr, will be

Operation	Time, hr
Cutting trees	= 0.89
Removing roots and stumps	= 0.92
Stacking material	= 0.68
Restacking material	= 0.35

Because of the nature of the work performed in clearing land, it is probable that the equipment will work not more than 40 min per hr. If the time per operation is adjusted to reflect this operating condition, the adjusted time per operation will be

1   Cutting trees, $0.89 \times \frac{60}{40}$ = 1.34 hr
2   Removing roots and stumps, $0.92 \times \frac{60}{40}$ = 1.38 hr
3   Stacking trees, $0.68 \times \frac{60}{40}$ = 1.02 hr
4   Restacking trees, $0.35 \times \frac{60}{40}$ = 0.53 hr

Operations 3 and 4 will use the same equipment, namely, a tractor-mounted rake. The time required to stack and restack the material will be 1.55 hr per acre. This should be the governing time for progress on the project, and this rate will be used for cutting and felling the trees and removing the roots and stumps.

The cost per acre, based on 1.55 hr per acre, will be

Moving in and out, $480.00 \div 46$ acres   = \$ 10.45
Tractors, $3 \times 1.55 = 4.65$ hr @ \$14.62   =   67.98
V blade, 1.55 hr @ \$2.75   =   4.26

Rake, 1.55 hr @ $2.50	=	3.87
Ripper, 1.55 hr @ $2.25	=	3.48
Ripper teeth, 1.55 hr @ $1.75	=	2.71
Hand tools and supplies	=	3.75
Fuel for burning, 30 gal @ $0.40	=	12.00
Tractor operators, 4.65 hr @ $6.25	=	29.06
Laborers, 4 men × 1.55 = 6.20 hr @ $3.25	=	20.15
Foreman, 1.55 hr @ $7.25	=	11.23
Pickup truck, 1.55 hr @ $2.10	=	3.25
Total cost		= $172.19

CONCRETE PAVEMENT

General information   The cost of concrete pavement in place includes the cost of fine-grading the subgrade; side forms; steel reinforcing, if required; aggregate; cement; mixing, placing, spreading, finishing, and curing concrete; expansion-joint material; and shaping the shoulders adjacent to the slab. If the subgrade is dry, it may be necessary to wet it before placing the concrete.

If the pavement is not uniformly thick, the average thickness may be determined in order to determine the area of the cross section. This area multiplied by the length will give the volume of concrete required, usually expressed in cubic yards. Payment for concrete pavement usually is at an agreed price per square yard of area.

Construction methods used   At least two methods are used in placing concrete pavement.

The older, and still-used, method is to bring the base under the pavement to the specified density, grade, and shape, and then to install side forms to confine the concrete until it sets and to control the thickness of the slab.

A newer but very satisfactory method is to use a slip-form paver to shape the slab. The two side forms, which confine the outer edges of the freshly placed concrete, are mounted on a self-propelled paver, which spreads, vibrates, and screeds the concrete to the specified thickness and surface shape as it moves along the job. Figure 5-6 illustrates a slip-form paver on a project.

Preparing the subgrade for concrete pavement   If the subgrade has been previously constructed, compacted, and shaped to the approximate final elevation, the preparation for the placing of concrete will usually include the installation of the side forms, fine-grading the subgrade to the exact shape and elevation, and possibly wetting and compacting after the fine-grading is completed.

The forms most commonly used are made of steel whose height is equal to the thickness of the concrete adjacent to the forms. Forms are manufactured in sections 10 ft long, with three holes per section for pins, which

Fig. 5-6 Placing concrete pavement with a slip-form paver.

are driven into the ground to maintain alignment and stability. As the tops of the forms are used to control the elevations of the subgrade and the concrete slab, they must be set to the exact elevation required.

Forms usually are left in place 8 to 12 hr after the concrete is placed, after which they are removed, cleaned, oiled, and reused. Handling may be done by hand or with a motor crane and a truck.

Fine-grading is usually accomplished with a self-propelled mechanical subgrader which rides on the forms. This machine loosens the earth to the required depth and removes it with conveyor blades which deposit it outside

Fig. 5-7 Subgrader for concrete pavement.

the forms. Self-propelled rollers or vibrating equipment may be used to compact the subgrade after the fine-grading is completed, with water added if necessary. See Fig. 5-7.

Handling, batching, and hauling materials A suitable site is needed for the storage and batching of materials. The equipment will include bulk cement bins, overhead bins for aggregate storage, equipped with weight batchers, and one or more clamshells to handle the aggregate. A cost study should be made prior to locating this plant to determine if more than one location is desirable. For a long paving project, more than one setup may be justified to reduce the length of haul for the batched materials.

Table 5-5 gives recommended sizes of roadbuilder's bins for storing and batching aggregate and sizes of clamshells for handling the aggregate.

If the concrete is to be mixed in a paver, dump trucks are used to haul the materials from the batching plant to the paver. Such trucks will haul two or more batches per load, depending on the size of the truck and the size of the batch. A batch includes the cement and aggregate required for one operation of the mixer. The weight of batched material will run about 3,850 lb per cu yd. If a size 34E paver, operating at 110 percent capacity, is used, the weight of a batch will be

$$\frac{34 \times 1.10 \times 3,850}{27} = 5,350 \text{ lb}$$

TABLE 5-5 Recommended Size of Bins and Clamshells for Handling and Storing Aggregate

Size mixer	Min. size bin, tons	Size clamshell bucket, cu yd	Size crane, cu yd	Boom length, ft	Operating radius, ft
1 27E single drum	75	¾	¾	45	40
1 34E single drum	75	1	1	45	40
1 16E dual drum	50	½	½	40	35
1 34E dual drum	100	1¾	1¾	50	42
2 34E dual drums	190	3	2½	60	50

For three batches, the total weight on the truck chassis will be about

Weight of 3 batches, 3 × 5,350 = 16,050 lb
Weight of truck body and hoist = 4,000 lb
_____
Total weight   = 20,050 lb

The number of trucks required to supply the aggregate may be obtained from the formula

$$N = \frac{60}{Kt} \times \left(\frac{60L}{S} + T\right) \tag{5-2}$$

where N = number of trucks
K = number of batches per truck
t = cycle time per batch, sec
L = round-trip distance, miles
S = average speed of trucks, mph
T = average time per truck at the batching plant and paver, plus reasonable delays, min

Mixing and placing concrete Concrete for pavements is usually mixed in paving mixers. The Mixer Manufacturer's Bureau of the Associated General Contractors of America lists three sizes as standard, namely, 16, 27, and 34 cu ft nominal capacities, when the mixers are operated on slopes not greater than 6 percent. Sizes 16E and 34E are available as single- and dual-drum units. The use of a dual-drum unit, which has a separate premixing and final mixing drum, will permit a mixer to produce a batch in less than a minute under favorable conditions.

Table 5-6 gives representative production rates for paving mixers based on the mixing time of 60 sec, when operating at 110 percent capacity. The

TABLE 5-6 Representative Production Rates for Paving Mixers for 60-sec Mixing Time

Type mixer	Single drum		Dual drum	
Size mixer	27E	34E	16E	34E
No. batches per hr	42	42	65	65
Vol. concrete per hr, cu yd	47	59	42	90
Vol. concrete per 8 hr, cu yd	376	472	336	720

rates are based on an operating factor of 0.85 for the single-drum mixers and 0.75 for the dual-drum mixers. For other operating conditions, the rates given in the table should be adjusted.

After a batch is mixed, it is discharged into a bucket, which moves along a boom attached to the mixer, to permit it to be distributed over the base. After concrete is placed across the full width of the pavement, it is spread and vibrated by a self-propelled spreader, which travels on the forms. Subsequent operations include screeding, belting, and finishing, all of which are accomplished by self-propelled equipment.

Recently, a modified paving method has been developed, which uses slip forms attached to the sides of a self-propelled machine, mounted on crawler tracks, which spreads, vibrates, and finishes the concrete pavement. The use of this machine is reported to reduce the cost of pavement $0.25 to $0.30 per square yard.

Joints In order to reduce the danger of irregular and unsightly joints across and along concrete pavement, it is common practice to install joints at regular intervals. Three types are installed, construction, transverse, and longitudinal.

Transverse joints are installed across the pavement at spacings varying from 15 to 30 ft, or sometimes more. Longitudinal center joints are installed along the length of the pavement, usually when the width exceeds 12 ft and the thickness is less than 12 in.

Joints may be constructed by grooving the freshly placed concrete, by installing premolded joint material or wood planks as the concrete is placed, or by sawing the grooves after the concrete has been placed. Self-propelled or hand-pushed machines, using abrasive or diamond blades, are used for this operation. The width of joints specified may vary from $\frac{1}{8}$ to $\frac{3}{8}$ in. Cutting speeds should vary from 2 to 10 fpm, depending on the width and depth of the joint, the type of blade used, and the kind of aggregate used. Figure 5-8 shows sections through several types of joints.

The grooved or sawed joints should be filled with a sealing compound, usually an asphaltic or rubber-base material, applied either hot or cold.

ESTIMATING CONSTRUCTION COSTS

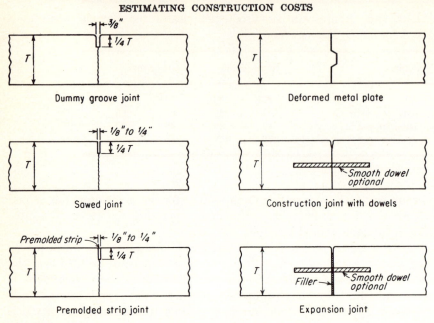

Fig. 5-8 Representative details of joints in concrete pavement.

Steel tie bars or rods may be installed across all joints, primarily to transfer shearing forces.

Curing concrete pavement Curing is accomplished by covering the fresh slab with burlap, cotton mats, waterproof paper, or an impervious membrane-producing compound, which is sprayed on the surface and sides soon after the concrete is placed.

The burlap and mats must be kept wet for the specified time, while the paper and membrane prevent or reduce the evaporation of initial water in the concrete. The membrane-producing compounds, frequently called curing compounds, will cover 30 to 50 sq yd per gal.

Example Estimate the total cost and the cost per square yard, for bid purposes, for placing a concrete pavement 30 ft wide, 9 in. average thickness, and 5.78 miles long.

The mix design specifies 524 lb of cement, 1,278 lb of fine aggregate, 2,096 lb of coarse aggregate, and 32.5 gal of water per cubic yard of concrete. A 34E dual-drum mixer, operating at 110 percent capacity will be used to mix the concrete.

Transverse joints, ⅛ in. wide and 2 in. deep, will be installed at 20-ft intervals across the slab. A longitudinal joint, ⅛ in. wide and 2 in. deep, will be installed along the centerline of the slab. All joints will be sawed within 24 hr after the concrete is placed. The joints will be sealed with hot asphaltic compound, applied with a pressure sealer.

No steel reinforcing or dowels will be placed in the pavement.

The batching plant for the aggregate will be set up near the midpoint of the project.

Fig. 5-9 Saw for cutting joints in concrete pavement. (*Clipper Manufacturing Co.*)

The plant will include a 500-bbl bulk-cement silo, equipped with a screw conveyor and a bucket elevator, a three-compartment 100-ton aggregate bin, and weight batchers, arranged for single-stop delivery to batch trucks. Trucks with a capacity of three batches each will haul the aggregate an average distance of 2.5 miles, at an average speed of 30 mph. A 1½-cu-yd crawler-mounted clamshell will be used to charge the aggregate into the bins.

Water will be obtained from a private pond, requiring a 3-in. gasoline-engine-operated pump having a capacity of 20,000 gal per hr. The water will be hauled an average distance of 3 miles, using trucks whose capacities are 2,500 gal each. The average haul speed is estimated to be 25 mph. It will require 10 min to fill a truck and also 10 min to empty a truck. The truck driver will operate the pump at the pond and also the pump to empty his truck tank into a wheel-mounted truck tank that will move along with the paver.

The calculations will be

$$\text{Volume of concrete, } \frac{5.78 \times 5{,}280 \times 30 \times 9}{27 \times 12} = 25{,}432 \text{ cu yd}$$

$$\text{Add for waste and overrun, } 2\% \times 25{,}432 = 509 \text{ cu yd}$$

$$\text{Total volume} = 25{,}941 \text{ cu yd}$$

$$\text{Area of pavement, } \frac{5.78 \times 5{,}280 \times 30}{9} = 101{,}728 \text{ sq yd}$$

$$\text{Mixer capacity, } 34 \times 1.10/27 = 1.39 \text{ cu yd per batch}$$

Mixer output, 65 × 1.39 = 90 cu yd per hr
Assume a 50-min working hour
Probable output of mixer, 90 × 50/60 = 75 cu yd per hr
Probable total time, 25,941 ÷ 75 = 346 hr

The costs will be based on 346 hr.
Use formula (5-2) to determine the number of trucks needed to haul the aggregate.

$$N = \frac{60}{Kt} \times \left(\frac{60L}{S} + T \right) = \frac{60}{3 \times 55} \times \left(\frac{60 \times 5}{30} + 12 \right) \tag{5-2}$$

$$= 8 \text{ trucks}$$

It may be desirable to have an extra truck available on a standby basis.
Quantity of water required for concrete only = 90 cu yd × 32.5 gal = 2,925 gal per hr
Note that the quantity of water is based on the maximum output of the mixer, because at times this amount of water will be needed.

Cycle time for water truck:
  Filling tank, 10 min   = 0.167 hr
  Emptying tank, 10 min   = 0.167 hr
  Traveling, 6 mi ÷ 25 mph = 0.240 hr
                                    ————————
  Total time   = 0.574 hr

Capacity of water truck, 2,500 gal ÷ 0.574 = 4,360 gal per hr
One water truck will be sufficient.
The costs will be

Materials:
  Cement, 25,941 cu yd × $\frac{524}{376}$ = 36,100 bbl @ $5.384   = $194,400.00

  Sand, 25,941 × $\frac{1,278}{2,000}$ = 16,576 tons @ $2.84   =   47,076.00

  Gravel, 25,941 × $\frac{2,096}{2,000}$ = 27,150 tons @ $3.60   =   97,767.00

  Water, 25,941 × 32.5 gal = 843 M gal @ $0.15   =   127.00
  Curing compound, 101,728 sq yd ÷ 45 sq yd per gal = 2,255 gal
    @ $2.28   =   5,145.00
  Sealing compound for joints, 8.5 tons @ $45.50   =   386.00
                                                                        ————————
  Subtotal cost of materials   = $344,901.00
Equipment:
  Moving to and from the project, excluding labor   = $ 2,850.00
  Batching plant, 346 hr @ $10.20   =   3,535.00
  Clamshell, 346 hr @ $24.05   =   8,312.00
  Mixer, 346 hr @ $22.92   =   7,930.00
  Dump trucks, 8 × 346 = 2,768 hr @ $9.20   =   25,467.00

Forms, 2 × 5.78 × 5,280 = 61,036 lin ft use @ $23.50 per 1,000 ft of
use .... = 1,435.00
Air compressor and hammer to drive form pins, 60 cfm, 346 hr @
$2.25 .... = 778.00
Fine-grader, 346 hr @ $7.25 .... = 2,515.00
Earth-roller, smooth wheel, 8 tons, 346 hr @ $1.05 .... = 363.00
Spreader vibrator, 346 hr @ $9.80 .... = 3,392.00
Finisher, 346 hr @ $4.75 .... = 1,646.00
Truck for hauling forms, 346 hr @ $4.08 .... = 1,411.00
Water truck, 346 hr @ $7.45 .... = 2,580.00
Water pump, 346 hr @ $1.65 .... = 571.00
Water tank at mixer, 346 hr @ $2.80 .... = 970.00
Motor grader, 346 hr @ $7.27 .... = 2,522.00
Concrete saws, including blades, 2 × 346 = 692 hr @ $9.85 ... = 6,816.00
Asphalt heater, 346 hr @ $4.20 .... = 1,455.00
Air compressor for applying joint sealer, 346 hr @ $2.25 ... = 778.00
Pickup trucks, ¾ ton, 2 × 346 = 692 hr @ $2.85 .... = 1,975.00
Sundry equipment .... = 375.00

Subtotal cost of equipment .... = $ 77,676.00
Labor:
Setting up and dismantling plant .... = $ 1,250.00
Batching plant operators, 2 men × 346 = 692 hr @ $4.80 ... = 3,322.00
Clamshell operator, 346 hr @ $6.75 .... = 2,335.00
Clamshell oiler, 346 hr @ $4.60 .... = 1,592.00
Truck drivers, 10 men × 346 hr = 3,460 hr @ $3.75 .... = 12,985.00
Form men, 4 × 346 = 1,384 hr @ $3.75 .... = 5,190.00
Fine-grader operator, 346 hr @ $6.25 .... = 2,163.00
Roller operator, 346 hr @ $4.25 .... = 1,474.00
Mixer operator, 346 hr @ $6.75 .... = 2,335.00
Spreader operator, 346 hr @ $6.25 .... = 2,163.00
Finisher operator, 346 hr @ $6.25 .... = 2,163.00
Grader operator, 346 hr @ $6.25 .... = 2,163.00
Saw operators, 2 men, 692 hr @ $6.25 .... = 4,325.00
Applying joint sealer, 2 men, 692 hr @ $5.25 .... = 3,625.00
Laborers, 6 men × 346 = 2,066 hr @ $3.25 .... = 6,715.00
Foreman, 346 hr @ $7.75 .... = 2,675.00

Subtotal labor cost .... = $ 56,475.00
Workman's compensation insurance, $56,475.00 × $4.37 per $100.00 = $ 2,467.96
Social security tax, $56,475.00 × 5.85% .... = 3,303.79
Unemployment tax, $56,475 × 3% .... = 1,694.25

Subtotal labor taxes .... = $ 7,466.00
Summary of costs:
Materials .... = $344,901.00
Equipment .... = 77,676.00
Labor .... = 56,475.00
Labor taxes .... − 7,466.00

Subtotal costs .... = $486,518.00

General overhead, 4% of $486,518.00   =   19,460.72
                                                            ──────────
Subtotal   = $505,978.72
Profit, 5% of $505,978.72   =   25,298.85
                                                            ──────────
Subtotal   = $531,277.57
Performance bond:
  $100,000.00 @ $7.50 per $1,000.00   =   750.00
  $431,277.57 @ $5.00 per $1,000.00   =   2,156.38
                                                            ──────────
Total cost, amount of bid   = $534,183.95
Cost per sq yd, $534,183.95 ÷ 101,728 sq yd   =   5.26

ASPHALT PAVEMENT

Asphalt pavements are obtained by mixing and placing one or more types of mineral aggregate and bitumen or asphaltic binders. It is beyond the scope of this book to cover all the methods used.

Aggregates The aggregates commonly used include sand, gravel, limestone, crushed iron ore, or other suitable crushed stone. Local materials should be used if they are of good quality. In order to produce the desired density, the aggregates should be proportioned by size from the largest to the smallest particles specified. Specifications usually designate the percentages passing and retained on screens having given size openings.

Asphalts Asphalts used for paving construction may be classified as road oils, or slow-curing (SC) asphalts; cutback asphalts, medium-curing (MC) asphalts; and rapid-curing (RC) asphalts; asphalt cements, either petroleum or natural asphalts; emulsified asphalts; and powdered asphalts. In addition to asphalts, coal tars and water-gas tars sometimes are used for paving construction.

Cold-mix asphaltic-concrete pavement If a cold-mix method is used to construct the pavement, the aggregate is mixed with the specified quantity of asphalt. The asphalt may or may not be heated, depending on the type of aggregate and asphalt used. Mixing is usually accomplished in a pugmill mixer for a specified time, about 1 min for most jobs. After the batch is thoroughly mixed, it is discharged into a truck and hauled to the job, where it is spread by hand or by a mechanical spreader in layers of specified thickness. Compaction to the desired density is obtained by rolling with smooth wheel or pneumatic rollers, producing a designated pressure. It may be necessary to place the material in two or more layers, with each layer rolled, in order to obtain uniform compaction throughout the full depth of the pavement.

If a travel plant is used to mix the material, the aggregate is placed in windrows along the road to be paved. The mixer picks up the aggregate with a bucket conveyor, mixes it with the asphalt, and deposits it in layers at the rear of the mixer. This method eliminates trucks for hauling the mixture and reduces spreading costs.

Hot-mix asphalt-concrete pavement If a hot-mix method is used to construct the pavement, the aggregate is heated in a rotary kiln or other suitable apparatus, permitting constant agitation, to a temperature of 300 to 375°F. After it is heated, it is screened and recombined to give the specified size grading, then discharged into a pugmill type of mixer. The asphalt cement is heated to the specified temperature, usually 275 to 375°F. The aggregate and the asphalt cement are mixed in proper proportions in the pugmill for the specified time or until a uniform mixture is obtained. The quantity of asphalt cement varies from 4 to 9 percent of the weight of the finished product. Both batch- and continuous-type mixers are used.

After the material is mixed, it is discharged into trucks and hauled to the job, where it is spread in uniformly thick layers by a mechanical spreader or by some other suitable method. After each layer is deposited, it is compacted by a wheel-type roller to the desired density. It may be necessary to cover the material with a canvas or tarpaulin during transit to prevent excessive loss of heat between the mixing plant and the job.

Cost of hot-mix asphaltic-concrete pavement The initial cost of the plant for producing hot-mix asphaltic-concrete pavement is relatively high. The cost of moving a plant to a location and setting it up for operation can run as high as $2,000 to $5,000 or more. Consequently, a single setup is generally made for a given job.

The cost of hot-mix asphaltic-concrete pavement will include the cost of aggregate, asphalt, moving the plant to the location, setting it up, operating the plant, hauling the material to the job, spreading, and rolling. The cost of operating the plant will include depreciation, interest, insurance and taxes, maintenance and repairs, and fuel. Additional costs may include the rental of a site for the plant and a railroad spur track for the delivery of aggregate and asphalt.

Equipment for hot-mix asphaltic-concrete pavement The total equipment for mixing and placing hot-mix asphaltic-concrete pavement will include such items as a clamshell to handle the aggregate, cold-aggregate storage and feeder bins, a cold-aggregate elevator, an aggregate drier, a dust collector, a hot-aggregate elevator, aggregate screens, hot-aggregate storage bins and batcher, a mixer, asphalt storage tanks, fuel-oil storage tank, steam boiler or hot oil heater to heat the asphalt, trucks to haul the asphalt mixture, an asphalt distributor, a mechanical spreader, and rollers to compact the pavement. If the subgrade or subbase is not already prepared, additional equipment will be required to do this work.

Example Estimate the total direct cost and the cost per ton for mixing and placing hot-mix asphaltic-concrete pavement 18 miles long, 24 ft wide, and 5 in. thick on a previously prepared subbase. Prior to placing a 3-in.-thick base, an asphalt prime coat will be applied to the top of the subbase at the rate of 0.3 gal per sq yd. A 2-in.-thick wearing course will be placed on top of the base with no tack coat required.

The paving mixtures shall meet the requirements for grading and mix composition shown in the accompanying table.

Course Thickness, in.	Base 3	Wearing 2
Sieve size	Combined aggregate including filler; per-cent passing, by weight	
2 in.	100	
1½ in.	95–100	
1 in.	......	100
¾ in.	70–85	95–100
½ in.	......	75–90
No. 4	35–50	45–60
No. 10	25–37	35–47
No. 40	15–25	23–33
No. 80	6–16	16–24
No. 200	2–6	6–12
Asphalt cement, percent of combined weight	6.0	6.0

The plant will require 60 percent coarse aggregate, larger than No. 10 sieve, and 40 percent fine aggregate and filler. Assume that 5 percent of the coarse aggregate, 10 percent of the fine aggregate, and 2 percent of the asphalt will be lost through waste or for other reasons. The combined material will weigh about 3,600 lb per cu yd when compacted.

The material will be mixed in a continuous mixer whose maximum capacity is 150 tons per hr. Based on a 50-min hour, the average output will be $0.833 \times 150 = 125$ tons per hr.

The mixing plant will include the units listed below with costs as indicated.

```
1 four-compartment bin and feeder
1 fuel-oil-fired drier
1 dust collector
1 hot elevator
1 screening unit
1 mixer, 150 tph continuous type
  Cost of above equipment       = $142,360.00
2 10,000-gal asphalt storage tanks =   11,640.00
1 2,000-gal fuel-oil storage tank  =      965.00
1 hot-oil heater                =    9,780.00
1 asphalt pump with engine      =    1,875.00
1 set of pipes for asphalt      =    4,860.00
900 gal of heat exchange oil    =      685.00
1 clamshell, size 1½ cu yd      =   61,270.00

  Subtotal cost                 = $233,435.00
```

Assume that this equipment has a useful life of 5 yr, with no salvage value at the end of its life, and that it will be used an average of 1,400 hr per year.

The average value will be 0.6 × $233,435.00 = $140,061.00

The annual cost will be:

Depreciation, $233,435.00 ÷ 5	= $46,687.00
Maintenance and repairs, 10% of $233,435.00	= 23,344.00
Investment, 14% × $140,061	= 19,609.00
Total annual fixed cost	= $89,640.00

Hourly costs:

Fixed cost, $89,640 ÷ 1,400	= $ 64.00
Hot-oil heater fuel, 16 gal @ $0.27	= 4.32
Drier fuel, 360 gal @ $0.26	= 93.60
Diesel fuel for mixer engine, 4.6 gal @ $0.32	= 1.48
Diesel fuel for drier engine, 8.2 gal @ $0.32	= 2.63
Lubricating oil, 0.5 quart @ $0.40	= 0.20
Other lubricants and grease	= 0.15
Total cost per hr	= $ 166.38
Cost per ton, $166.38 ÷ 125 tph	= 1.33

The total quantities will be

Area, 18 miles × 5,280 ft × 24 ft/9	= 254,000 sq yd
Material for base course, $\dfrac{254,000 \times 3 \times 3,600}{36 \times 2,000}$	= 38,200 tons
Material for wearing course, $\dfrac{254,000 \times 2 \times 3,600}{36 \times 2,000}$	= 25,400 tons
Total weight	= 63,600 tons
Total weight of aggregate, 0.94 × 63,600	= 59,800 tons

Assume that the cost of moving the plant will be charged to this job, and that the cost of moving out will be charged to the next job. If this cannot be done, the entire cost of moving to and from this job should be charged to this job.

The cost will be

Materials:

Coarse aggregate, 0.6 × 59,800 × 1.05 = 37,674 tons @ $3.15	= $118,673.00
Fine aggregate, 0.4 × 59,800 × 1.10 = 26,312 tons @ $2.65	= 69,695.00
Asphalt, 0.6 × 63,600 × 1.02 = 3,880 tons @ $36.80	= 142,784.00
Priming oil, 254,000 sq yd × 0.3 gal = 76,200 gal @ $0.16	= 12,192.00
Total cost of materials	= $343,344.00
Cost per ton, $343,344.00 ÷ 63,600	= 5.38

Equipment at mixing plant only:

Moving to job and setting up the plant	= $ 3,460.00
Plant cost, 63,600 tons/125 tph = 508 hr @ $166.38 per hr	= 84,524.00
Total cost of mixing plant	= $ 87,984.00

Labor at mixing plant:

Foreman, 508 hr @ $7.75	=	$	3,938.00
Crane operator, 508 hr @ $6.75	=		3,427.00
Crane oiler, 508 hr @ $4.75	=		2,415.00
Weighers, 2 × 508 = 1,016 hr @ $4.75	=		4,830.00
Batching plant operator, 508 hr @ $6.75	=		3,427.00
Plant oiler-mechanic, 508 hr @ $5.25	=		2,665.00
Laborers, 3 × 508 = 1,524 hr @ $3.25	=		4,950.00
Total cost of labor at mixing plant	=	$	25,652.00

Hauling mixture from mixing plant to roadway, subcontract, 63,600 tons @ $0.52 = $ 33,072.00

Equipment placing pavement:

Asphalt distributor, 1,500 gal cap at 60% of 508 = 305 hr @ $9.53	=	$	2,905.00
Laydown paver, 125 tph, 508 hr @ $8.45	=		4,300.00
Smooth-wheel roller, 12 ton, 508 hr @ $4.27	=		2,169.00
Pneumatic roller, self-propelled, 12 ton, 508 hr @ $6.65	=		3,378.00
Total equipment cost placing	=	$	12,752.00

Labor placing pavement:

Foreman, 508 hr @ $7.75	=	$	3,942.00
Distributor operator, 305 hr @ $5.75	=		1,753.00
Helper, 305 hr @ $3.75	=		1,145.00
Paver operator, 508 hr @ $6.25	=		3,192.00
Roller operators, 2 men, 1,016 hr @ $5.75	=		5,870.00
Laborers, 3 × 508 = 1,524 hr @ $3.25	=		4,965.00
Total labor cost placing	=	$	20,867.00

Other direct costs:

Land-site rental, 4 months @ $75.00	=	$	300.00
Plant and road tools	=		650.00
Service trucks, 2 × 508 = 1,016 hr @ $2.44	=		2,480.00
Barricades, flares, signals	=		650.00
Truck driver, 508 hr @ $3.75	=		1,905.00
Helper, 508 hr @ $3.25	=		1,654.00
Superintendent, 4 months @ $1,500.00	=		6,000.00
Clerk-timekeeper, 4 months @ $650.00	=		2,600.00
Total other costs	=	$	16,239.00

Summary of direct costs:

Materials	=	$	343,344.00
Equipment at mixing plant	=		87,984.00
Labor at mixing plant	=		25,652.00
Hauling from plant to roadway	=		33,072.00
Equipment placing pavement	=		12,752.00
Labor placing pavement	=		20,867.00
Other direct costs	=		16,239.00
Total direct cost	=	$	539,910.00
Cost per ton, $539,910 ÷ 63,600 tons	=		8.50

Factors affecting the cost of asphalt pavement There are a number of factors which affect the cost of asphalt pavement in place. These factors include the size of the job, weather conditions, the cost of maintaining traffic, if necessary, the ability to keep the plant operating at its rated output, the distance the mixed aggregate must be hauled, the distance from the plant to a railroad siding if aggregate is delivered by rail, etc. Each of these factors should be analyzed with respect to the particular job.

Flexible-base and double-surface-treatment asphalt pavement For many secondary roads which are not subjected to heavy traffic or heavy truck loads, a pavement constructed of 6 to 12 in. of flexible base and an asphalt and aggregate surface about ½ in. thick, more or less, is frequently used. Such pavements have relatively low initial costs and give satisfactory service for the use to which they are subjected.

Material The base material may be caliche, shell, bank-run gravel, processed gravel, crushed iron ore, crushed stone, or other suitable local material.

The aggregate for the wearing surface may be graded gravel and sand, shell, crushed stone, or crushed limestone rock asphalt. The surface treatment may be applied in one, two, or three separate operations.

Preparing the subgrade The customary operations in constructing a pavement by this method begin with the preparation of the subgrade. This operation consists in grading, sprinkling, if necessary, and compacting the subgrade to the desired shape and density. The equipment required includes motor graders; sprinkler trucks; and tamping-type, smooth-wheel, or pneumatic rollers.

Preparing the flexible base After the subgrade has been properly prepared, the aggregate for the flexible base is placed to give the desired thickness. The aggregate should be graded from the smallest to the largest acceptable sizes in such proportions that a dense mass will be produced. The aggregate should contain sufficient binder material to solidify the base. It may be necessary to add water as the base is shaped and compacted. The material is spread in 3- to 6-in. layers with a grader and compacted with multiwheel pneumatic or smooth-wheel rollers to the desired shape and density. It is frequently specified that the base shall be opened to traffic for several weeks for further compaction and seasoning.

Placing the wearing surface After the base has been brought to the desired condition by traffic, the surface is reshaped and compacted if necessary. It is swept or broomed clean, and a surface treatment of hot asphalt is applied at a designated rate, using a self-propelled pressure distributor. The rates of application vary from about 0.15 to 0.40 gal per sq yd of surface. Immediately following the application of the asphalt, the aggregate is spread uniformly over the surface by a mechanical spreader or by a grader. For small jobs the aggregate may be spread by hand. The rates of application are generally expressed in terms of the number of square yards of surface coverage per cubic yard of aggregate, such as 1 cu yd per 75 sq yd of surface. One cubic yard per 35 sq yd will give a surface thickness of approximately 1 in. After the aggregate is spread uniformly, it is rolled or subjected to traffic in order to combine it with the asphalt. If a second surface treatment is to be applied, it is customary to roll the first layer immediately with a smooth-wheel roller. This operation is followed with a second application of asphalt at the specified rate, which is covered uniformly with another layer of aggregate. The maximum-size aggregate in the final surface treatment is usually smaller than that used in the first treatment. Following the application of the aggregate for the final surface treatment, the entire surface should be broomed and thoroughly rolled with a self-propelled smooth-wheel roller. After the work is completed, there should be a slight excess of aggregate on the surface. The pavement may now be opened to traffic.

Example   Estimate the total direct cost and the cost per square yard for the base course and the surface treatment for a flexible-base and double-surface-treatment asphalt pavement for the following project.

Length of highway	6.24 miles
Width of flexible base	22 ft
Thickness of base	8 in.
Width of asphalt pavement	22 ft
Gal asphalt per sq yd, first application	0.25
Gal asphalt per sq yd, second application	0.20
Cu yd aggregate per sq yd, first application	1:75
Cu yd aggregate per sq yd, second application	1:150

The base course is to be bank-run gravel available locally, with an average haul distance of 4.8 miles.

The aggregate for surface treatment will be durable particles of gravel or crushed stone graded by size as follows.

For the first application:	Percent
Retained on $\frac{1}{2}$-in. screen	0
Retained on $\frac{1}{4}$-in. screen	60–80
Retained on 10-mesh screen	90–100
For the second application:	
Retained on $\frac{3}{8}$-in. screen	0
Retained on $\frac{1}{4}$-in. screen	5–20
Retained on 10-mesh screen	70–95
Retained on 20-mesh screen	95–100

The required quantities of materials are determined as follows:

$$\text{Surface area of base course,} \quad \frac{6.24 \text{ miles} \times 5{,}280 \text{ ft} \times 22 \text{ ft}}{9 \text{ sq ft}} \quad = 80{,}540 \text{ sq yd}$$

$$\text{Quantity of aggregate required for base course,} \quad \frac{80{,}540 \times 8}{3 \times 12} \quad = 17{,}920 \text{ cu yd}$$

Quantity of aggregate required for first surface application, 80,540 ÷ 75 sq yd per cu yd   = 1,075 cu yd

Quantity of aggregate required for second surface application, 80,540 ÷ 150 sq yd per cu yd   =   538 cu yd

Quantity of asphalt required for first application, 80,540 sq yd × 0.25 gal   = 20,135 gal

Quantity of asphalt required for second application, 80,540 sq yd × 0.20 gal   = 16,108 gal

The base will be constructed, using gravel placed along the subbase under a contract with a local supplier, then spread over the subbase with a motor grader with a 12-ft blade. The material will be compacted with a self-propelled 5 by 5-ft dual drum sheep's-foot compactor, sprinkled with water, then rolled with a 12-ton three-wheel smooth roller.

The aggregate will be placed at a rate of 60 cu yd per hr. It is estimated that sprinkling will require an average of 24 gal of water per cu yd of aggregate. This is equal to 1,440 gal of water per hr.

The costs will be

Base course:

Aggregate, 17,920 cu yd @ 2.25	= $40,320
Grader, 17,920 cu yd ÷ 60 cu yd per hr, = 299 hr @ $8.65	= 2,587
Sheep's-foot roller, 299 hr @ $12.48	= 3,731
Smooth-wheel roller, 299 hr @ $4.09	= 1,223
Water truck, 299 hr @ $6.80	= 2,033
Water, 17,920 cu yd × 24 gal per cu yd = 430 M gal @ $0.25	= 108
Truck driver, 299 hr @ $3.75	= 1,121
Grader operator, 299 hr @ $7.25	= 2,168
Roller operators, 2 men, 598 hr @ $7.25	= 4,336
Laborers, 3 men, 897 hr @ $3.25	= 2,915
Foreman, 299 hr @ $7.75	= 2,317
	————
Total cost of base	= $62,859
Cost per cu yd, $62,859 ÷ 17,920 cu yd	= 3.50

Aggregate for the two surface treatments:
    Rate of placing first application, 25 cu yd per hr
    Time required to place aggregate, 1,075 ÷ 25 = 43 hr

Cost of first application:

Aggregate, delivered to roadway, 1,075 cu yd @ $5.25	= $ 5,644
Spreader, unmounted, 43 hr @ $0.75	= 32
Spreader operators, 2 men, 86 hr @ $3.25	= 279
Rotary broom, 9-ft self-propelled, 43 hr @ $2.18	= 94
Broom operator, 43 hr @ $5.75	= 247

Roller, 12-ton smooth-wheel, 43 hr @ $4.09	=	175
Roller operator, 43 hr @ $7.25	=	312
Laborers, 2 men, 86 hr @ $3.25	=	279
Foreman, 43 hr @ $7.75	=	333

Subtotal cost of first application = $ 7,395

Rate of placing aggregate for second application, 12.5 cu yd per hr.

Time required to place aggregate, 538 cu yd ÷ 12.5 cu yd per hr = 43 hr.

Cost of second application:

Aggregate, delivered to roadway, 538 cu yd @ $5.75	=	$ 3,094
Spreader, unmounted, 43 hr @ $0.75	=	32
Spreader operators, 2 men, 86 hr @ $3.25	=	279
Rotary broom, 9-ft self-propelled, 43 hr @ $2.18	=	94
Broom operator, 43 hr @ $5.75	=	247
Roller, 12-ton smooth-wheel, 43 hr @ $4.09	=	175
Roller operator, 43 hr @ $7.25	=	312
Laborers, 2 men, 86 hr @ $3.25	=	279
Foreman, 43 hr @ $7.75	=	333

Subtotal cost of second application = $ 4,845

Cost of asphalt in place for 2 applications:

Asphalt delivered to storage tank at job, 20,135 + 16,108 = 36,243 ×

1.02 for waste = 36,968 gal @ $0.22	=	$ 8,133
Distributor, 2 × 43 = 86 hr @ $6.98	=	598
Distributor operator, 86 hr @ $4.75	=	408
Heating asphalt, fuel, 86 hr @ $0.36	=	31
Fireman, 86 hr @ $5.25	=	451
Asphalt storage tanks and pumps, 86 hr @ $1.85	=	159
Laborers, 2 men, 172 hr @ $3.25	=	558

Foreman is paid under aggregate placing

Total cost of asphalt = $10,338

Cost of moving plant to job, setting it up, and returning to the storage yard = $ 1,650

Summary of costs:

Base course	=	$62,859
Aggregate for first application	=	7,395
Aggregate for second application	=	4,845
Asphalt	=	10,338
Moving in and out	=	1,650

Total cost	=	$87,087
Cost per sq yd, $87,087 ÷ 80,540	=	1.08

PROBLEMS

5-1 Estimate the cost per acre for clearing and grubbing 56 acres of sandy land. All trees and shrubs and roots larger than 1 in. in diameter in the top 18 in. of soil are to be removed, stacked, and burned on the site. There are no water ponds or streams to delay progress on the operations.

Use the information appearing in this book as a guide in determining the production rates, and for the cost of equipment and wage rates.

The average tree count per acre is as follows:

Elm trees, 24 to 36 in. diameter	4
Elm trees, 12 to 24 in. diameter	26
Oak trees, 24 to 36 in. diameter	8
Oak trees, 12 to 24 in. diameter	16
Smaller trees	64

The cost of moving to the job and back to the storage yard will be $380.00.

5-2 An area of 280 acres is to be cleared of all trees and shrubs above the surface of the ground. Stumps and roots below the surface of the ground need not be removed.

The trees will be cut off even with the surface of the ground using one or more tractor-mounted V blades and then stacked for burning on the site using one or more tractor-mounted rakes such as the one illustrated in Fig. 5-4. The soil is sandy clay.

Use the wage rates and equipment costs from the example beginning on page 85 of this book.

The average tree count per acre is

Hardwood trees, 24 to 36 in. diameter	16
Hardwood trees, 12 to 24 in. diameter	48
Smaller trees	56

The cost of moving to the job and back to the storage yard will be $575.

5-3 Estimate the total cost and the cost per square yard of concrete, for bid purposes, for furnishing the materials and constructing a pavement 24 ft wide, 9 in. thick, and 6.40 miles long. Use the same types of equipment, equipment costs, wage rates, material costs, and overhead costs as those used in the example beginning on page 92 of this book.

The batched aggregate will be hauled to the mixer in trucks that hold three batches each. The average haul distance will be 2 miles.

Use the same batch mix as in the example.

No reinforcing steel or dowels will be placed in the pavement.

Joints will be sawed as specified in the example.

The cost of moving the plant to the project and back to the storage yard will be $3,280.00, excluding the cost of labor in setting it up and taking it down.

5-4 Estimate the total direct cost and the cost per ton for furnishing materials and for mixing and placing hot-mix asphaltic-concrete pavement 12.0 miles long, 28 ft wide, and 4 in. thick on a previously prepared base. The pavement will be placed in two layers, each 2 in. thick after compaction.

Use the same methods, material costs, equipment, equipment costs, wage rates, and production rates as those used in the example beginning on page 97 of this book.

Change the asphaltic cement to 7 percent of the combined weight of the concrete.

REFERENCES

1   Latin-American Land Development Seminar, Program and Proceedings, Rome Plow Company, Cedartown, Georgia, September 25, 1966.
2   Cost of Clearing Land, Agricultural Experiment Station of Auburn University, Auburn, Alabama, Circular 133, June, 1959.

6

Piling and Bracing

General information When a project requires excavation into earth which is so unstable that the walls must be supported to prevent them from caving into the pit or trench, it will be necessary to install a system of shores, braces, or solid sheeting along the walls to hold the earth in position. If groundwater is present, it may be necessary to install semiwatertight sheeting around the walls to exclude or reduce the flow of water into the pit. Timber and steel are used for braces and sheeting.

Load-bearing piles are installed under a structure to transmit the loads from the structure into a deeper soil which has sufficient strength to support the loads. Timber, concrete, and steel are used for load-bearing piles.

Shoring trenches If a trench is excavated deeper than 5 ft into reasonably firm earth, shoring is generally required. Rough lumber, such as 2- by 10-in. or 2- by 12-in. planks (as long as the deepest portion of the trench), may be installed on opposite sides of the trench in a vertical position to prevent the earth from caving in. The spacing of shores should be about 6 to 8 ft along the trench. Wood or metal trench braces, spaced about 3 to 4 ft apart, one above the other, may be used to hold the shores against the walls of the trench. The shores and braces are removed as the trench is back-filled. The shores may be used several times, and the metal trench braces should last several years. A truck will be needed to haul the shores and braces to different locations along the trenches. Two or more men will be required to install and remove the shores and braces. A method of estimating the cost of shoring a trench is illustrated in the following example.

Example Estimate the cost of providing shores for 100 ft of trench with depth varying from 8 to 10 ft, using 2- by 12-in. planks and trench braces. The planks will be spaced 6 ft apart along the trench. Trench braces will be placed not more than 2 ft above the bottom of the trench and spaced not more than 3 ft apart vertically to within 2 ft below the surface of the ground.

Assume that the shore lumber will be used 30 times.
The estimated cost will be determined as follows.

No. of shores, $100 \div 6 = 17$
No. of planks, $2 \times 17 = 34$

Lumber, 34 pc, 2 in $\times$ 12 in. $\times$ 12 ft 0 in. $= \dfrac{816 \text{ fbm} \times \$150.00 \text{ per M fbm}}{1,000 \times 30 \text{ uses}}$ $= \$ 4.08$

Trench braces, 3 per shore, $= 3 \times 17 = 51$ @ \$0.06 $= 3.06$
Labor installing, 17×0.40 hr $= 6.8$ hr @ \$3.50 $= 23.80$
Labor removing, 17×0.30 hr $= 5.1$ hr @ \$3.50 $= 17.85$
Truck, part time, 3 hr @ \$3.12 $= 9.36$

     Total cost $= \$58.15$

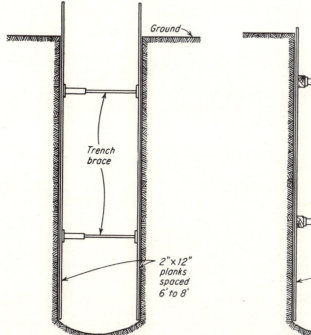

Fig. 6-1   Trench braces and shores in a sewer trench.

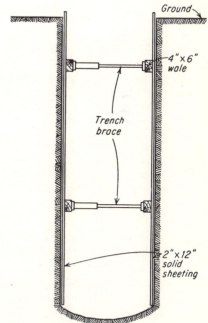

Fig. 6-2   Trench braces, wales, and solid sheeting.

Sheeting trenches If the earth is so unstable that it must be restrained for the full areas of the wall of a trench, it will be necessary to install solid sheeting to the full depth of the trench, as illustrated in Fig. 6-2. For depths up to about 12 ft, 2-in.-thick lumber may be used for sheeting with 4- by 6-in. lumber wales and braces, while for depths of 12 to 20 ft, 3-in.-thick lumber should be used for sheeting with 6- by 8-in. or 8- by 8-in. lumber for wales and braces. The lumber may be rough sawed or *S4S*.

Depending on the stability of the earth, it may be necessary to drive the sheeting and install some of the braces ahead of excavating. The sheeting may be driven with a maul or with a pneumatic hammer, such as a paving breaker equipped with a suitable driving head.

Example Estimate the cost of installing and removing solid sheeting and braces for 100 ft of trench 7 ft deep. Use 2- by 12-in. lumber 8 ft 0 in. long for sheeting with two horizontal rows of 4- $\times$ 6-in. wales on each side of the trench for the full length of the trench. Trench braces will be placed 4 ft apart along each row of wales.

Assume that the sheeting can be used four times and the wales ten times.

The sheeting will be driven one plank at a time using a 75-cfm air compressor and a pneumatic hammer.

The estimated cost will be determined as follows.

Sheeting, $\dfrac{200 \text{ ft} \times 12}{11.25} = 214 \text{ pc} = \dfrac{214 \times 2 \times 8 \times 12}{12 \times 1,000} = 3.424$ M fbm @

$70.00 per use .... = \$239.70

Wales, $\dfrac{200 \text{ ft} \times 4 \times 6}{12 \times 1,000} = 0.4$ M fbm @ $28.00 per use .... = 11.20

Trench braces, $2 \times 100 \div 4 = 50$ @ $0.06 .... = 3.00

Labor-driving sheeting, 2 men, 214 pc $\div$ 5 pc per hr = 42.8 hr $\times$ 2 = 85.6 hr @ $3.50 .... = 299.60

Labor-installing wales and braces, 50 each $\div$ 5 per hr = 10 hr $\times$ 2 men = 20 hr @ $3.50 .... = 70.00

Labor removing sheeting, wales, and braces, one fourth of time to install = 26.4 hr @ $3.50 .... = 92.50

Air compressor, 85.6 $\div$ 2 = 42.8 hr @ $2.78 .... = 119.20

Air hammer, hose, etc., 42.8 hr @ $0.40 .... = 17.15

Total cost .... = \$852.35

Pile-driving equipment Equipment used to drive piles on land usually consists of a skid rig or a truck-mounted or a crawler-mounted crane, leads, a hammer, and a source of steam or compressed air to drive the hammer. When piles are driven in water, the driving rig is usually mounted on a barge.

Pile-driving hammers The actual driving of piles usually is accomplished with hammers. Several types are available, including the drop hammer, single-acting steam hammer, double-acting steam hammer, differential-acting steam hammer, self-contained diesel-operated hammer, and vibratory hammer.

Fig. 6-3 A single-acting steam hammer driving a steel shell for a cast-in-place concrete pile. (*Municipal Construction.*)

The size of a drop hammer is indicated by the weight of the hammer, whereas the size of a steam or diesel hammer is indicated by the theoretical foot-pounds of energy delivered each blow. Steam hammers can be operated by compressed air at pressures of 80 lb to 100 psi.

Table 6-1 gives recommended sizes of hammers for different types and sizes of piles, and driving conditions. Table 6-2 gives information for various sizes and types of hammers.

Steel-sheet piling The cost of steel-sheet piling in place will include the cost of the piling, driving equipment, and labor. If the piling is to be salvaged, there will be an additional cost for extracting it. It is common practice to drive two piles simultaneously.

TABLE 6-1   Recommended Sizes of Hammers for Driving Piles, in Theoretical Foot-pounds of Energy per Blow

Length of piles, ft	Depth of penetration	Weight of piles, lb per lin ft											
		Steel sheet*			Timber			Concrete		Steel		
		20	30	40	30	40	60	150	400	40	80	120
Driving through ordinary earth, moist clay, and loose gravel, normal frictional resistance												
25	½	2,000	2,000	3,600	3,600	3,600	7,000	7,500	15,000	3,600	7,000	7,500
	Full	3,600	3,600	6,000	3,600	6,000	7,000	7,500	15,000	4,000	7,500	7,500
50	½	6,000	6,000	7,000	7,000	7,000	7,500	15,000	20,000	7,000	7,500	12,000
	Full	7,000	7,000	7,500	7,500	7,500	12,000	15,000	20,000	7,500	12,000	15,000
75	½	......	7,000	7,500	......	......	15,000	......	30,000	7,500	15,000	15,000
	Full	......	......	12,000	......	......	15,000	......	30,000	12,000	15,000	20,000
Driving through stiff clay, compacted sand, and gravel, high frictional resistance												
25	½	3,600	3,600	3,600	7,500	7,000	7,500	7,500	15,000	5,000	9,000	12,000
	Full	3,600	7,000	7,000	7,500	7,000	7,500	12,000	15,000	7,000	10,000	12,000
50	½	7,000	7,500	7,500	12,000	7,500	12,000	15,000	25,000	9,000	15,000	15,000
	Full	......	7,500	7,500	......	7,500	15,000	......	30,000	12,000	15,000	20,000
75	½	......	7,500	12,000	......	12,000	15,000	......	36,000	12,000	20,000	25,000
	Full	......	......	15,000	......	15,000	20,000	......	50,000	15,000	20,000	30,000

* The indicated energy is based on driving two steel-sheet piles, simultaneously. When driving single piles, use approximately two-thirds of the indicated energy.

TABLE 6-2   Data on Pile-driving Hammers

| Hammer | Hammer model | Weight, lb | | Strokes per min | Length of stroke, in. | Theoretical energy, ft-lb per blow |
		Complete unit	Ram			
Vulcan, single-acting	2	6,700	3,000	70	29	7,260
	1	9,600	5,000	60	36	15,000
	0	16,250	7,500	50	39	24,375
	OR	18,050	9,300	50	39	30,225
McKiernan-Terry, single-acting	S5	12,460	5,000	60	39	16,250
	S8	18,300	8,000	55	39	26,000
	S10	22,380	10,000	55	39	32,500
	S14	31,700	14,000	60	33	37,500
	S20	38,650	20,000	60	36	60,000
Vulcan, differential-acting	30C	7,250	3,000	133	12.5	7,260
	50C	12,140	5,000	120	15.5	15,100
	140C	27,980	14,000	103	15.5	36,000
	200C	39,050	20,000	98	15.5	50,200
McKiernan-Terry, double-acting	9B3	7,000	1,600	145	17	8,750
	10B3	10,850	3,000	105	19	13,100
	11B3	14,000	5,000	95	19	19,150
Raymond International, hydraulic	65CH	14,615	6,500	130	16	19,500
	80CH	17,782	8,000	130	16	24,450
McKiernan-Terry, diesel	DE-10	3,518	1,100	48	108	6,600
	DE-20	6,325	2,000	48	113	12,000
	DE-30	9,075	2,800	48	129	16,800
	DE-40	11,275	4,000	48	129	24,000
Link-belt Speeder, diesel	180	4,550	1,725	90		8,100
	440	10,300	4,000	86		18,200
	520	12,545	5,070	80		26,300

Table 6-3 gives information for sheet piles manufactured by the United States Steel Company.

Table 6-4 gives representative rates for driving steel-sheet piles, when they are driven in pairs, using the size hammer recommended in Table 6-1.

Example   Estimate the cost of furnishing and driving steel-sheet piling for a cofferdam to enclose a rectangular area 60 by 100 ft. The piles will be 16 in. wide, weighing 30.7 lb per

TABLE 6-3 Properties of U.S. Steel Company Steel-sheet Piles

Section number	Width, in.	Web thickness, in.	Weight, lb Per lin ft of pile	Per sq ft of wall
MP-101	15	$3/8$	35.0	28.0
MP-102	15	$1/2$	40.0	32.0
MP-117	15	$3/8$	38.8	31.0
MP-110	16	$31/64$	42.7	32.0
MP-112	16	$3/8$	30.7	23.0
MP-113	16	$1/2$	37.3	28.0
MP-116	16	$3/8$	36.0	27.0
MP-115	$19 5/8$	$3/8$	36.0	22.0
MP-27	18	$3/8$	40.5	27.0
MP-32	21	$3/8$	56.0	32.0
MP-38	18	$3/8$	57.0	38.0

TABLE 6-4 Representative Number of Steel-sheet Piles Driven per Hour

Length of pile, ft	Depth of penetration	Weight of pile, lb per lin ft 20	30	40
20	$1/2$	6	$5 3/4$	$5 1/2$
	Full	$5 1/2$	$5 1/4$	$5 1/4$
25	$1/2$	5	$4 3/4$	$4 1/2$
	Full	$4 1/4$	4	4
30	$1/2$	$4 1/2$	$4 1/4$	4
	Full	4	$3 3/4$	$3 1/2$
35	$1/2$	4	$3 3/4$	$3 1/2$
	Full	$3 1/2$	$3 1/4$	3
40	$1/2$	. . .	$3 1/4$	3
	Full	. . .	3	$2 3/4$
45	$1/2$	. . .	. . .	$2 3/4$
	Full	. . .	. . .	$2 1/2$
50	$1/2$	. . .	. . .	$2 1/2$
	Full	. . .	. . .	$2 1/4$

lin ft, section number MP112, Table 6-3, 24 ft long. The piles will be driven in pairs to full penetration into soil having normal frictional resistance.

A hammer delivering approximately 6,000 ft-lb of energy per blow should be used. A suitable steel cap should be placed on top of the piles to protect them from damage during driving. The hammer will be suspended from an 8-ton 12-ft-radius gasoline-engine-powered crawler-mounted crane. Steam will be supplied by a 25-hp boiler, fired with fuel oil.

The cost will be determined as follows.

$$\text{No. piles required, } \frac{320 \text{ ft} \times 12 \text{ in.}}{16 \text{ in.}} = 240$$

Corner piles will be of the same section, bent at right angles.
The cost will be

Piling, 240 × 24 × 30.7 = 176,832 lb @ $0.16	=	$28,293.00
Moving equipment to the job and back to the storage yard	=	320.00
Crane, 240 piles ÷ 4 per hr = 60 hr @ $10.88	=	652.80
Hammer, 60 hr @ $3.04	=	182.40
Boiler, fuel, and water, 60 hr @ $3.60	=	216.00
Other equipment and supplies, 60 hr @ $2.60	=	156.00
Allow 8 hr for the crew to set up and take down the equipment		
Foreman, 60 + 8 = 68 hr @ $7.75	=	526.00
Crane operator, 68 hr @ $6.75	=	458.00
Crane oiler, 68 hr @ $4.75	=	323.00
Fireman, 68 hr @ $5.75	=	391.00
Man on hammer, 68 hr @ $6.75	=	458.00
Helpers, 2 men, 136 hr @ $3.75	=	510.00
Total cost	=	$32,487.20

Wood piles The cost of wood load-bearing piles in place will include the cost of the piles delivered to the job; the cost of moving the pile-driving equipment to the job, setting it up, taking it down, and moving out; and the cost of equipment and labor driving the piles. As the cost of moving in, setting up, taking down, and moving out is the same regardless of the number of piles driven, the cost per pile will be lower for a greater number of piles.

The cost of piles is usually based on the length, size, quality, treatment, and location of the job. An estimator should obtain prices for the sizes to be used before preparing the estimate. As piles are sometimes broken in driving them, it may be desirable to include a few extras in the estimate. If the contractor is paid for the total number of linear feet of piles driven, any piles which cannot be driven to full penetration will result in some wastage. When it is necessary to cut the tops of piles off to a fixed elevation, the estimate should include the cost of cutting.

Driving wood piles A relatively fixed amount of energy is required to drive a pile of a given length into a given soil, regardless of the frequency of the blows. Consequently, a hammer delivering a given amount of energy

TABLE 6-5 Representative Number of Wood Piles Driven per Hour, Full Penetration

Length of piles, ft	Piles driven per hr	
	Normal friction	High friction
20	5	4
28	4¼	3
32	3½	2½
36	3	2¼
40	2¾	2
50	2½	1¾
60	2	1½

per blow will reduce the driving time if it strikes more blows per minute. This is particularly true in driving piles into materials where skin friction and not point resistance is to be overcome, as the skin friction will not have as much time to develop between blows when the blows are struck more frequently.

Table 6-5 gives the approximate number of wood piles that should be driven per hour for various lengths and driving conditions. The rates are based on using a hammer of the proper size, as indicated in Table 6-1.

Example Estimate the cost of furnishing and driving 160 wood piles 36 ft long into a soil having normal frictional resistance. The piles will be approximately 14 in. in diameter at the butt and 7 in. at the tip. The piles will be treated with creosote preservative at the rate of 8 lb per cu ft.

The average weight of the piles will be about 37 lb per lin ft. Reference to Table 6-1 indicates that the hammer should deliver approximately 7,000 theoretical ft lb of energy per blow. Use a Vulcan size 2 single-acting hammer. The driving rate is estimated to be 3 piles per hour.

Include 5 extra piles for possible breakage or damage.

The cost will be

Piles for the job, 165 × 36 = 5,940 lin ft @ $1.85	=	$10,989.00
Moving equipment to the job and back to the storage yard	=	420.00
Crane, 8 ton, 12 ft-radius, 160 ÷ 3 = 53 hr @ $10.88	=	576.50
Pile hammer, 53 hr @ $3.04	=	161.00
Boiler, fuel, water, etc., 53 hr @ $3.60	=	190.80
Leads, hose, accessories, 53 hr @ $3.25	=	172.00
Allow 16 hr for the crew to set up and dismantle the equipment		
Foreman, 53 + 16 = 69 hr @ $7.75	=	535.00
Crane operator, 69 hr @ $6.75	=	466.00
Crane oiler, 69 hr @ $4.75	=	327.80

Fireman, 69 hr @ $5.75   =   397.00
Man on hammer, 69 hr @ $6.75   =   466.00
Helpers, 2 men, 138 hr @ $3.75   =   517.50
                                                              _____
Total cost   = $15,218.60
Cost per pile, $15,218.60 ÷ 160   =   95.40
Cost per lin ft, $95.40 ÷ 36   =   2.65

Concrete piles   There are many kinds of concrete piles. The more commonly
used types are precast, cast in place in metal tubes, and cast in place in
drilled holes. Because of the varied conditions under which concrete piles
are cast and driven, it is not possible to prepare a general estimate that will
apply to a particular project. In order to estimate the cost, the estimator
must understand all the factors which affect the cost.

Precast concrete piles   Precast concrete piles are usually square or octagonal in cross section. Round piles must be cast in vertical tubes, which
makes handling difficult. The cross section may vary uniformly from the
butt to the tip, or it may be constant except for a short length near the tip,
which has a sharp taper to facilitate driving.

The reinforcing steel consists of longitudinal bars enclosed in a steel-
wire spiral, according to the specifications of the designer. For square and
octagonal piles, which are cast in horizontal forms, the reinforcing should be
fabricated into a single unit for each pile, before placing it into the form.

If piles meeting the specifications cannot be purchased already cast,
it will be necessary to cast them at the job. To reduce handling costs, they
should be cast near the driving site. A level area is selected, if available,
for the casting bed. An adequately supported wood floor is built on the
ground, and wood side forms are erected. The side forms may be reused
several times. The reinforcing is installed, and the concrete is poured into
the forms. Piles should be cast sufficiently early to gain the necessary
strength before they are driven. It will be necessary to keep them wet while
they are curing. Burlap, straw, or sand may be used for this purpose. If the
piles are close enough, they can be pulled to driving position by the driving
rig with a wire rope.

Cast-in-place concrete piles   Several methods of casting concrete piles
in place are used. In general, they involve driving tapered steel shells or
steel pipes which are later filled with concrete. While the shells are left in
place, the pipes sometimes are withdrawn as the concrete is deposited.

Cast-in-place piles are especially suitable for use on projects where
the soil conditions are such that the depth of penetration is not known in
advance and the depth varies among the piles driven.

Monotube piles   The monotube piles is obtained by driving a fluted
tapered-steel tube, closed at the tip, to the desired penetration. Additional
sections may be welded to the original tube as driving progresses to increase

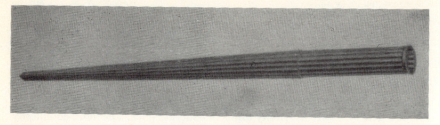

Fig. 6-4 Monotube pile. (*Union Metal Manufacturing Company.*)

the length, if required. After driving is completed in a given area, the condition of each tube is examined by lowering a light into the tube. A damaged tube can be removed and replaced with another. Concrete is then poured into the tubes. No additional reinforcing is required. If steel dowels are specified, they are driven into the tops of the piles before the concrete hardens (Fig. 6-4).

Several advantages are claimed for this type of pile. Damaged or broken piles will not result from driving. Driving subsequent piles will not damage piles previously driven as the concrete is not poured until driving has progressed beyond the area affected by driving. There is no loss of materials, as any excess tube length can be cut off and used on another tube. Cutting off of piles that cannot be driven to the expected penetration is eliminated.

Table 6-6 gives dimensions and other information on type F tubes.

TABLE 6-6 Data on Monotube Piles, Type F, Taper 1 in. in 7 ft, Tip Diameter 8 in.

Length, ft	Butt diam, in.	Weight of shells, lb			Volume of concrete per pile, cu ft
		No. 11 gauge	No. 9 gauge	No. 7 gauge	
10	9.4	137	166	195	3.5
15	10.1	202	248	293	5.8
20	10.8	271	334	396	8.4
25	11.5	348	430	511	11.4
30	12.2	425	526	627	14.3
35	12.9	512	635	757	18.1
40	13.6	600	744	988	22.1
45	13.9	700	865	1,030	26.6
50	14.6	794	983	1,172	31.0
55	15.3	894	1,111	1,328	36.5
60	16.0	1,004	1,244	1,438	42.2
70	17.4	1,231	1,529	1,826	55.5
75	18.1	1,356	1,685	2,013	62.8

Raymond steel-encased concrete
piles are installed by driving
the required length of steel shell
and internal steel mandrel;
withdrawing the mandrel, leaving
the steel shell in place; inspecting
the driven shell internally; and
filling the shell with concrete.
Numbered shell sections are made
in 4-, 8-, 12-, and 16-ft. lengths.
Longer lengths can be furnished
for special conditions. The point
section is closed at the bottom
by a flat steel plate welded to the
boot ring. Shell sections are screw-
connected.

8	18 3/8 in.
7	17 3/8 in.
6	16 3/8 in.
5	15 3/8 in.
4	14 3/8 in.
3	13 3/8 in.
2	12 3/8 in.
1	11 3/8 in.
0	10 3/8 in.
00	9 1/2 in.
000	8 5/0 in.

Fig. 6-5 Raymond step-taper pile. (*Raymond International Inc.*).

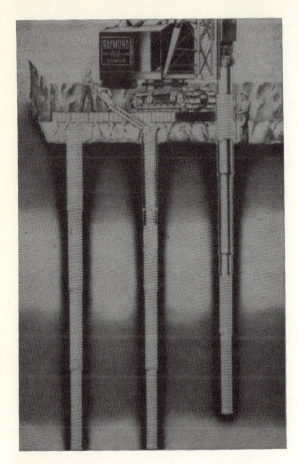

Fig. 6-6 Driving Raymond step-taper piles. (*Raymond International Inc.*)

Raymond step-tapered concrete piles This pile is obtained by driving a corrugated-steel shell, made by joining sections 4-, 8-, 12-, or 16-ft long to produce a total length to fit the particular need of a project. A shell may be assembled using more than one length, if desired. Figure 6-5 illustrates a pile shell after the sections are assembled. The lower end of the bottom shell is closed with a steel plate or a hemispherical boot. If shells having tip diameters larger than the one shown in Fig. 6-5 are desired, the sections below the desired diameter are omitted.

A steel mandrel, which has stepped outer surfaces to correspond with the inside diameters of the shell sections, is inserted into the shell before it is driven. This mandrel bears on each of the stepped shoulders of the shell to distribute the driving energy throughout the shell. When the desired penetration is obtained, the mandrel is withdrawn and the shell is inspected for possible damage, after which it is filled with concrete.

Information furnished by the manufacturer gives the volume of concrete required to fill the shells.

Cost of cast-in-place concrete piles In estimating the cost of cast-in-place concrete piles, it is necessary to determine the cost of the shells delivered to the project, the cost of equipment and labor required to drive the shells, and the cost of the concrete placed in the shells. All these costs except the cost of driving the shells are easy to obtain. The rate of driving varies with the length of the piles, the class of soil into which they are driven, the type of driving equipment used, the spacing of the piles, the topography of the site, and weather conditions. If a project requires the driving of a substantial number of piles, it may be desirable to make subsoil tests in order to obtain reliable information related to the ease or difficulty of driving piles, and the lengths of piles required (Fig. 6-6).

Steel piles Steel piles are used when they seem more suitable or economical than wood or concrete piles. They are especially suited to projects where it is necessary to drive piles through considerable depths of poor soil in order to reach solid rock or other formation having high load-supporting properties. Steel H sections or wide-flange beams are most frequently used, although fabricated sections are sometimes used for special conditions. Rolled sections are less expensive than fabricated sections.

Driving operations are similar to those for other piles. Steel piles do not require the care in handling that must be observed in handling precast-concrete piles. The danger of damaging a pile is reduced by using a driving cap which fits the particular pile.

TABLE 6-7 Data on Selected Steel-pile Sections

Size, in.	Area, sq in. H sections	Weight, lb per lin ft
14 × 14½	34.44	117
	30.01	102
	26.19	89
	21.46	73
12 × 12	21.76	74
	15.58	53
10 × 10	16.76	57
	12.95	44
	12.35	42
8 × 8	10.60	36

TABLE 6-8   Approximate Number of Steel Piles
Driven to Full Penetration,* per Hour

Length of pile, ft	Weight, lb per lin ft			
	102	74	57	36
30	3	3¼	3½	3½
35	2½	2¾	3	3
40	2¼	2¼	2½	2½
45	1½	1¾	1¾	
50	1½	1½	1¾	
55	1¼	1½		
60	1	1		
70	¾			
80	½			

* The rates are based on piles being delivered to the job in
lengths up to 40 ft. For lengths greater than 40 ft, a field
weld will be required for each pile.

Table 6-7 gives the weights in pounds per linear foot for various steel
piles.

Table 6-8 gives approximate rates for driving steel piles, using a hammer
of the recommended size.

Example   Estimate the cost for bid purposes for furnishing and driving 180 12- by 12-in.
H-section steel piles 40 ft long, weighing 74 lb per lin ft. The piles will be driven to full
penetration into soil having high frictional resistance.

Reference to Table 6-1 indicates that a 15,000-ft-lb single-acting steam hammer will be
satisfactory.

A diesel-engine-powered crawler crane will be used to lift the piles into position, support the leads, and lift the hammer. The size crane required may be determined as follows:

Item	Weight on crane, lb
Piles, 40 ft × 74 lb	= 2,960
Hammer	= 9,600
Leads	= 5,000
Pile cap	= 440
Total weight	= 18,000 lb
	= 9 tons

Assume that during the driving of the piles the maximum radius required will be 25
ft for the crane. This radius and the load on the crane will give a product equal to $9 \times 25 =$
225 ton ft. Use a 20-ton 12-ft-radius crane to give a load-radius product of 240 ton ft.

Reference to Table 6-8 indicates a probable driving rate of 2-¼ piles per hour. The time required to drive the piles should be about $180 \div 2\text{-}\frac{1}{4} = 80$ hr. This time will be used. The cost will be

Material:

Piles, $180 \times 40 \times 74 = 532{,}800$ lb @ $0.14	= $74,592.00

Equipment:

Moving to and from the job	= 825.00
Crane, 80 hr @ $14.50	= 1,160.00
Hammer, 80 hr @ $2.76	= 220.80
Boiler, fuel, water, hose, etc., 80 hr @ $8.50	= 680.00
Leads, 80 hr @ $2.25	= 180.00
Pick-up truck, ¾ ton, 96 hr @ $2.86	= 274.00
Utility truck, 2 ton, 96 hr @ $3.81	= 366.40
Sundry equipment, tools, and supplies, 80 hr @ $2.50	= 200.00
Subtotal cost of equipment	= $ 2,906.20

Labor:

Allow 16 hr to set up and dismantle the equipment	
Foreman, $80 + 16 = 96$ hr @ $7.75	= $ 744.00
Crane operator, 96 hr @ $6.75	= 648.00
Crane oiler, 96 hr @ $4.75	= 456.00
Fireman, 96 hr @ $5.50	= 527.50
Guidemen, $2 \times 96 = 192$ hr @ $4.75	= 911.50
Laborers, $2 \times 96 = 192$ hr @ $3.25	= 623.00
Superintendent, three-fourth month @ $1,500.00	= 1,125.00
Subtotal cost of labor	= $ 5,035.00

Summary of direct costs:

Materials	= $74,592.00
Equipment	= 2,906.20
Labor	= 5,035.00
Subtotal	= $82,533.20
Workman's compensation insurance, $5,035.00 \times $8.14 per $100.00	= 410.06
Social security tax, 5.85% of $5,035.00	= 294.55
Unemployment tax, 3% of $5,035.00	= 151.05
Subtotal	= $83,388.86
General overhead, 5% of $83,388.86	= 4,169.45
Subtotal	= $87,558.31
Profit, 8% of $87,588,31	= 7,004.66
Subtotal	= $94,562.97
Performance bond, 1% of $94,562.97	= 945.63
Total cost, amount of bid	= $95,508.60
Cost per lin ft, $95,508.60 \div 7,200$ ft	= 13.27

Jetting piles into position If the formation into which the piles are driven contains considerable sand, the rate of driving may be increased by jetting with water. To accomplish this with concrete piles, steel pipes may be placed in the piles when they are cast. The pipe, 1½ to 2 in. in diameter, should extend from the tip to within about 2 ft of the top, where it protrudes from the side by the use of an ell. It may require 100 to 400 gal of water per min or more to jet a concrete pile into position. Required pressures will be 100 to 200 psi. For suitable soil conditions this method will sink piles much more quickly than with pile-driving hammers.

Drilled shafts and underreamed footings   A method which is used to produce modified cast-in-place concrete piles is to drill holes to the desired depths with rotary-drilling rigs. The diameters of the holes vary from about 10 to 72 in. or more. Depths up to approximately 100 ft are possible. Near the bottom, the diameter of the hole is gradually increased with an underreaming bit to form a footing resembling a cone. In the event that there is danger of the soil caving into the hole, an open steel cylinder, having an outside diameter slightly less than that of the hole, is temporarily placed

TABLE 6-9   Volume of the Shaft and the Underream for Various Diameters of Holes

Volume of shaft per foot of depth

Diam, in.	Volume, cu ft	Diam, in.	Volume, cu ft
12	0.78	30	4.90
14	1.07	36	7.07
16	1.40	48	12.60
18	1.77	60	19.60
24	3.14	72	28.30

Volume of underream*

Diam bottom, in.	Volume, cu ft	Diam bottom, in.	Volume, cu ft
18	0.9	56	26.4
20	1.2	60	32.7
24	2.0	72	56.0
30	4.1	84	86.4
36	7.2	96	125.0
42	11.2	108	217.6
48	16.7	120	301.4

* Varies with the shape of the underream.

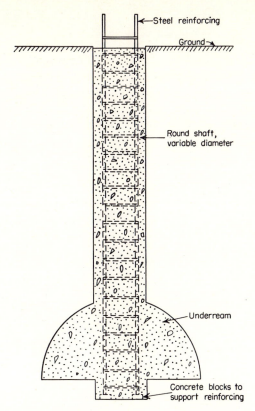

Fig. 6-7 Drilled hole for cast-in-place concrete pile.

in the hole until it is filled with concrete. If reinforcing steel is required, it is prefabricated into proper units and lowered into position. Lugs, attached to the units, hold them in position at the center of the piles.

Cost of concrete piles in drilled holes The cost of piles in place, using this method, will include the cost of drilling the holes, the cost of reinforcing, and the cost of concrete in place.

Table 6-9 gives the volume of the shaft and the underream for the type of hole illustrated in Fig. 6-7.

Example Estimate the cost of 50 concrete piles cast in drilled shafts and undereamed footings, similar to the one in Fig. 6-7. The shafts will be 20 ft long, with 18-in. diameters. The bottom diameter of the underreams will be 60 in. The soil is tight sandy clay, firm enough to stand without support in the drilled holes.

Reinforcing steel will consist of four ⅞-in. round bars 25 ft long and a ¼-in.-diameter spiral with a 12-in. pitch.

Ready-mixed concrete will be used, with the delivery trucks discharging directly into the holes.

The quantities will be

Volume of shafts, 50 × 20 × 1.77	= 1,770 cu ft
Volume of underreams, 50 × 32.7	= 1,635 cu ft
Total volume	= 3,405 cu ft
	= 126.5 cu yd
Reinforcing steel:	
Longitudinal bars, 50 × 25 × 2.044 lb	= 10,220 lb
Spiral steel, 4,580 ft × 0.167	= 765 lb
Total weight	= 10,985 lb
Weight of tie wire, 26 lb	

The cost will be

Reinforcing steel in place, 10,985 lb @ $0.21	= $2,306.85
Tie wire, 26 lb @ $0.25	= 6.50
Concrete in place, 126.5 cu yd @ $22.75	= 2,877.87
Moving the equipment to and from the job	= 130.00
Drilling rig, 50 holes × 1½ hr per hole, = 75 hr @ $10.50	= 787.50
Add 8 hr for labor to move in and out	
Foreman, 75 + 8 = 83 hr @ $7.75	= 643.00
Driller, 83 hr @ $6.75	= 560.00
Helpers, 2 men, 166 hr @ $3.75	= 622.50
Total cost	= $7,934.22
Cost per pile, $7,934.22 ÷ 50	= 158.68

PROBLEMS

6-1 Estimate the total direct cost and the cost per linear foot for furnishing and driving 96 creosote-treated wood piles to full penetration into soil having high frictional resistance. The piles will be 40 ft long, with 14 in. minimum diameter at the butt and 6 in. minimum diameter at the tip. The tip of each pile will be fitted with a metal point to facilitate driving.

The crane used on the job will be a 12-ton 12-ft-radius diesel-engine-powered crawler-mounted unit.

The cost of moving the equipment to the job and back to the storage yard will be $420.00

Use the information appearing in the example beginning on page 114 and in Appendix B to determine the cost of the job.

The piles will cost $2.25 per lin ft delivered to the job. The pile points will cost $3.80 each.

It will not be necessary to cut off the tops of the piles.

6-2 Estimate the total direct cost and the cost per linear foot for furnishing and driving the piles of Prob. 6-1 to full penetration into soil having normal frictional resistance. Use a metal point on the tip of each pile.

The cost of moving to the job and back to the storage yard will be $360.00.

6-3 Estimate the total direct cost and the cost per linear foot for furnishing and driving 84 12- by 12-in. steel piles weighing 53 lb per lin ft to full penetration into soil having high frictional resistance. The piles will be 40 ft long.

The piles will cost $0.18 per lb delivered to the job.

The cost of moving the equipment to the job and back to the storage yard will be $540.00.

Use a 12-ton 12-ft-radius diesel-engine-powered crawler-mounted crane to handle the piles and hammer.

Use cost information appearing in the example on page 120 and in Appendix B to estimate the cost of the project.

6-4 Estimate the total direct cost and the cost per linear foot for furnishing and driving 66 14- by 14-in. steel piles weighing 117 lb per lin ft into soil having high frictional resistance. The piles will be driven to full penetration. The piles will be 40 ft long.

Assume that the leads will weigh 6,000 lb. Select the size hammer suitable for driving the piles and then select the size crane required to handle the piles, hammer, and leads. Assume a 12-ft radius for the crane.

Material costs, wage rates, and other conditions will be the same as for the example on page 120.

7

Concrete Structures

Costs of concrete structures The items which govern the total costs of concrete structures include the following:

1 Forms
2 Reinforcing steel
3 Concrete
4 Finishing, if required
5 Curing

When preparing an estimate, it is suggested that the cost of each of these items be determined separately and that a code number be assigned to each item. The cost of an item should include all materials, equipment, and labor required for that item. Following this procedure will simplify the preparation of an estimate.

FORMS

Forms for concrete structures Concrete structures may be constructed in any shape for which it is possible to build forms. However, the cost of forms for complicated shapes is considerably greater than for simple shapes because of the extra material and labor costs required to build them and the low salvage value after they are used.

Since the major cost of a concrete structure frequently results from the cost of forms, the designer of a structure should consider the effect which the shape of a structure will have on the cost of forms.

Forms for concrete are fabricated from lumber, plywood, steel, aluminum, and various composition materials, either separately or in combination. If the form material will be used only a few times, lumber will usually be more economical than steel or aluminum. However, if the forms can be fabricated into panel sections or other shapes, such as round column forms, that will be used many times, the greater number of uses obtainable with steel and aluminum may produce a lower cost per use than with lumber.

That material should be selected for forms which will give the lowest total cost of a structure, considering the cost of the forms plus the cost of finishing the surface of the concrete, if required, after the froms are removed. The use of wood planks usually will leave form marks which may have to be removed at considerable cost, whereas the use of plywood, pressed wood, or metal forms might eliminate the cost of removing form marks. It is good practice to spend an extra 5 cents per sq ft on forms if by so doing it is possible to eliminate a finishing cost of 10 cents per sq ft.

The cost of forms will include the cost of materials, such as lumber, nails, bolts, form ties, and the cost of labor making, erecting, and removing the forms. Frequently there will be a cost for power equipment, such as saws and drills, and hand tools. If it is possible to reuse the forms, an appropriate allowance should be made for salvage value. If the forms are treated with oil prior to each use, the cost of the oil should be included.

The cost of materials for forms   The cost of materials for forms should include an allowance for waste.

The net dimensions of lumber are given in Chap. 11. Since the net dimensions are less than the nominal dimensions, it is necessary to use the former when determining the quantity required. Standard lengths are available in multiples of 2 ft, with the longer lengths carrying an extra charge. If an odd length, such as 9 ft, is desired, it may be obtained by cutting an 18-ft-long plank into two equal lengths, or it may be obtained by cutting 1 ft off a 10-ft-long plank, with a resulting waste.

Plywood used for forms should be made with waterproof glue, which will not dissolve when wet. It is available in thicknesses varying from $\frac{1}{4}$ to 1 in. or more and in sheets 4 ft wide and 8, 10, or 12 ft long. Other dimensions may be obtained on special order.

Nails required for forms   The quantity of nails required for forms usually will vary from 10 to 20 lb per M fbm of lumber for the first use of the lumber and from 5 to 10 lb per M fbm for additional uses if the forms can be reused without completely refabricating them.

Form oil   When form oil is used to treat the surfaces that will come in contact with concrete, it should be applied at a rate of 300 to 500 sq ft per gal. It may be applied with mops, brushes, or pressure sprayers.

Labor required to build forms   The factors which determine the amount of labor required to build forms for concrete structures include the following:

1   Size of the forms.

2   Kind of materials used. Large sheets of plywood require less labor than planks.

3   Shape of the structure. Complicated shapes require more labor than simple shapes.

4   Location of the forms. Forms built above the ground require more labor than forms built on the ground or on a floor.

5   The extent to which prefabricated form panels or sections may be used.

6   Rigidity of dimension requirements.

7   The extent to which power equipment is used to fabricate the forms.

If forms are prefabricated into panels or sections, then assembled, used, removed, and reused, it is desirable to estimate separately the labor required to make, assemble, and remove them. Since making is required only once, for additional uses it will only be necessary to assemble and remove the forms.

The production rates given in this book are based on using power saws and other equipment as much as possible. If the fabricating is done with hand tools, the labor required should be increased.

The tables on production rates give the minimum and maximum number of man-hours required to do a specified amount of work, both for carpenters and helpers. If union regulations specify the ratio of helpers to carpenters for certain locations, it may be necessary to transfer some of the helper time to carpenter time. The tables include an allowance for lost time by using a 45- to 50-min hr.

Forms for footings   Many soils are sufficiently rigid to permit their use as forms for concrete footings poured into holes dug for the footings. When this is possible, there will be no costs for lumber or carpenter's building forms.

If the soil is not sufficiently rigid, it will be necessary to install forms. Forms for most footings consist of 1-in.-thick lumber planks, which are

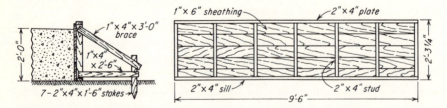

Fig. 7-1   Forms for a concrete footing.

cleated together, side by side, in the correct width. These planks are designated as sheathing. The sheathing is held in place by 2- by 4-in. studs and braces, spaced about 18 in. apart around the outside of the forms. Figure 7-1 shows the details of a typical form for a footing.

The quantity of lumber will be

Sheathing, 5 pc, 1 in. × 6 in. × 10 ft 0 in.	= 25 fbm
Plates, 2 pc, 2 in. × 4 in. × 10 ft 0 in.	= 14 fbm
Studs, 7 pc, 2 in. × 4 in. × 2 ft 0 in.	= 9 fbm
Stakes, 7 pc, 2 in. × 4 in. × 1 ft 6 in.	= 7 fbm
Braces, 7 pc, 1 in. × 4 in. × 3 ft 0 in.	= 7 fbm
Braces, 7 pc, 1 in. × 4 in. × 2 ft 6 in.	= 6 fbm
Total lumber	= 68 fbm

Lumber per sq ft of surface, 68 ÷ 19 = 3.6 fbm
Nails, 68 fbm @ 10 lb per M fbm = 0.7 lb

The labor-hours required to build and install forms for 100 sq ft of surface-contact area for footings should be approximately as given in the accompanying table.

	Laborer	
Carpenter	Erecting	Removing
3½	2	1

Forms for foundation walls Foundation walls include basement walls for buildings, retaining walls, concrete cores for earth-filled dams, vertical walls for water reservoirs, etc. The forms for such walls consist of 1-in.-thick sheathing, 2- by 4-in. or 2- by 6-in. studs, double 2- by 4-in. or 2- by 6-in. wales, and 2- by 4-in. or 2- by 6-in. braces. The sheathing, studs, and wales are erected on both sides of the wall, while the braces are spaced about 6 to 8 ft apart on one side of the wall only.

Design of forms for foundation walls In order to estimate correctly the quantity of materials for wall forms, it is necessary to design the forms for the particular wall. Proper design will assure adequate safety with economy.

The factors which affect the design are

1 The consistency and proportions of the concrete
2 The rate of filling the form
3 The temperature of the concrete
4 The method of placing the concrete
5 The depth of drop and distribution of reinforcing steel

These factors affect the pressure at any given depth below the surface of the freshly placed concrete. The rate of filling and the temperature determine the effective depth of liquid concrete, that is, the depth which has not set long enough to possess internal rigidity. As the concrete sets, it reduces the pressure on the forms. Concrete sets more slowly at low temperatures than at high ones. Lower maximum pressures will be exerted by concrete having a low slump than one having a high slump.

The maximum pressure of concrete on wall and column forms may be determined from the following formulas, which were developed by the American Concrete Institute, and published in 1963 [1].

For walls:

$$P_m = 150 + \frac{9,000R}{T} \qquad \text{for } R \text{ up to 7 ft per hr} \tag{7-1}$$

$$P_m = 150 + \frac{43,400}{T} + \frac{2,800R}{T} \qquad \text{for } R \text{ more than 7 ft per hr} \tag{7-2}$$

For columns:

$$P_m = 150 + \frac{9,000R}{T} \tag{7-3}$$

where P_m = maximum pressure on forms, lb per sq ft
  R = rate of filling forms, ft per hr
  T = temperature of concrete, °F

The maximum pressure obtained from formula (7-2) is limited to 2,000 lb per sq ft, and for formula (7-3) it is limited to 3,000 lb per sq ft. The pressures are for concrete that is consolidated by internal vibration as it is placed.

Table 7-1 gives the maximum pressures resulting from placing concrete in wall forms. Table 7-2 gives the maximum pressures resulting from placing concrete in column forms.

Regardless of the pressures given in Tables 7-1 and 7-2, the maximum pressures on wall and column forms will not exceed the weight of 1 cu ft of concrete times the depth to the given area in feet.

The pressures given in Table 7-1 and Table 7-2 are effective for concrete weighing 150 lb per cu ft. If forms are filled with concrete whose weight is more or less than 150 lb per cu ft, the maximum pressures appearing in the two tables should be multiplied by a factor equal to the weight of 1 cu ft of the concrete in question divided by 150 [2].

Design of forms for concrete walls using tables   A number of publications contain tables that may be used when designing or building forms for concrete walls. In general, these tables have been useful and satisfactory.

TABLE 7-1 Relation between the Rate of Filling Wall Forms, Maximum Pressure, and Temperature (ACI)

Rate of filling forms, ft per hr	Maximum concrete pressure, lb per sq ft						
	Temperature, °F						
	40	50	60	70	80	90	100
1	375	330	300	279	262	250	240
2	600	510	450	409	375	350	330
3	825	690	600	536	487	450	420
4	1,050	870	750	664	600	550	510
5	1,275	1,050	900	793	712	650	600
6	1,500	1,230	1,050	921	825	750	690
7	1,725	1,410	1,200	1,050	933	850	780
8	1,793	1,466	1,246	1,090	972	877	808
9	1,865	1,522	1,293	1,130	1,007	912	836
10	1,935	1,578	1,340	1,170	1,042	943	864
15	2,185*	1,858	1,573	1,370	1,217	1,099	1,004
20	2,635*	2,138*	1,806	1,570	1,392	1,254	1,144

* These values are limited to 2,000 lb per sq ft.

TABLE 7-2 Relation between the Rate of Filling Column Forms, Maximum Pressure, and Temperature (ACI)

Rate of filling forms, ft per hr	Maximum concrete pressure, lb per sq ft						
	Temperature, °F						
	40	50	60	70	80	90	100
1	375	330	300	279	262	250	240
2	600	510	450	409	375	350	330
3	825	690	600	536	487	450	420
4	1,050	870	750	664	600	550	510
5	1,275	1,050	900	793	712	650	600
6	1,500	1,230	1,050	921	825	750	690
7	1,725	1,410	1,200	1,050	937	850	780
8	1,950	1,590	1,350	1,179	1,050	950	870
9	2,175	1,770	1,500	1,307	1,162	1,050	960
10	2,400	1,950	1,650	1,436	1,275	1,150	1,050
12	2,850	2,310	1,950	1,693	1,500	1,350	1,230
15	3,525*	2,850	2,400	2,093	1,837	1,650	1,500
20	4,650*	3,750*	3,150*	2,721	2,400	2,150	1,950

* These values are limited to 3,000 lb per sq ft.

TABLE 7-3   Design of Forms for Concrete Walls

Minimum temperature of concrete °F	50			70			90		
Rate of filling forms, ft per hr	2	4	6	2	4	6	2	4	6
Maximum pressure, lb per sq ft	510	870	1,230	409	664	921	350	550	750
Maximum spacing of studs for safe value of sheathing, in.									
For 1 in. sheathing	22	17	14	24	19	16	26	21	18
For 2 in. sheathing	38	29	24	42	33	28	45	36	31
Maximum spacing of wales for safe value of studs, in.									
2 × 4 studs 1 in. sheathing	26	23	21	28	25	23	29	26	24
4 × 4 studs 1 in. sheathing	40	35	33	43	38	35	45	40	37
2 × 6 studs 1 in. sheathing	41	36	33	44	39	36	46	41	38
2 × 6 studs 2 in. sheathing	31	27	25	33	29	27	35	31	29
4 × 4 studs 2 in. sheathing	31	27	25	33	29	27	34	30	28
3 × 6 studs 2 in. sheathing	41	36	33	43	38	35	45	41	37
Maximum spacing of form ties for safe values of wales, in.									
Double 2 × 4 wale 2 × 4 stud 1S	34	28	24	37	31	27	39	33	29
Double 2 × 4 wale 4 × 4 stud 1S	30	24	21	32	27	24	34	29	26
Double 2 × 4 wale 2 × 6 stud 1S	27	22	20	29	24	22	31	26	23
Double 2 × 6 wale 2 × 6 stud 1S	43	35	31	46	38	34	49	41	37
Double 2 × 6 wale 3 × 6 stud 2S	43	35	31	48	39	35	50	41	37

Table 7-3 contains information which should enable one to select a design that is suitable for most projects and conditions. The following conditions apply to the information contained in the table.

1 All lumber used is *S4S* stress-grade pine or fir as specified by the National Forest Products Association publication 06, National Design Specifications for Stress-grade Lumber and its Fastenings, 1973 edition, Washington, D.C.

2 Current dimensions of lumber, effective on January 1, 1973, were used in determining the values appearing in the table.

3 The listed maximum pressures of concrete were determined by using the American Concrete Institute formulas [formulas (7-1) and (7-3) in this book]. These formulas provide for the increased pressures resulting from compacting the concrete with interval vibrators as it is placed.

4 The listed maximum spacings for form ties are based on limiting the stresses in the lumber to safe values. When these spacings are used, the designer or form builder should be certain that he selects form ties that will be strong enough to resist the total stresses in the ties. Or, if it is desirable to use ties having lower strengths, they may be used with safety if the spacing of the ties is reduced sufficiently. For example, if a given design from Table 7-3 permits a maximum safe spacing of 48 in. but requires that 6,000-lb form ties be used for this spacing, the spacing may be reduced to 24 in., which will permit the use of 3,000-lb/ties.

5 The values appearing in Table 7-3 limit the deflections of all members, including sheathing, to not more than l/360.

6 Most of the values appearing in Table 7-3 are limited by the allowable unit fiber stress in bending, which, because of the short duration, permits stresses up to 1,800 psi in stress-grade lumber.

7 If concrete is placed in forms at rates other than those listed in Table 7-3, or if it is placed at different temperatures from those listed, the table may still be used to design forms. For example, assume that the concrete is placed at a temperature of 60 deg and the forms are filled at a rate of 8 ft per hr. Table 7-1 indicates a maximum pressure of 1,246 lb per sq ft. Table 7-3 indicates that if forms are filled at a rate of 6 ft per hr at a temperature of 50 deg, the maximum pressure will be 1,230 lb per sq ft. If the forms are designed to resist this pressure, they should be safe for the pressure of 1,246 lb per sq ft.

Materials required for forms for foundation walls Wall forms are frequently built as indicated in Fig. 7-2. Assume that the rate of filling the forms will be 4 ft per hr at a temperature of 70 deg.

The sheathing will be 1- by 6-in. *D* and *M* lumber, usually called center

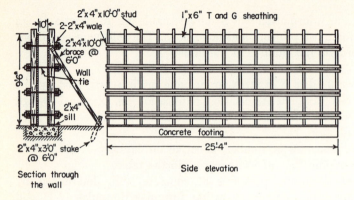

Fig. 7-2 Forms for a concrete wall.

match, whose actual dimensions are ¾ in. thick and 5⅛ in. wide. Studs will be 2- by 4-in., and wales will be double 2- by 4-in. planks.

Reference to Table 7-3 indicates the following pressure and spacings:

Maximum pressure, 664 psf
Spacing of studs, 19 in.
Spacing of wales, 25 in.
Spacing of form ties, 31 in.

The wall is 9 ft 6 in. high and 25 ft 4 in. long.

The sheathing will be supplied in commercial lengths of 12 ft 0 in. and 14 ft 0 in. The number of pieces of each length for each side of the forms will be $9 \times 12 + 6 = 114$ in. $\div$ 5⅛ in. $= 22.3$ pc (use 23 pc).

The total quantity of sheathing will be:
$2 \times 23 = 46$ pc, 1 in. $\times$ 6 in. $\times$ 12 ft 0 in. $= 276$ fbm
$2 \times 23 = 46$ pc, 1 in. $\times$ 6 in. $\times$ 14 ft 0 in. $= 322$ fbm
Studs required:
Length of wall, $25 \times 12 + 4 = 304$ in.
Spacing of studs, 19 in.
No. studs required per side, $(304/19) + 1 = 17$
No. studs required for wall, $2 \times 17 = 34$
Lumber required, 34 pc, 2 in. $\times$ 4 in. $\times$ 10 ft 0 in. $= 227$ fbm
Wales required:
Height of wall, 114 in.
Spacing of wales, 25 in.
No. required per side, $114 \div 25 = 4.55$. Use 5 wales*
For each wale use 2 pc of 2 $\times$ 4 in. $\times$ 12 ft 0 in. and 2 pc of 2 $\times$ 4 in. $\times$ 14 ft 0 in. lumber
Lumber required:
20 pc, 2 $\times$ 4 in. $\times$ 12 ft 0 in. $=$ 160 fbm
20 pc, 2 $\times$ 4 in. $\times$ 14 ft 0 in. $=$ 182 fbm

* It is possible that the reduced pressure near the top of the wall may permit a safe increase in the spacing of the wales sufficient to reduce the number of wales to four per side.

Sills required:

2 pc, 2 × 4 in. × 12 ft 0 in.		=	16 fbm
2 pc, 2 × 4 in. × 14 ft 0 in.		=	18 fbm

Scab splices required:

No. required per side, 2 × 5 = 10

Total number, 2 × 10 = 20

Lumber required:

20 pc, 2 × 4 in. × 2 ft 0 in.   =   27 fbm

Braces required, one side only:

Spacing, 6 ft 0 in., with one at each end

No. required, 5

Lumber, 5 pc, 2 × 4 in. × 10 ft 0 in.   =   33 fbm

Stakes:

5 pc, 2 × 4 in. × 3 ft 0 in.   =   10 fbm

        Total quantity of lumber   = 1,271 fbm

Area of the wall, 2 × 9.5 × 25.33 = 481 sq ft

Quantity of lumber required per sq ft of area, 1,271 ÷ 481 = 2.60 fbm

Quantity of nails required, 1,271 × 10 lb per M fbm = 13 lb

Form ties required:

No. per wale, 304 in. ÷ 31 in. = 10

No. per side, 5 × 10 = 50

Maximum load on a tie, 25 × 31/144 × 664 = 3,574 lb

Use 4,000-lb ties, or, if 3,000-lb ties are to be used, the spacing should be reduced to
31 × 3,000/3,574 = 26 in.

Form ties   In order to resist the internal pressure resulting from the con-

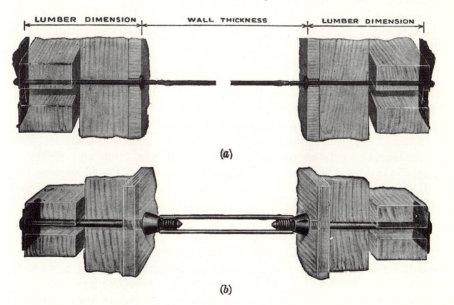

Fig. 7-3 Form ties for concrete wall: (*a*) snap tie; (*b*) coil tie. (*Superior Concrete Accessories, Inc.*)

crete, some positive method of holding the forms in position is necessary. Although wire may be used, this method is not so satisfactory or so dependable as the use of commercial form ties. A great many varieties are available. Most of them serve two purposes. They hold the forms apart prior to placing the concrete, then resist the bursting pressure of the concrete after it is placed.

The types available consist of narrow steel bands, plain rods, rods with hooks or buttons on the ends, and threaded rods, with suitable clamps or nuts to hold them in position. While most ties are designed to break off inside the concrete, there are some which can be pulled out of the wall after the forms are removed and the holes filled with concrete. Manufacturers specify the safe working stresses for their ties.

In ordering form ties it is necessary to specify the thickness of the wall, sheathing, studs, and wales.

As a guide to the cost of form ties, a popular type, having a safe working strength of 3,000 lb, is priced as shown in the accompanying table.

For wall thickness, in.	Cost per 100
From 4 to 12	$15.75
Over 12 to 18	17.85
Over 18 to 24	20.00
Over 24 to 30	21.10
Over 30 to 36	23.10

Two end clamps are required for each tie. The cost will be about $20.00 per 100 clamps, with a credit of $17.50 per 100 usable clamps if returned to the supplier within a specified time, usually about 4 months.

Labor building wood forms for foundation walls   For walls up to about 12 ft high, 2- by 4-in. studs and double 2- by 4-in. wales are adequate. For greater heights, up to approximately 20 ft, 2- by 6-in. studs and double 2- by 6-in. wales may be desirable. Double rows of braces should be used for walls over 12 ft high, one near the middle and the other near the top of the wall.

Table 7-4 gives the quantities of lumber and form ties and the labor-hours for erecting forms for 100 sq ft of walls for various heights. The quantity of form ties should be based on the area of only one side of the wall.

Plywood forms   If the specifications require a wall having a smooth finish or if the wall is to be rubbed with a stone to remove form marks, it may be economical to use plywood as a sheathing to reduce finishing costs. A special plywood with waterproof glue is used. The plywood is made in

TABLE 7-4   Quantities of Lumber and Form Ties and the Labor-hours Required for 100 Sq Ft of Wall Forms

Height of wall, ft	Lumber, fbm	Form ties*	Labor-hour building and removing	
			Carpenter	Helper
4	175	16	3.0–4.0	2.0–3.0
6	200	14	3.5–5.5	2.0–3.0
8	225	14	4.0–5.0	2.5–3.5
10	250	16	4.5–5.5	3.0–4.0
12	275	16	5.0–6.0	3.5–4.5
14	325	10	5.5–6.5	4.0–5.0
16	350	10	6.5–7.5	4.5–5.5
18	375	10	7.0–8.0	5.0–6.0

* The form ties must be strong enough to withstand the stresses for the particular uses. The lumber and form ties are based on filling the forms at a rate of 4 ft per hr at a temperature of 70 deg.

various thicknesses. Those commonly used are ¼, ⅜, ½, ⅝, and ¾ in. If a 1-in. wood sheathing is used, a ¼-in.-thick plywood inside the sheathing will be satisfactory. If the plywood serves as the sheathing, the thicker sections should be used. It is necessary to install the plywood with the outer layers perpendicular to the studs.

Plywood is available in 4- by 4-ft, 4- by 8-ft, 4- by 10-ft, and 4- by 12-ft sizes. Studs should be spaced so that there will be a stud at the end of each plywood sheet. In order to reduce the damage to plywood to a minimum, a standard pattern for holes for wall ties should be adopted. The adoption of such a practice will reduce the need for drilling additional holes for subsequent uses. The use of small nails will reduce the damage while removing. With reasonable care, plywood can be used many times.

Table 7-5 gives the quantities of plywood, lumber, and form ties and the labor-hours required for 100 sq ft of concrete wall using ¾-in.-thick plywood for various heights. For wall heights other than multiples of 2 ft, there will be side wastage. Also door and window openings, wall columns, and irregular-shaped walls will usually cause end wastage. The estimator should consider the extent of wastage from the nature of the structure.

Prefabricated form panels Contractors frequently make prefabricated form panels using 2- by 4-in. lumber for frames and 1-in.-thick planks or ¾-in.-thick plywood for sheathing. If planks are used for the sheathing,

TABLE 7-5   Quantities of ¾-in. Plywood, Lumber, and Form Ties and the Labor-hours Required for 100 Sq Ft of Wall

Height of wall, ft	Plywood, sq ft	Lumber, fbm	Form ties*	Labor-hours building, erecting, and removing	
				Carpenter	Helper
4	100	100	16	3.0–3.5	1.5–2.0
6	100	125	14	3.5–4.0	1.5–2.0
8	100	160	14	3.5–4.5	2.0–2.5
10	100	200	16	4.0–5.0	2.5–3.0
12	100	240	16	4.5–5.5	3.5–4.0
14	100	300	10	6.0–7.0	4.0–4.5
16	100	325	10	6.5–7.5	4.5–5.0
18	100	350	10	7.0–8.0	5.0–5.5

* The form ties must be strong enough to resist the stress for the particular use. The lumber and form ties are based on a rate of fill of approximately 4 ft per hr at a temperature of 70 deg.

the panels may be made to any desired sizes. However, if plywood sheathing is used, it is good practice to make panels 2 ft 0 in. wide and 4, 6, 8, 10, or 12 ft long. An assortment of lengths will permit considerable flexibility in fitting the lengths to variable length walls. When the panels are erected, one on top of the other, the 2-ft width will permit the use of form ties at the top and bottom of adjacent panels, thus eliminating the need for holes through the plywood.

Figure 7-4 shows a method of constructing a typical panel, using ¾-in.-thick plywood for the sheathing and 2- by 4-in. planks for the frame. The total weight of this panel will be about 140 lb, which will require two men to handle it.

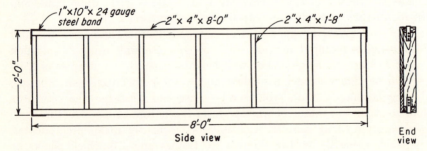

Fig. 7-4   Prefabricated panel form.

The quantity of lumber required for the panel in Fig. 7-4 will be

Plywood, $2 \times 8 = 16$ sq ft
    2 pc, 2 in. $\times$ 4 in. $\times$ 8 ft 0 in.   = 11 fbm
    7 pc, 2 in. $\times$ 4 in. $\times$ 1 ft 8 in.   = 8 fbm
                                        ———
         Total lumber   = 19 fbm
    Quantity of lumber per sq ft, $19 \div 16$ = 1.2 fbm

The labor required to build the panel should be about as follows:

Carpenter building frame, 19 fbm $\div$ 75 fbm per hr = 0.25 hr
Carpenter ripping and installing plywood   = 0.15 hr
                                        ———
         Total carpenter time   = 0.40 hr
Helper   = 0.20 hr
         Carpenter time per 100 sq ft, $^{100}\!/_{16} \times 0.4$ = 2.5 hr

Figure 7-5 shows a typical set of forms for a foundation wall using prefabricated panels. The bottoms of the forms are held together by steel

ESTIMATING CONSTRUCTION COSTS

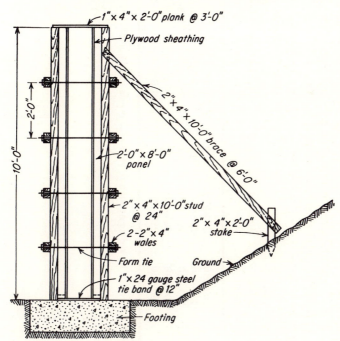

Fig. 7-5 Forms for concrete wall.

TABLE 7-6   Quantities of Plywood, Lumber, and Form Ties and the Labor-hours Required for 100 Sq Ft of Wall Area, Using Prefabricated Panels

Height of wall, ft	Ply-wood, sq ft	Lumber, fbm	Form ties*	Labor-hours making panels		Labor-hours erecting and removing forms	
				Carpenter	Helper	Carpenter	Helper
4	100–110	200–225	11	2.0–2.5	1.5–2.0	2.5–3.0	2.0–2.5
6	100–110	210–230	15	2.0–2.5	1.5–2.0	3.0–3.5	2.5–3.0
8	100–110	220–240	16	2.0–2.5	1.5–2.0	3.5–4.0	3.0–3.5
10	100–110	225–250	17	2.0–2.5	1.5–2.0	4.0–4.5	3.5–4.0
12	100–110	230–260	18	2.0–2.5	1.5–2.0	4.5–5.0	4.0–4.5

* The form ties must be strong enough to resist the stress for the particular use.

tie bands, which pass under the concrete wall and are nailed to the bottom members of the panel frames. The tops of the forms may be held together with planks, as shown in Fig. 7-5, or with steel tie bands. The top and bottom edges of the panels may be grooved slightly to receive the form ties.

Table 7-6 gives the quantities of plywood, lumber, and form ties and the labor-hours required for 100 sq ft of contact area for concrete walls using prefabricated panels, based on a rate of fill not exceeding 4 ft per hr at a temperature of 70 deg.

The increase in the number of uses of the lumber, and the reduction in the labor costs after the panels are made, will frequently permit a saving in form costs, when compared with other types of forms.

Commercial prefabricated forms Plywood sheathing attached to steel or aluminum frames in various sizes may be purchased or rented from the manufacturers. Figure 7-6 illustrates a set of wall forms in place.

Fig. 7-6 Universal wall forms. (*Universal Form Clamp Co.*)

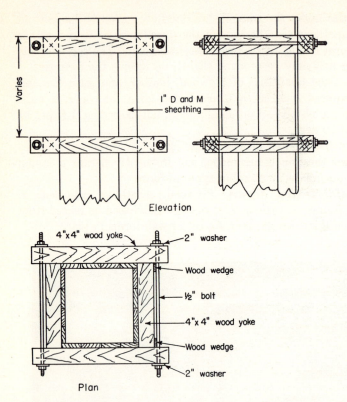

Elevation

Plan

Fig. 7-7 Column clamp made of wood yokes and bolts.

Forms consisting of steel sheets attached to steel frames may also be purchased or rented.

All these forms can be used with straight or circular walls or for walls which vary in thickness, either uniformly or in steps.

Forms for concrete columns Forms for concrete columns are made of lumber or steel. Steel forms are economical if the savings in cost of finishing the concrete and erecting the forms together with the number of uses justify the higher initial cost than for forms made from lumber.

Forms should be designed to resist the high pressures which result from quick filling with concrete. If the forms are filled in 30 min or less, the concrete will exert the full hydrostatic pressure based on a weight of 150 lb per cu ft.

If the forms are to be reused, they should be designed so that they can be removed with the least possible dismantling, usually by removal in two symmetrical sections. For square or rectangular columns, the sides are made of tongue-and-groove planks, prefabricated to the correct widths and lengths, or of plywood. The planks for a side are fastened together with 1- by 4-in.

or 1- by 6-in. wood cleats, spaced 3 to 4 ft apart along the form sides. The sides are assembled for the form. In order to facilitate use and reuse, forms should be identified with numbers or letters written on them.

Clamps are installed around the forms prior to filling them with concrete. The clamps are made of lumber, lumber and bolts, or steel. The steel clamps are adjustable for a wide range of sizes and may be reused several hundred times. The spacing should be such that the column forms will safely resist the maximum possible internal pressure.

For small columns, approximately 12 in. wide, 1- by 4-in. wood clamps, with 4-in. side perpendicular to the form, are sometimes used. Each clamp will require four pieces, which are lapped where they contact each other. Each lap should be nailed with five to six 8d nails. The clamps should be spaced about 12 to 16 in. apart.

For columns larger than 12 in., clamps should consist of 4- by 4-in. yokes and bolts, with two washers per bolt. The yokes are drilled for the bolts. Figure 7-7 shows the details of this clamp.

If it is expected that column clamps will be needed for many uses, it will be more economical and satisfactory to purchase or rent adjustable steel clamps. Each clamp is made of four arms, about $\frac{5}{16}$ in. thick and varying in width from 2 to 3 in., depending on the length. The arms are fastened together in pairs with rivet hinges. Slots spaced along the arms permit their use on columns varying in size from approximately 10 by 10 in. to 48 by 48 in. Figure 7-8 shows how the clamps are assembled around the forms. For most columns using 1-in. sheathing, it is sufficiently accurate to assume one steel clamp for each foot of height.

Table 7-7 shows the approximate safe spacing of steel clamps and yoke and bolt clamps for various column heights.

Fig. 7-8   Adjustable column clamps. (*Symons Corporation.*)

TABLE 7-7

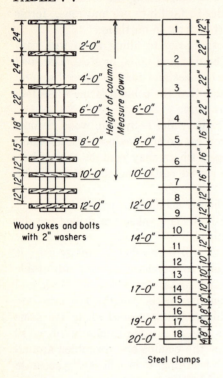

Wood yokes and bolts
with 2" washers

Steel clamps

Material required for forms for concrete columns The sheathing or sides for forms may be 1-in.-thick *D* and *M* lumber or plywood. *S4S* lumber is available in net widths $3\frac{1}{2}$-, $5\frac{1}{2}$-, $7\frac{1}{4}$-, $9\frac{1}{4}$-, and $11\frac{1}{4}$-in. *D* and *M* lumber, frequently called tongue-and-groove lumber, is available in net widths $3\frac{1}{8}$, $5\frac{1}{8}$, $6\frac{7}{8}$, and $8\frac{7}{8}$ in. For quantity purposes these widths are designated as 4, 6, 8, and 10 in. wide.

Example Determine the quantity of lumber required for forms for a concrete column 16 by 16 in. by 11 ft 6 in. long, using 1-in.-thick *D* and *M* lumber, 1- by 6-in. wood cleats and steel-column clamps.

The sheathing for two sides will be 16 in. wide; for the other two sides, it will be $17\frac{1}{2}$ in. wide. The 16-in. sides can be made with the following pieces:

Item	Net width, in.
2 pc 5-$\frac{1}{8}$ in.	= $10\frac{1}{4}$
2 pc 3-$\frac{1}{8}$ in.	= $6\frac{1}{4}$
Total	= $16\frac{1}{2}$

The 17½-in. side can be made with the following pieces:

Item	Net width, in.
3 pc 5-⅛ in.	= 15-⅜
1 pc 3-⅛ in.	= 3-⅛
Total	= 18½

It will be necessary to rip a small quantity of lumber from the side of each panel. Five 1- by 6-in. cleats will be used for each side panel.

The quantity of lumber required will be

For side panels:
10 pc, 1 × 6 in. × 12 ft 0 in.	= 60 fbm
6 pc, 1 × 4 in. × 12 ft 0 in.	= 24 fbm
For cleats, 20 pc, 1 × 6 in. × 1 ft 6 in.	= 15 fbm
For floor template, 4 pc, 2 × 4 in. × 3 ft 0 in.	= 8 fbm
Total quantity	= 107 fbm

Area of column surface, $4 \times 16/12 \times 11.5 = 61.3$ sq ft
Quantity of lumber per sq ft of area, $107 \div 61.3$ = 1.75 fbm

If $S4S$ lumber is used for the sheathing the quantity will be reduced to about 1.6 fbm per sq ft of area.

Because the lumber used as braces for column forms can be removed, generally without damage, and reused elsewhere, it should not be necessary to charge all, if any, of this lumber to the column forms.

Cost of lumber for forms   The cost per use for lumber for forms should be based on its initial cost at the job divided by a realistic number of times that it can be used. For example, in this book the specified cost of common grade lumber is $220.00 per M fbm. For some special grades and sizes the

TABLE 7-8   Materials Required for Column Forms

Kind of lumber	Kind of clamps	
	Yokes and bolts	Steel clamps
D and M	3.0 fbm	1.7 fbm
$S4S$	2.75 fbm	1.5 fbm
Bolts and washers	0.43 lb	None

cost may be higher, as it is in actual practice. If form lumber can be used five times, the cost per use will be $44.00 per M fbm.

Quantity of nails required for forms If *D* and *M* or *S4S* lumber is used for forms, the quantity of nails required should be about 10 to 15 lb per M fbm of lumber.

If plywood is used for sheathing, the quantity of nails should be less, usually about 7 to 12 lb per M fbm of lumber, including the plywood.

Cost of adjustable steel column clamps Steel clamps should cost from $0.15 to $0.30 per use, the cost varying with the size of the columns.

Labor making and erecting forms for concrete columns Column forms are usually prefabricated at the job prior to erecting in place. The lumber is cut and ripped to the proper sizes, using power saws, the sides are drawn together on a table, using carpenter's clamps, and the cleats are attached with 8d common nails, which are clinched after they are driven. The four sides are assembled and nailed together or clamped together using two clamps. After the forms are assembled, they are carried to the place of use and set into templates that have been accurately and securely located. It may be necessary to support the forms above the floor or column base until the column reinforcing is fastened to the dowels. Some contractors prefer to place the column reinforcing first and assemble the forms around it.

If it is not necessary to rebuild the forms for each use, the estimator should separate the labor into two operations, making the forms and erecting and removing them. For subsequent uses, if the sizes are not altered, only the labor for erecting and removing the forms should be considered.

Example Estimate the cost of forms for 20 columns, 16 by 16 in. by 11 ft 6 in. long. The sheathing will be 1-in. *D* and *M* lumber. Adjustable steel clamps will be used with the columns. Assume that the lumber will be used five times. The lumber will cost $200.00 per M fbm at the job.

TABLE 7-9 Approximate Labor-hours Required to Make, Erect, and Remove 100 Sq Ft of Wood Column Forms

Labor	Wood yokes and bolts	Adjustable steel clamps
Carpenter making	3.5–4.5	2.5–3.5
Carpenter erecting	6.0–7.0	5.5–6.5
Helper making	1.0–1.5	0.5–1.0
Helper erecting and removing	3.5–4.5	4.0–5.0

The cost for the first use will be

Lumber, 1,227 sq ft × 1.8 fbm per sq ft = 2,210 fbm @ $0.04	= $	88.40
Nails, 2,210 fbm × 15 lb per M fbm = 33 lb @ $0.25	=	8.25
Form oil, 1,227 sq ft ÷ 400 sq ft per gal = 3.1 gal @ $1.20	=	3.72
Power saws, 2,210 fbm ÷ 400 fbm per hr = 5.5 hr @ $0.35	=	1.93
Column clamps, 20 columns × 8 clamps = 160 @ $0.20	=	32.00
Carpenter making, 1,227 sq ft × 3 hr per 100 sq ft = 36.8 hr @ $7.67	=	282.26
Helper making, 1,227 sq ft × 1.0 hr per 100 sq ft = 12.3 hr @ $5.79	=	71.22
Carpenter erecting, 1,227 sq ft × 6.0 hr per 100 sq ft = 73.6 hr @ $7.67	=	564.52
Helper erecting and removing, 1,227 sq ft × 4.5 hr per 100 sq ft = 55.2 hr @ $5.79	=	319.61
Foreman, based on 5 carpenters, 22 hr @ $8.50	=	187.00
Total cost	=	$1,558.91
Cost per sq ft, $1,558.91 ÷ 1,227 sq ft	=	1.27

The cost for additional uses will be

Lumber, 2,210 fbm @ $0.04	= $	88.40
Nails, 2,210 fbm × 5 lb per M fbm = 11 lb @ $0.25	=	2.75
Form oil, 3.1 gal @ $1.20	=	3.72
Column clamps, 160 @ $0.20	=	32.00
Carpenter erecting, 73.6 hr @ $7.67	=	564.52
Helper erecting and removing, 55.2 hr @ $5.79	=	319.61
Foreman, 73.6 hr ÷ 5 = 14.7 hr @ $8.50	=	124.95
Total cost	=	$1,135.95
Cost per sq ft, $1,135.95 ÷ 1,227 sq ft		0.93

Economy of reducing the size of concrete columns The previous example demonstrated the economy effected by reusing the same-size column forms. As noted, the cost per square foot of surface area for the additional uses of the forms, without remaking to different sizes, was reduced from $1.27 to $0.93 for a saving of $0.34 per square foot. The saving on forms for a column in the example was

$$\frac{\$1,558.91 - \$1,135.95}{20} = \$21.15$$

If these columns are erected in a multistory building whose load conditions will permit the sizes of the columns for higher floors to be reduced in size to 14 by 14 in., thereby requiring that different forms be made, or that the forms from the larger columns be remade for use as forms for the smaller columns, the additional cost of making or remaking the smaller forms may exceed the saving in the cost of concrete required for the smaller columns. Thus, reducing the sizes of concrete columns in order to reduce the cost of concrete may not be economical. Also, if a larger-size column is used for the higher floors, it is probable that the quantity of reinforcing

TABLE 7-10 Maximum Safe Loads in Pounds from T-head or Stringers on Wood Shores*

Size shore, in.	Size T head or stringer, in.		
	4 × 4 rough	4 × 4 S4S	2 in. S4S
4 × 4 rough	8,000	7,000	3,000
4 × 4 S4S	7,000	6,125	2,625

* The loads are limited to 500 lb for each square inch of contact area between the members.

steel required in a column may be less than would be required in a smaller column. This practice could result in additional savings.

Shores Shores are members of form work whose function is to support loads from the forms for concrete beams, girders, slabs, etc. Shores may be made from commercial lumber or they may be obtained from commercial manufacturers, by purchase or rental.

Wood shores Wood shores are frequently made at the project from lumber in sizes 4 by 4 in., 4 by 6 in., 6 by 6 in., or larger. Either rough-sawed or S4S lumber may be used. Figure 7-11 illustrates typical shores in use.

If the vertical posts of wood shores are adequately braced to prevent horizontal movement or buckling under loading conditions, the allowable compressive stress may be as high as 1,600 lb per sq in. However, if a wood T head or other wood member rests on the top of the post, the maximum safe load may be limited by the allowable unit stress between the post and the member resting on the post. The maximum compressive stress acting perpendicular to the grain of wood should be limited to 500 lb per sq in.

TABLE 7-11 Load Capacities for Symons Shores, Pounds*

Length, ft		Height extended to, ft						
From	To	7.5	9.0	10.0	11.0	12.0	13.0	14.0
4.5	7.5	6,000						
6.5	11.5	6,000	6,000	5,500	4,500			
7.25	13.0	6,000	6,000	6,000	5,500	3,500	3,000	
8.25	15.0	.....	6,000	6,000	6,000	4,000	3,500	2,500

* These loads may be increased 100 percent if two-way horizontal braces are installed at the midpoints provided the maximum load does not exceed 6,000 lb.

Consider a shore consisting of a 4- by 4-in. *S4S* post and a T head of the same size and material. The actual sizes of the lumber will be 3½ by 3½ in. The area of contact between the top of the post and the T head will be 12.25 sq in. The maximum safe load on the post will be 500 × 12.25 = 6,125 lb. Even though the post may be capable of withstanding a load of 1,600 × 12.25 = 19,600 lb, the maximum load for this design should not exceed 6,125 lb.

Table 7-10 gives the safe loads that may be transmitted to the tops of wood shores from horizontal wood members resting on the tops of the shores. **Adjustable shores** Figure 7-9 illustrates an adjustable shore made by one manufacturer. Table 7-11 gives information for this shore.

Quantity of lumber required for concrete beams using wood shores Figure 7-11 illustrates the details of constructing wood forms and shores for a typical beam-and-slab floor. Two plans for constructing beam forms are shown. Plan A requires that the 4- by 4-in. T heads be long enough to permit one 2- by 4-in. ledger to be nailed on each side to resist lateral move-

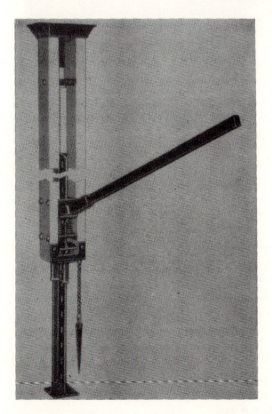

Fig. 7-9 Adjustable shore. (*Symons Corporation.*)

Fig. 7-10   Adjustable shores supporting floor forms. (*Symons Corporation.*)

ment of the 2- by 4-in. studs. For plan B the bottoms of the studs extend below the beam soffit about 6 in. to permit a 1- by 6-in. tie to be installed for each pair of studs. Plan B permits the use of shorter T heads than those required for plan A.

If the concrete beams of Fig. 7-11 are spaced 12 ft 0 in. apart, with an intermediate stringer midway between the beams, as shown, the beam shores will support the weight of the beam plus a slab 6 ft wide. The volume of concrete per linear foot of beam will be $6 \times 1 \times 0.5$ plus $1 \times 1.33 =$

TABLE   7-11A   Quantity   of   Lumber Required for Forms for Concrete Beams

Type of shores	Lumber required, fbm per sq ft of surface	
	For plan in Fig. 7-11	
	A	B
Wood to 9 ft long	3.50–4.0	3.2–3.7
Adjustable	. . . .	2.3–2.8

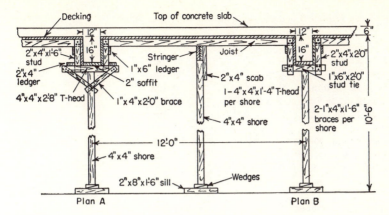

Fig. 7-11 Wood forms and shores for concrete beams.

4.33 cu ft. The weight will be 650 lb. Add 40 psf for live load, for the storage of form lumber and other materials during construction, equal to 6 × 40 = 240 lb per lin ft. The total load will be 890 lb per lin ft. If 4- by 4-in. *S4S* shores and T heads are used, the maximum safe load per shore will be 6,500 lb. The maximum safe spacing of shores will be 6,500 ÷ 890 = 7.3 ft. In order to eliminate the danger of objectionable deflection between shores, the spacing should not exceed 5 ft.

Example Determine the quantity of lumber required for a beam 12 in. wide, 16 in. deep, and 18 ft 10 in. long, based on plan A of Fig. 7-11. With a shore at each end of the beam, five shores will be required.

The quantity of lumber will be

    Soffit:
      1 pc, 2 in. × 8 in. × 20 ft 0 in.   =   27fbm
      1 pc, 2 in. × 6 in. × 20 ft 0 in.   =   20 fbm
    Sides:
      Use 3½ pc, 1 × 6 in. per side
      7 pc, 1 in. × 6 in. × 20 ft 0 in.   =   70 fbm
    Studs, 22 pc, 2 in. × 4 in. × 1 ft 6 in. =   22 fbm
    Ledgers, 2 pc, 2 in. × 4 in. × 20 ft 0 in. =   27 fbm
    Shores:
      5 pc, 4 in. × 4 in. × 9 ft 0 in.   =   60 fbm
      5 pc, 4 in. × 4 in. × 3 ft 0 in.   =   20 fbm
      10 pc, 1 in. × 4 in. × 2 ft 0 in.   =   7 fbm
    Sills, 5 pc, 2 in. × 8 in. × 1 ft 6 in.   =   10 fbm
                                              ——
    Total quantity of lumber   = 263 fbm
    Area of the beam, 69 sq ft
    Quantity of lumber per sq ft, 263 ÷ 69 =   3.8 fbm

If the method shown in plan B is used, the quantity of lumber will be

Soffit:
1 pc, 2 in. × 8 in. × 20 ft 0 in.	=	27 fbm
1 pc, 2 in. × 6 in. × 20 ft 0 in.	=	20 fbm
Sides, 7 pc, 1 in. × 6 in. × 20 ft 0 in.	=	70 fbm
Studs, 22 pc, 2 in. × 4 in. × 2 ft 0 in.	=	30 fbm
Cross braces, 11 pc, 1 in. × 6 in. × 2 ft 0 in.	=	11 fbm

Shores:
5 pc, 4 in. × 4 in. × 9 ft 0 in.	=	60 fbm
5 pc, 4 in. × 4 in. × 1 ft 4 in.	=	9 fbm
10 pc, 1 in. × 4 in. × 1 ft 6 in.	=	5 fbm
Sills, 5 pc, 2 in. × 8 in. × 1 ft 6 in.	=	10 fbm

Total quantity of lumber	=	242 fbm
Quantity of lumber per sq ft, 242 ÷ 69	=	3.5 fbm

Quantity of lumber required for forms for concrete beams using adjustable shores   If adjustable shores instead of wood shores are used,

TABLE 7-12   Labor-hours Required to Build 100 Sq Ft of Forms for Concrete Beams

Operation	Using 4- by 4-in. wood shores	Using adjustable shores
For inside beams and girders		
Making soffits, sides, and wood shores:		
Carpenter	3.5–4.5	3.0–3.5
Helper	1.0–1.5	1.0–1.5
Erecting soffits, sides, and shores:		
Carpenter	5.0–6.0	5.0–5.5
Helper	1.5–2.0	1.5–2.0
Removing forms:		
Helper	1.5–2.0	1.5–2.0
For outside beams and girders		
Making soffits, sides, and wood shores:		
Carpenter	3.5–4.5	3.0–4.0
Helper	1.0–1.5	1.0–1.5
Erecting soffits, sides, and shores:		
Carpenter	7.0–8.0	6.7–7.0
Helper	3.0–3.5	3.0–3.5
Removing forms:		
Helper	3.0–3.5	3.0–3.5

the maximum safe load will be 6,000 lb. The maximum safe spacing, based on the safe load, will be about 6 ft 6 in. However, the spacing will be limited to 4 ft 0 in, with no horizontal braces required.

The quantity of lumber required using plan B of Fig. 7-11 and adjustable shores will be 242 − 84 = 158 fbm. The quantities are determined from the example in the previous article.

The quantity of lumber per square foot will be 158 ÷ 69 = 2.3 fbm.

Labor required to build forms for concrete beams In building and erecting forms for a concrete beam it is customary to make the soffit and then, starting at a column form, erect the soffit on the shores. The shores are wedged to the correct elevation and temporarily braced. The side forms for the beams are made and cleated with the 2- by 4-in. studs, to the full length of the beam. Then they are lifted to position, nailed to the soffit, and secured against lateral failure by the method shown in plan A or B (Fig. 7-11). After the tops of the forms are brought to a straight line, the joists for the decking are installed, and then the decking is installed.

If the side forms are to be used on other beams, they should be assembled and erected in a manner which will permit them to be removed intact. The estimator should show separately the labor required to make and to erect the forms, as this practice will permit him to include only erection costs for additional uses.

Table 7-12 gives the labor-hours required to build forms for concrete beams.

Example Estimate the cost of forms for 12 inside concrete beams, 13 in. wide, 19 in. deep, and 19 ft 2 in. long, using 4- by 4-in. wood shores 9 ft 0 in. long. The forms will be used six times on the job.

The area of the beams will be $\dfrac{12 \times (19 + 19 + 13)}{12} \times$ 19 ft 2 in. = 980 sq ft.

Assume that the lumber will cost $200.00 per M fbm at the job.
The cost for the first use will be

Lumber, 980 sq ft × 3.5 fbm per sq ft = 3,430 fbm @ $0.033	= $	113.19
Nails, 3, 430 fbm × 15 lb per M fbm = 51 lb @ $0.25	= $	12.80
Form oil, 980 sq ft ÷ 400 sq ft per gal = 2.5 gal @ $1.20	=	3.00
Power saws, 3,430 fbm ÷ 400 fbm per hr = 8.6 hr @ $0.35	=	3.00
Carpenters, 980 sq ft × 10 hr per 100 sq ft = 98 hr @ $7.67	=	751.66
Helpers, 980 sq ft × 5.75 hr per 100 sq ft = 56.3 hr @ $5.79	=	325.98
Foreman, based on using 6 carpenters, = 16.3 hr @ $8.50	=	138.55
Total cost	=	$1,348.18
Cost per sq ft, $1,348.18 ÷ 980	=	1.38

The cost for additional uses will be

Lumber, 3,430 fbm @ $0.033	= $	113.19
Nails, 3,430 fbm × 7 lb per M fbm = 24 lb @ $0.25	=	6.00

Form oil, 2.5 gal @ $1.20	=	3.00
Power saws, no cost		
Carpenters, 980 sq ft × 6 hr per 100 sq ft = 58.8 hr @ $7.67	=	451.00
Helpers, 980 sq ft × 4.25 hr per 100 = 41.5 hr @ $5.79	=	240.28
Foreman, 58.8 ÷ 6 = 9.8 hr @ $8.50	=	83.30
Total cost	= $	896.77
Cost per sq ft, $896.77 ÷ 980	=	0.92

Forms for flat-slab-type concrete floors   The flat-slab-type concrete floor is sometimes used when column heads support the floors, without beams or girders. Since there are no beam forms to assist in supporting the slab forms, it is necessary to classify all forms as slab forms. The cost of forms will vary with the thickness of the slab and the height of the slab above the lower floor.

Figure 7-12 illustrates the method commonly used in building forms for this type of slab.

In order to determine the quantity of lumber required, it is necessary to know the thickness of the slab, the probable live load on the slab while the forms are in place, and the height of the slab above the supporting floor. The live load includes form lumber, brick, tile, and other materials temporarily stored on the slab. The live load will usually amount to about 40 to 50 psf of area.

The form builder has considerable freedom in selecting the size and spacing of joists and stringers. Joist sizes commonly used are 2 by 4 in., 2 by 6 in., 2 by 8 in., and 2 by 10 in. *S4S*. Stringer sizes are 2 by 8 in., 2 by 10 in., 4 by 4 in., and 4 by 6 in. *S4S*. The 2-in.-thick stringers give good economy, but the bearing stresses where they rest on the shores may be excessive and should be checked before the final design is selected. The

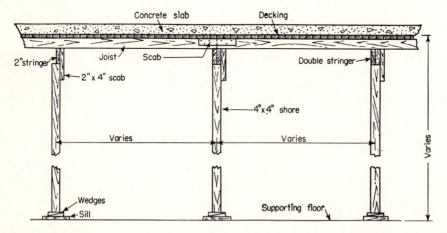

Fig. 7-12   Wood forms for flat-slab concrete floor.

TABLE 7-13 Safe Spans for Forms for Concrete Slabs Based on Using 1-in. Decking and Joists Spaced 2 ft 0 in. on Centers* All Lumber is S4S.

Thickness of slab, in.		4	5	6	8	10
Total load, psi		90	103	115	140	165
Size joist, in.		Safe span for joists				
2 × 4		4 ft 9 in.	4 ft 9 in.	4 ft 3 in.	3 ft 9 in.	3 ft 9 in.
2 × 6		6 ft 6 in.	6 ft 6 in.	6 ft 0 in.	5 ft 6 in.	5 ft 0 in.
2 × 8		8 ft 9 in.	8 ft 3 in.	8 ft 0 in.	7 ft 6 in.	7 ft 0 in.
2 × 10		10 ft 0 in.	10 ft 0 in.	9 ft 6 in.	8 ft 9 in.	8 ft 3 in.
Size stringer, in.	Stringer spacing	Safe span for stringers				
2 × 8	6 ft 0 in.	5 ft 6 in.	4 ft 6 in.	4 ft 0 in.	3 ft 6 in.	3 ft 0 in.
2 × 10	7 ft 0 in.	5 ft 6 in.	4 ft 6 in.	4 ft 0 in.	3 ft 6 in.	3 ft 0 in.
4 × 4	5 ft 0 in.	4 ft 3 in.	3 ft 9 in.	3 ft 9 in.	3 ft 3 in.	
4 × 6	6 ft 0 in.	5 ft 6 in.	5 ft 6 in.	5 ft 0 in.	4 ft 9 in.	4 ft 3 in.

* The spans are based on a fiber stress in bending of 1,800 psi of net size for the joist or stringer. Wood capable of withstanding this stress should be used, or the spans should be reduced.

exercise of care in designing forms will result in economical forms, considering the cost of materials and labor.

Design of forms for flat-slab type concrete floors   Table 7-13 gives information that may be used to design forms for flat-slab concrete floors.

The following example illustrates how the information in Table 7-13 may be used to design forms.

Example   Determine the quantity of lumber per square foot of floor area required for a 6-in.-thick concrete slab whose clear height above the lower floor is 11 ft 2 in. Assume a live load of 40 psf. The decking will be 1 by 6 in. S4S, whose net width is 5½ in.

The joists will be spaced 2 ft 0 in. on centers.

If the selected spacings for joists, stringers, and shores will permit the use of commercial lumber lengths, which vary in 2-ft steps, with little or no end wastage, this will generally reduce the quantity and the cost of form materials. For example, if the spacing of shores is 4 ft, the stringers should be 12, 16, or 20 ft long.

Reference to Table 7-13 indicates that 2- by 8-in. joists will permit a safe span of 8 ft 0 in. Thus 2- by 8-in. joists 16 ft 0 in. long, spaced 2 ft 0 in. on centers, will be used.

Investigate the desirability of using 4- by 6-in. S4S or 2- by 10-in. S4S stringers.

If 4- by 6-in. stringers are used, the maximum safe span will be 5 ft 0 in. The area supported by a shore will be 5 ft × 8 ft = 40 sq ft. The total weight on a shore from this area will be 40 × 115 = 4,600 lb. The area in compression between a stringer and the top of a 4- by 4-in. shore will be 3.5 × 3.5 = 12.25 sq in. The unit compressive stress on the stringer will be 4,600 ÷ 12.25 = 375 lb per sq in., which is satisfactory.

If 2- by 10-in. stringers are used, the maximum safe span will be 4 ft 0 in., which will be the maximum safe spacing of shores. The area supported by a shore will be 4 × 8 ft = 32 sq ft. The total weight on a shore will be 32 × 115 = 3,680 lb. The area in compression between a stringer and the top of a shore will be 1.5 × 3.5 = 5.25 sq in. The unit compressive stress on the stringer will be 3,680 ÷ 5.25 = 700 lb per sq in. This stress exceeds the allowable unit stress. If 2- by 10-in. stringers are used, the spacing of the shores must be reduced to less than 4 ft., to limit the stress on the stringer to not more than 500 lb per sq in. The maximum safe spacing will be 4 × 500/700 = 2.86 ft

Use 4- by 6-in. S4S stringers, with shores spaced 4 ft 6 in. apart to limit deflection.

The forms will consist of the following materials:

Decking, 1 × 6 in. S4S, 16 ft 0 in. long
Joists, 2 × 8 in. S4S, 16 ft 0 in. long
Stringers, 4 × 6 in. S4S, 18 ft 0 in. long
Shores, 4 × 4 in. S4S, 10 ft 0 in. long

In order to determine the quantity of lumber required per square foot of slab, consider an area of slab 8 ft 0 in. wide and 18 ft 0 in. long = 144 sq ft.
The quantity of lumber required will be

Decking, 8 × 12 ÷ 5.5 = 17.5 pc of 1 × 6 in. × 18 ft 0 in.	= 157 fbm
Joists, 18 ÷ 2 = 9 pc of 2 × 8 in. × 8 ft × 0 in.	= 96 fbm
Stringer, 1 pc of 4 in. × 6 in. × 18 ft 0 in.	= 36 fbm
Shores, 18 ÷ 4.5 = 4 pc of 4 in. × 4 in. × 10 ft 0 in.	= 53 fbm
Scabs, 4 pc of 2 × 4 in. × 2 ft 0 in.	= 6 fbm
Sills for shores, 4 pc of 2 × 8 in. × 1 ft 6 in.	= 8 fbm
Two-way braces for shores, 4 × 8 ft 0 in. = 32 ft + 18 ft = 50 ft, 1 × 6 in.	= 25 fbm
	—————
Total quantity	= 381 fbm
Quantity of lumber per sq ft, 381 ÷ 144 sq ft	= 2.7 fbm

TABLE 7-14 Quantities of Lumber Required per Square Foot of Flat-slab Concrete Floor in FBM*

Ceiling height, ft	Thickness of slab					
	6 in.	8 in.	10 in.	6 in.	8 in.	10 in.
	4- × 4-in. wood shores			Adjustable shores		
10	2.7	2.8	3.0	2.3	2.4	2.7
12	2.8	2.9	3.1	2.3	2.4	2.7
14	3.0	3.1	3.4	2.5	2.6	3.0
16	3.1	3.2	3.6	2.5	2.6	3.0

* The values in the table are based on a live load of 40 lb per sq ft on the slab. The values do not include lumber for column heads, stairs, elevator shafts, etc.

If adjustable shores are used instead of wood shores, the lumber for shores and scabs will not be needed. This will reduce the quantity of lumber to 381 − 59 = 322 fbm. The quantity of lumber per square foot will be 322 ÷ 144 = 2.24 fbm.

Quantity of lumber required for forms for flat-slab-type concrete floors Table 7-14 gives the approximate quantities of lumber required per square foot of slab area for various thicknesses of slab and heights of ceilings.

Labor required to build forms for flat-slab-type concrete floors Table 7-15 gives the number of labor-hours required to build and remove forms for flat-slab concrete floors.

Example Estimate the direct cost of furnishing, building, and removing forms for a flat-slab concrete floor whose size is 56 ft 8 in. by 88 ft 6 in. The slab will be 6 in. thick. There will

TABLE 7-15 Labor-hours Required to Build and Remove 100 Sq Ft of Forms for Flat-slab Concrete Floors and Column Heads.

Ceiling height, ft	Labor	4- × 4-in. wood shores	Adjustable shores
10–14	Carpenter	5.0 6.0	4.5–5.5
	Helper	3.5–4.0	3.0–3.5
14–16	Carpenter	5.5–6.5	5.0–6.0
	Helper	4.0–4.5	3.5–4.0

be 16 column heads size 6 by 6 ft each. The ceiling height is 11 ft 8 in. Assume that the lumber will be used four times. The lumber will cost $220.00 per M fbm at the job.

The quantities are

Gross area of slab, 56.67 × 88.5	= 5,015 sq ft
Area of column heads, 16 × 36	= −576 sq ft
Net area of slab	= 4,439 sq ft

If adjustable shores are used, the cost will be

Lumber for slab, 4,439 sq ft × 2.3 fbm per sq ft = 10,210 fbm @ $0.055	=	$ 561.55
Lumber for column heads, 576 sq ft × 4.0 fbm per sq ft = 2,304 fbm @ $0.055	=	126.72
Nails, 12,514 fbm × 12 lb per M fbm = 150 lb @ $0.25	=	37.50
Form oil, 5,015 sq ft ÷ 400 sq ft per gal = 12.5 gal @ $1.20	=	15.00
Shores, 4,439 sq ft ÷ 24 sq ft per shore, 185 each @ $0.45	=	83.25
Power saws, 12,514 fbm ÷ 400 fbm per hr = 31.3 hr @ $0.35	=	10.95
Carpenters, 5,015 sq ft × 5.0 hr per 100 sq ft = 250.7 hr @ $7.67	=	1,922.87
Helpers, 5,015 sq ft × 3.0 hr per 100 sq ft = 150.4 hr @ $5.79	=	870.82
Foreman, based on using 5 carpenters, 250.7 ÷ 5 = 50.1 hr @ $8.50	=	425.85
Total cost	=	$4,054.51
Cost per sq ft, $4,054.51 ÷ 5,015	=	0.81

If 4- × 4-in. wood shores are used, the cost will be

Lumber for slab, 4,439 sq ft × 2.7 fbm per sq ft = 11,985 fbm @ $0.055	=	$ 659.17
Lumber for column heads, 576 sq ft × 4.25 fbm per sq ft = 2,448 fbm @ $0.055	=	134.64
Nails, 14,433 fbm × 12 lb per M fbm = 173 lb @ $0.25	=	43.25
Form oil, 5,015 sq ft ÷ 400 sq ft per gal = 12.5 gal @ $1.20	=	15.00
Power saws, 14,433 fbm ÷ 400 fbm per hr = 36.0 hr @ $0.37	=	13.35
Carpenters, 5,015 sq ft × 5.5 hr per 100 sq ft = 275.8 hr @ $7.67	=	2,115.39
Helpers, 5,015 sq ft × 3.5 hr per 100 sq ft = 175.5 hr @ $5.79	=	1,016.15
Foreman, based on using 5 carpenters, 275.8 ÷ 5 = 55.2 hr @ $8.50	=	469.20
Total cost	=	$4,466.15
Cost per sq ft, $4,466.15 ÷ 5,015	=	0.89

Thus the use of adjustable shores is cheaper than using wood shores.

Forms for slabs for beam-and slab-type concrete floors The forms for the slab for beam-and-slab-type concrete floors consist of the decking, joists, stringers, shores, and braces, if required. The ends of the joists are usually supported by the beam forms. If the beams are spaced more than approximately 8 ft 0 in. on centers, it will usually be economical to install a stringer between the beams to assist in supporting the decking. The use of a stringer will reduce the load on the beam forms and will permit the use of smaller joists.

Figure 7-13 illustrates the types of forms used for slab floors. Plan A does not require stringers to support the forms, while plan B requires stringers and shores.

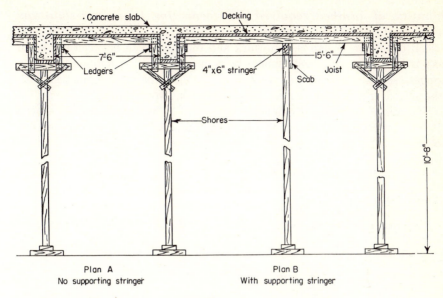

Fig. 7-13 Wood forms for beam-and-slab concrete floor.

Quantity of lumber required for forms for beam-and-slab-type concrete floors This determination is based on using plan A of Fig. 7-13. If the concrete slab is 6 in. thick and the clear span between the beam faces is 7 ft 6 in., Table 7-13 indicates that it will be necessary to use 2- by 8-in. *S4S* lumber for joists, spaced 2 ft 0 in. on centers. Decking will be 1- by 6-in. *S4S* lumber, whose actual width is 5½ in. Ledgers will be 1- by 6-in. *S4S* lumber. Assume that the bay will be 24 ft 6 in. long.

The quantities are determined as follows:

Area of slab, $7.5 \times 24.5 = 184$ sq ft
No. pieces of decking required for 7.5 ft of width, $7.5 \times 12 \div 5.5 = 16.35$ (use 17 pc)
For the length of 24.5 ft use
   1 pc, 14 ft 0 in. long
   1 pc, 12 ft 0 in. long

The quantity of lumber will be

Decking, 17 pc, 1×6 in. $\times$ 12 ft 0 in.	= 102 fbm
17 pc, 1×6 in. $\times$ 14 ft 0 in.	= 119 fbm
Joists, no. $24.5 \div 2 = 12.25 + 1 = 14$ pc, 2×8 in. $\times$ 8 ft 0 in.	= 149 fbm
Ledgers, 2 pc, 1×6 in. $\times$ 12 ft 0 in.	= 12 fbm
2 pc, 1×6 in. $\times$ 14 ft 0 in.	= 14 fbm
Total quantity	= 396 fbm
Quantity per sq ft, $396 \div 184$	= 2.15 fbm

Quantity of lumber required for forms for beam-and-slab-type concrete floors This determination is based on using plan B of Fig. 7-13. If the concrete slab is 6 in. thick and the clear span between the beam faces is 15 ft 6 in., the joists should be supported at their midpoints with a stringer, as indicated in Fig. 7-13. The stringer may be supported with either wood or adjustable shores. Use 4- by 4-in. wood shores.

Reference to Table 7-13 indicates the use of 2- by 8-in. *S4S* joists spaced 2 ft 0 in. on centers. Stringers will be 4 by 6 in. *S4S* with shores spaced not more than 5 ft 0 in. apart.

The quantities are determined as follows:

Length of bay, 24 ft 6 in.
Area of bay, 15.5 × 24.5 = 380 sq ft
No. pieces of decking required for 15.5 ft of width, 15.5 × 12 ÷ 5.5 = 33.9 (use 34 pc)

The quantity of lumber will be

Decking, 34 pc, 1 × 6 in. × 12 ft 0 in.	= 204 fbm
34 pc, 1 × 6 in. × 14 ft 0 in.	= 238 fbm
Joists, no., 24.5 ÷ 2 = 12.25 + 1 = 14 pc, 2 × 8 in. × 16 ft 0 in.	= 299 fbm
Stringer, 1 pc, 4 × 6 in. × 12 ft 0 in.	= 24 fbm
1 pc, 4 × 6 in. × 14 ft 0 in.	= 28 fbm
Shores, use 4 ft 0 in. spacing under the 12 ft 0 in. length, and 4 ft 8 in. under the 14 ft 0 in. length of stringer for a total of 7 shores. 7 pc, 4 in. × 4 in. × 10 ft 0 in.	= 93 fbm
Scabs, 7 pc, 2 × 4 in. × 2 ft 0 in.	= 10 fbm
Sills, 7 pc, 2 × 8 in. × 1 ft 6 in.	= 14 fbm
Horizontal braces:	
2 pc, 1 × 6 in. × 14 ft 0 in.	= 14 fbm
7 pc, 1 × 6 in. × 8 ft 0 in.	= 28 fbm
Total quantity	= 952 fbm
Quantity per sq ft, 952 ÷ 380	= 2.5 fbm

Table 7-16 gives the quantities of lumber required for forms for the slab only of beam-and-slab-type concrete floors. The width of the slab is the clear distance between the faces of the beams.

Labor required to build forms for beam-and-slab-type concrete floors The labor required to build forms for the slab only for beam-and-slab-type concrete floors will be slightly less for plan A than for plan B, since stringers and shores are not installed. The beam forms will be in place when the slab forms are started. Ordinarily the 1- by 6-in. ledgers on which the joists rest will be nailed to the beam forms before erecting them. The joists will be cut to length, sized if necessary, and nailed in place. Then the decking will be installed.

If supporting stringers and shores are required, they will be installed after some or all of the joists are in place, but prior to installing the decking.

TABLE 7-16   FBM of Lumber Required per Square Foot for Forms for the Slab Only of Beam-and-slab-type Concrete Floors

With no supporting stringers				
Height of ceiling, ft	Thickness of slab, in.			
	4	6	8	10
All heights	2.0	2.2	2.3	2.4

With supporting stringers

Height of ceiling, ft	4- by 4-in. wood shores				Adjustable shores			
	Thickness of slab, in.				Thickness of slab, in.			
	4	6	8	10	4	6	8	10
10	2.3	2.4	2.5	2.6	2.1	2.2	2.3	2.4
12	2.4	2.5	2.6	2.7	2.2	2.3	2.4	2.5
14	2.6	2.7	2.8	2.9	2.4	2.5	2.6	2.7
16	2.7	2.8	2.9	3.0	2.5	2.6	2.7	2.8

TABLE 7-17   Labor-hours Required to Build and Remove the Slab Forms for 100 Sq Ft of Beam-and-slab-type Concrete Floors

With no supporting stringers

Height of ceiling, ft	Labor	Labor-hours
All heights	Carpenter	3.0–4.0
	Helper	2.0–2.5

With supporting stringers

Height of ceiling, ft	Labor	Labor-hours	
		Wood shores	Adjustable shores
10–14	Carpenter	4.0–5.0	4.0–4.5
	Helper	3.0–3.5	2.5–3.0
14–16	Carpenter	4.5–5.5	4.0–4.5
	Helper	3.5–4.0	3.5–4.0

Table 7-17 gives the labor-hours required to build and remove the forms for 100 sq ft of beam-and-slab-type concrete floors.

Example Estimate the cost of building and removing forms for a beam-and-slab-type concrete floor using adjustable shores. The slab will be 15 ft 4 in. wide, 38 ft 6 in. long, and 7 in. thick. The ceiling height will be 12 ft 2 in. Assume that the lumber will cost $220.00 per M fbm at the job, and that it will be used four times.
The area of the slab will be 15.33 × 38.5 = 590 sq ft.
The cost will be

Lumber, 590 sq ft × 2.4 fbm per sq ft = 1,416 fbm @ $0.055	=	$ 77.85
Nails, 1,416 fbm × 12 lb per M fbm = 17 lb @ $0.25	=	4.25
Form oil, 590 sq ft ÷ 400 sq ft per gal 1.5 gal @ $1.20	=	1.80
Shores, 11 @ $0.45	=	4.95
Power saws, 1,416 fbm ÷ 400 fbm per hr = 3.5 hr @ $0.35	=	1.23
Carpenters, 590 sq ft × 4.2 hr per 100 sq ft = 24.8 hr @ $7.67	=	190.22
Helper, 590 sq ft × 3.0 hr per 100 sq ft = 17.7 hr @ $5.79	=	102.48
Foreman, based on using five carpenters, 24.5 ÷ 5 = 5 hr @ $8.50	=	42.50
Total cost	=	$425.28
Cost per sq ft, $425.28 ÷ 590 sq ft	=	0.72

Forms for metal-pan and concrete-joist-type concrete floors There is an increasing use of metal-pan- and concrete-joist-type concrete floors. This type of construction has a number of advantages over other types of forms for concrete floors. The total quantity of concrete is less than for slab floors, which reduces the cost of materials and also permits the use of smaller beams, columns, and footings. If the pans are used enough times, the cost per square foot of use, including the labor cost of installing and

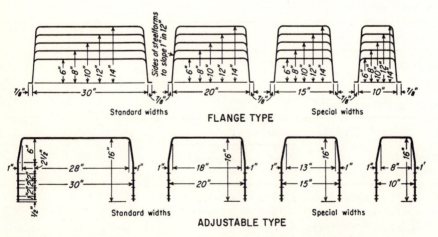

Fig. 7-14 Dimensions of Meyer Steelforms. (*Ceco Steel Products Corp.*)

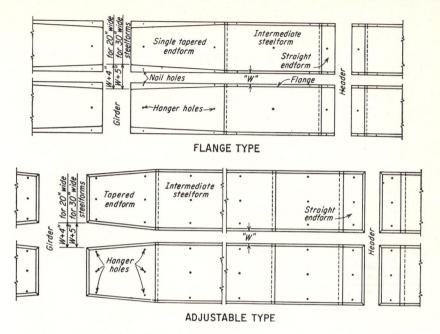

Fig. 7-15 Plan view of Meyer Steelforms. (*Ceco Steel Products Corp.*)

removing, should compare favorably with forms built entirely of lumber.
Some types of pans permit removal within about a week.

This type of construction has several disadvantages. Few building
contractors can afford to purchase the pans, as they cannot be reused unless
a similar project is constructed. The structure must be designed for the use
of a designated size and type of pan. It is difficult to form around irregular
structural units and openings.

Several plans are used for the installation of the metal pans. Contractors
can rent the pans at an agreed price per square foot per use, with a minimum number of uses. Also, they can subcontract the pans only in place and
removed, or they can subcontract the furnishing of the pans and supporting
forms completely.

Figure 7-14 gives details of the widths and depths of flange and adjustable types of Meyer Steelforms. Figure 7-15 shows plan views of the
installation of flange and adjustable types of Meyer Steelforms. Figure
7-16 shows the details of supporting forms for the flange and adjustable
types of metal-pan forms. The stringers may be supported by wood or adjustable shores. Figure 7-17 shows various stages in the installation and
use of metal-pan forms.

Metal pans are available which are designed to be self-supporting for
spans up to 12 ft or more. These forms do not require intermediate stringers
and shores.

ESTIMATING CONSTRUCTION COSTS

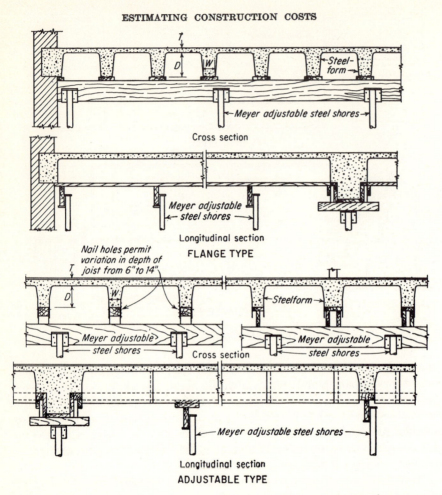

Fig. 7-16 Supporting forms for metal pans. (*Ceco Steel Products Corp.*)

Table 7-18 gives the quantities of concrete required for Meyer Steelform construction. Manufacturers of other forms will furnish similar information for their forms.

Lumber required for metal-pan and concrete-joist construction
The lumber required for metal-pan and concrete-joist construction consists of centering strips or planks, 2 by 6 in. or 2 by 8 in., stringers, shores, and braces. For irregularities around structural units or openings, lumber is sometimes used instead of metal pans.

The spacing of the centering strips will equal the width of the pan plus the thickness of the joist. By the use of pans of special widths it is usually possible to install pans for any width floor. Also, the pans are available in

Fig. 7-17 Installing metal pans. (*Ceco Steel Products Corp.*)

various lengths, usually in 1-ft intervals, which permits the pans to be used for any length floor.

Stringers may be 2 by 8 in., 2 by 10 in., or 4 by 6 in. They should be spaced, in so far as possible, an exact number of feet apart in order to reduce the end wastage of lumber in centering strips. A spacing of 4 ft 0 in. is frequently used, which permits the use of centering strips in lengths 8 ft 0 in., 12 ft 0 in., 16 ft 0 in., or 20 ft 0 in. A 2-in. stringer is not thick enough to permit safe end-butting joints of centering strips. If butt joints of centering strips are necessary, the thickness of 2-in. stringers should be increased by nailing a lumber strip to the side of each stringer.

Wood or adjustable shores may be used to support the stringers. The spacing should be 4, 5, 6, 7, or 8 ft, depending on the load and the length of the shores. The use of such spacing will permit the use of commercial lengths of stringers, with a minimum end wastage. Spacings 6, 7, and 8 ft are generally possible.

Example Determine the quantity of lumber required for 1 sq ft of floor, using metal pans of the flange type. The thickness of the floor will be 12½ in. This floor thickness is obtained with a 10-in.-deep pan and a slab thickness of 2½ in. over the pan. Joists will be 6 in. thick, and the pans will be 30 in. wide. The size of the bay will be 16 ft 0 in. by 24 ft 0 in. The height of the ceiling will be 12 ft 3 in.

TABLE 7-18 Concrete Required for Meyer Steelform Construction

Flange type

For 20-in. widths

Depth of Steelform, in.	Width of joist, in.	Cu ft of concrete per sq ft of floor with slab thickness over Steelforms of			Cu ft per lin ft of beam using tapered ends	
		2 in.	2½ in.	3 in.	2 in.	4 in.
6	4	0.262	0.304	0.346	0.08	0.16
	5	0.279	0.321	0.363	0.08	0.16
	6	0.293	0.335	0.377	0.08	0.16
8	4	0.298	0.340	0.382	0.12	0.24
	5	0.320	0.362	0.404	0.12	0.24
	6	0.339	0.381	0.423	0.11	0.22
10	4	0.336	0.378	0.420	0.16	0.32
	5	0.363	0.405	0.447	0.16	0.32
	6	0.387	0.429	0.471	0.15	0.30
12	4	0.377	0.419	0.461	0.21	0.42
	5	0.408	0.450	0.492	0.20	0.40
	6	0.437	0.479	0.521	0.20	0.40
14	4	0.420	0.462	0.504	0.26	0.52
	5	0.456	0.498	0.540	0.25	0.50
	6	0.490	0.532	0.574	0.24	0.48

For 30-in. widths

Depth of Steelform, in.	Width of joist, in.	Cu ft of concrete per sq ft of floor with slab thickness over Steelforms of			Cu ft per lin ft of beam using tapered ends	
		2½ in.	3 in.	3½ in.	4 in.	5 in.
6	5	0.293	0.334	0.374	....	0.13
	6	0.304	0.346	0.387	....	0.13
	7	0.315	0.357	0.398	....	0.13
8	5	0.322	0.364	0.405	....	0.18
	6	0.338	0.380	0.421	....	0.18
	7	0.351	0.392	0.434	....	0.17
10	5	0.353	0.395	0.436	....	0.24
	6	0.372	0.414	0.455	....	0.23
	7	0.389	0.430	0.472	....	0.23
12	5	0.386	0.427	0.469	....	0.30
	6	0.408	0.450	0.491	....	0.30
	7	0.430	0.470	0.512	....	0.29
14	5	0.420	0.460	0.493	....	0.37
	6	0.447	0.488	0.530	....	0.36
	7	0.470	0.511	0.553	....	0.35

Adjustable type

For 20-in. widths

Depth of Steelform, in.	Width of joist, in.	Cu ft of concrete per sq ft of floor with slab thickness over Steelforms of			Cu ft per lin ft of beam using tapered ends	
		2 in.	2½ in.	3 in.	2 in.	4 in.
6	3½	0.262	0.304	0.345	....	0.14
	4½	0.279	0.321	0.362	....	0.13
	5½	0.295	0.337	0.378	....	0.13
8	3½	0.289	0.329	0.370	....	0.18
	4½	0.309	0.350	0.393	....	0.18
	5½	0.331	0.372	0.414	....	0.17
10	3½	0.312	0.353	0.395	....	0.23
	4½	0.340	0.381	0.424	....	0.22
	5½	0.367	0.408	0.450	....	0.21
12	3½	0.336	0.378	0.420	....	0.27
	4½	0.371	0.412	0.455	....	0.26
	5½	0.403	0.444	0.486	....	0.25
14	3½	0.361	0.403	0.444	....	0.31
	4½	0.402	0.443	0.485	....	0.30
	5½	0.438	0.480	0.522	....	0.29

For 30-in. widths

Depth of Steelform, in.	Width of joist, in.	Cu ft of concrete per sq ft of floor with slab thickness over Steelforms of			Cu ft per lin ft of beam using tapered ends	
		2½ in.	3 in.	3½ in.	4 in.	5 in.
6	5½	0.300	0.341	0.383	0.10	
	6½	0.311	0.353	0.395	0.10	
	7½	0.322	0.364	0.406	0.09	
8	5½	0.326	0.368	0.408	0.13	
	6½	0.341	0.383	0.424	0.12	
	7½	0.355	0.397	0.438	0.12	
10	5½	0.352	0.394	0.434	0.16	
	6½	0.371	0.413	0.454	0.15	
	7½	0.389	0.430	0.472	0.15	
12	5½	0.378	0.420	0.461	0.18	
	6½	0.401	0.443	0.484	0.18	
	7½	0.422	0.464	0.505	0.17	
14	5½	0.404	0.446	0.486	0.21	
	6½	0.430	0.472	0.514	0.21	
	7½	0.456	0.497	0.539	0.20	

Amount of concrete given for tapered endforms is for one side of beam only.

The volume of concrete per square foot of floor will average 0.372 cu ft, the weight of which is 56 lb. Add 40 psf for temporary additional load. This gives a total average load of 96 psf on the forms.

The 6-in.-wide joist requires 2- by 8-in. centering strips, 3 ft 0 in. on centers. A 2- by 8-in. *S4S* will safely support the load with a span of 4 ft 0 in. Accordingly, the stringers will be spaced 4 ft 0 in. on centers. The load per linear foot of stringer will be 4 by 96 = 384 lb. For a 2- by 10-in. *S4S* stringer the safe span, based on an allowable fiber stress of 1,800 psi, is 8.7 ft. A shore spacing of 8 ft 0 in. will be used. A shore will be required at the end of each stringer. The ends of the centering strips at the edges of the floor bay will be supported by the beam forms.

The area of floor will be 16 × 24 = 384 sq ft

The quantity of lumber will be

Centering strips, 24 ÷ 3 ft = 8 + 1 = 9 pc, 2 × 8 in. × 16 ft 0 in.	= 192 fbm
Ledgers at beams, 2 pc, 1 × 6 in. × 24 ft 0 in.	= 24 fbm
Stringers, 3 pc, 2 × 10 in. × 24 ft 0 in.	= 120 fbm
Shores, 12 pc, 4 × 4 in. × 10 ft 0 in.	= 160 fbm
Scabs, 12 pc, 2 × 4 in. × 2 ft 0 in.	= 16 fbm
Sills, 12 pc, 2 × 8 in. × 1 ft 6 in.	= 24 fbm
Braces:	
6 pc, 1 × 6 in. × 12 ft 0 in.	= 36 fbm
4 pc, 1 × 6 in. × 16 ft 0 in.	= 32 fbm
Total quantity of lumber	= 604 fbm
Lumber per sq ft of floor, 604 ÷ 384	= 1.6 fbm

The preceding quantity of lumber is based on an ideal bay, with no wastage because of end sawing. On most jobs this condition will not exist, and end wastage of centering strips and stringers will probably average about 10 percent. Adding this amount to the foregoing quantities will increase the total quantity of lumber by 10 percent of 312 = 31 fbm. The resulting quantity of lumber will be 635 fbm, giving 1.7 fbm per square foot of floor area.

If adjustable shores are used instead of wood shores, the quantity of lumber will be reduced by about 160 fbm. The elimination of scabs and sills will be offset by the use of more braces. The total quantity of lumber for the bay will be about 635 − 160 = 475 fbm. The quantity of lumber per square foot of floor area will be 475 ÷ 384 = 1.3 fbm. Thus the reduction in the quantity of lumber will be about 0.3 fbm per sq ft.

Table 7-19 gives the quantities of lumber in fbm per square foot of floor area for metal-pan and concrete-joist types of concrete floors, using wood shores and adjustable shores.

Labor required to build forms and install metal pans for concrete floors The labor required to build the forms and install the pans for concrete floors may be divided into three operations. The first operation consists in cutting the centering strips, stringers, and shores to the correct lengths. The second operation consists in erecting the lumber in place. The third operation consists in installing the pans on the centering strips and securing them in place with nails driven through holes in the flanges. The labor required to remove the lumber and the pans should be included. If

TABLE 7-19   FBM of Lumber per Square Foot of Floor for Metal-pan- and Concrete-joist-type Concrete Floors*

| Height of ceiling, ft | 4- by 4-in. wood shores | | | | Adjustable shores | | | |
| | Thickness of floor, † in. | | | | Thickness of floor, † in. | | | |
	8½	10½	12½	14½	8½	10½	12½	14½
10	1.5	1.5	1.5	1.6	1.2	1.3	1.3	1.4
12	1.6	1.6	1.7	1.8	1.2	1.3	1.3	1.4
14	1.7	1.8	1.8	1.9	1.3	1.4	1.4	1.5
16	1.8	1.9	1.9	2.0	1.4	1.5	1.5	1.6
18	2.0	2.1	2.1	2.2	1.5	1.6	1.7	1.8

* The quantities are for construction where the ends of the centering strips are supported by beam forms.
† The thickness of floor equals the depth of the pan plus a slab thickness of 2½ in. over the top of the pan.

TABLE 7-20   Labor-hours* Required to Build and Remove Forms and to Install and Remove Metal Pans for 100 Sq Ft of Floor

| Labor | 4- by 4-in. wood shores | | | | Adjustable shores | | | |
| | Thickness of floor, in. | | | | Thickness of floor, in. | | | |
	8½	10½	12½	14½	8½	10½	12½	14½
Cutting forms:								
Carpenter	1.2	1.2	1.25	1.25	1.1	1.1	1.2	1.2
Helper	0.7	0.7	0.8	0.08	0.6	0.6	0.7	0.7
Erecting forms:								
Carpenter	2.4	2.4	2.6	2.6	2.2	2.2	2.4	2.4
Helper	1.0	1.0	1.2	1.2	0.9	0.9	1.0	1.0
Removing forms:								
Helper	1.2	1.2	1.3	1.3	1.2	1.2	1.3	1.3
Installing pans:								
Mechanic	0.35	0.40	0.45	0.50	0.35	0.40	0.45	0.50
Helper†	0.50	0.55	0.60	0.65	0.50	0.55	0.60	0.65
Removing pans:								
Helper	0.90	0.95	1.00	1.05	0.90	0.95	1.00	1.05

* The labor-hours are based on reasonably large simple areas, using skilled mechanics to install the pans. For complex jobs the labor-hours should be increased. If carpenters are used to install the pans, all labor-hours installing pans should be increased about 10 percent.
† If helpers are not used, add the time to that given for mechanics.

the lumber can be reused without additional cutting, only the labor for the last two operations should be considered for additional uses. The lumber is usually cut with a power saw set up at the job.

If the pans are installed by carpenters instead of by experienced mechanics, the rates of installation will not be so rapid as by mechanics. Metal pan subcontractors use skilled mechanics, while contractors who rent the pans use carpenters to install them.

Table 7-20 gives the quantities of labor required to build forms and install metal pans for 100 sq ft of floor.

Example Estimate the cost of form lumber and metal pans in place and removed for a total floor area 39 ft 6 in. wide by 96 ft 6 in. long, of which 526 sq ft will be beams. The floor thickness will be $10\frac{1}{2}$ in. The joints will run one way only. The ceiling height will be 10 ft 8 in. The structure will be a dormitory, with corresponding irregularities in the floor areas. Carpenters will install the pans. Adjustable shores will be used.

The gross area of the floor will be $39.5 \times 96.5 = 3,812$ sq ft. Deducting the area of the beams leaves a net floor area of 3,286 sq ft for form purposes.

Assume that the lumber will cost $220.00 per M fbm at the job, and that it will be used eight times.

The metal pans will be rented, with the renter to pay the cost of freight to the job and back to the supplier.

The cost in place will be

Form lumber, 3,286 sq ft $\times$ 1.3 fbm per sq ft = 4,272 fbm @ $0.0275	= $	117.48
Nails, 4,272 fbm $\times$ 12 lb per M fbm = 51.5 lb @ $0.25	=	12.90
Shores, 3,286 sq ft $\div$ 24 sq ft per shore = 137 @ $0.45	=	61.65
Pan rental, 3,286 sq ft @ $0.23	=	755.78
Freight on pans	=	96.00
Form oil, 3,286 sq ft $\div$ 400 sq ft per gal = 8 gal @ $1.20	=	9.60
Power saws, 4,272 fbm $\div$ 600 fbm per hr = 7 hr @ $0.35	=	2.45
Cutting and erecting forms:		
Carpenters, 3,286 sq ft $\times$ 3.3 hr per 100 sq ft = 108.4 hr @ $7.67	=	831.43
Helper, 3,286 sq ft $\times$ 1.5 hr per 100 sq ft = 49.3 hr @ $5.79	=	285.45
Removing forms:		
Helper, 3,286 sq ft $\times$ 1.2 hr per 100 sq ft = 39.4 hr @ $5.79	=	228.13
Installing pans:		
Carpenter, 3,286 sq ft $\times$ 0.45 hr per 100 sq ft = 14.8 hr @ $7.67	= $	113.52
Helper, 3,286 sq ft $\times$ 0.55 hr per 100 sq ft = 18.1 hr @ $5.79	=	104.80
Removing pans:		
Helper, 3,286 sq ft $\times$ 0.95 hr per 100 sq ft = 31.2 hr @ $5.79	=	180.65
Foreman, based on using four carpenters, 123.2 hr $\div$ 4 = 30.8 hr @ $8.50	=	261.80
Total cost		$3,061.64
Cost per sq ft, $3,061.64 $\div$ 3,286	=	0.932

For pans of types and sizes different from the ones used in this example, the estimator should make a similar determination of the costs.

Concrete stairs The construction of concrete stairways is complicated and presents a problem for the estimator. The costs per unit of volume or area vary a great deal, depending on the length of the tread, the width of the tread, the height of the riser, the shape of the supporting floor for the shores, whether the treads are square or rounded, and whether the ends of the treads and risers are open or closed with curbs. Some stairs are straight run, while others have an intermediate landing. Some stairs are completely in the open, while others have a wall on one or both sides. All these conditions affect the cost of stairs.

The width of treads varies from 8 to 12 in., and the height of risers varies from 6 to 8 in. for different stairs.

Steel dowels should be set in the concrete floor and the beam supporting the landing prior to placing the concrete in the floor and landing. These dowels are tied to the reinforcing steel for the stairs as it is placed.

Lumber required for forms for concrete stairs Figure 7-18 shows the details of one method of constructing forms for concrete stairs. The height from floor to floor is 10 ft 6 in. Using a riser height of 7 in., 18 risers and 17 treads will be required. The treads will be 12 in. wide, which gives a total horizontal length of 17 ft 0 in. The supporting floor is horizontal, which requires shores of variable lengths to support the forms.

Example   Estimate the cost of materials and labor for building forms for the stairs illustrated in Fig. 7-18.

The cost will be

Sheathing, 16 × 6 + 20% waste = 115 fbm @ $0.22	= $ 25.30
Joists, 5 pc, 2 in. × 8 in. × 16 ft 0 in. = 107 fbm @ $0.22	= 23.54
Side stringers:	
2 pc, 2 in. × 12 in. × 20 ft 0 in. = 80 fbm @ $0.26	= 20.80
2 pc, 2 in. × 4 in. × 22 ft 0 in. = 29 fbm @ $0.22	= 6.38
Riser planks, 18 pc, 2 in. × 8 in. × 6 ft 0 in. = 144 fbm @ $0.22	= 31.68
Riser cleats, 36 pc, 2 in. × 4 in. × 1 ft 0 in. = 24 fbm @ $0.22	= 5.28
Riser stringer, 1 pc, 2 in. × 4 in. × 20 ft 0 in. = 13 fbm @ $0.22	= 2.86
Riser braces, 18 pc, 2 in. × 4 in. × 1 ft 0 in. = 12 fbm @ $0.22	= 2.64
Stringers, 2 pc, 4 in. × 6 in. × 6 ft 0 in. = 24 fbm @ $0.22	= 5.28
Shores:	
2 pc, 4 in. × 4 in. × 6 ft 0 in. = 16 fbm @ $0.22	= 3.52
2 pc, 4 in. × 4 in. × 2 ft 0 in. = 6 fbm @ $0.22	= 1.32
Sills, 4 pc, 2 in. × 8 in. × 1 ft 6 in. = 8 fbm @ $0.22	= 1.76
Wedges, 5 fbm @ $0.22	= 1.10
Headers, 1 pc, 1 in. × 8 in. × 6 ft 0 in. = 4 fbm @ $0.22	= 0.88
Braces, 3 pc, 1 in. × 6 in. × 12 ft 0 in. = 18 fbm @ $0.22	= 3.96
Nails, 10 lb @ $0.25	= $ 2.50
Total cost	= $138.80

The cost just given is for a single use. If the forms can be removed and used again on stairs of the same design, the cost per use will be reduced by approximately the number of uses. Frequently the forms for stairs are sup-

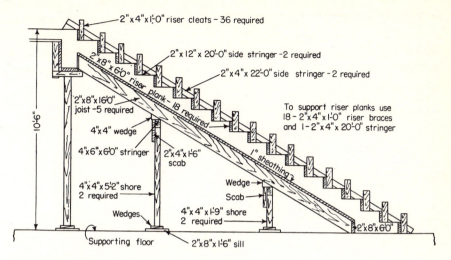

Fig. 7-18 Wood forms for concrete stairs.

ported by shores resting on existing lower stairs. If this construction is used, the lengths of all shores will be increased to 9 ft 0 in. for Fig. 7-18.

Labor required to build forms for concrete stairs Because of the irregularities of the forms for concrete stairs, it is difficult to estimate accurately the cost of building forms. It is necessary to cut the shores to length, attach the 4- by 6-in. stringers, set the shores and stringers in place, and brace them securely. The 4- by 4-in. wedges are nailed to the tops of the stringers and the joists before the sheathing is installed. The sheathing is cut and nailed to the joists. The side stringers, consisting of a 2 in. by 12 in. by 20 ft 0 in. and a 2 in. by 4 in. by 22 ft 0 in., are cleated together, and the positions for the 2- by 8-in. riser planks are marked on the stringers. The stringers and the riser planks are installed. Because of the length of the riser planks, 6 ft 0 in., a center support should be used to hold them against deflection. A 2- by 4-in. riser stringer is placed at the center of the riser planks, and each riser plank is secured to it with a 2-in. by 4-in. by 1-ft 0-in. riser brace.

Two carpenters should be used to build the forms for the stairs. Each carpenter should fabricate and install about 25 fbm of lumber per hour.

REINFORCING STEEL

Types and sources of reinforcing steel Reinforcing for concrete may consist of steel bars or welded-wire fabric, used separately or together. The cost of bars may be estimated by the pound, hundred-weight, or ton, while

TABLE 7-21 Sizes, Areas, and Weights of Reinforcing Bars

Bar No.	Size, diam, in.	Area, sq in.	Weight, lb per ft
2	¼	0.05	0.167
3	⅜	0.11	0.376
4	½	0.20	0.668
5	⅝	0.31	1.043
6	¾	0.44	1.502
7	⅞	0.60	2.044
8	1.0	0.79	2.670
9	1.128	1.00	3.400
10	1.270	1.27	4.303
11	1.410	1.56	5.313

the cost of welded-wire fabric may be estimated by the pound or square foot.

Usually bars are fabricated to the required lengths and shapes by commercial shops prior to delivery to a project. Such shops are equipped with machines that will perform the fabricating operations more economically than when fabricating is performed on the job. Upon request, these shops will furnish quotations covering the supplying and fabricating of all rein-

TABLE 7-22 Quantity of Reinforcing Steel

Bar mark	No. required	Bar size	Length	Weight, lb per ft	Total weight, lb
A	120	4	30 ft 0 in.	0.668	2,405
B	56	4	20 ft 0 in.	0.668	749
C	116	5	24 ft 0 in.	1.043	2,900
D	42	6	12 ft 4 in.	1.502	780
E	36	6	14 ft 8 in.	1.502	794
F	28	6	19 ft 8 in.	1.502	826
G	16	7	18 ft 6 in.	2.044	604
H	72	7	22 ft 3 in.	2.044	3,280
I	84	7	18 ft 8 in.	2.044	3,200
J	24	8	24 ft 0 in.	2.670	1,535
K	18	8	21 ft 6 in.	2.670	1,037
Total weight					18,110

forcing for a given project. Estimators frequently request such quotations before preparing estimates.

Properties of reinforcing bars Table 7-21 gives the sizes, areas, and weights of reinforcing bars.

Estimating the quantity of reinforcing steel When the reinforcing steel consists of bars of different sizes and lengths, each size and length should be listed separately. A form such as the one used in Table 7-22 will simplify the listing and reduce the danger of errors. Each size and length should be assigned a number or a letter of the alphabet.

Cost of reinforcing steel The items which determine the cost of reinforcing steel delivered to a project are:

1 The base cost of the bars at the fabricating shop
2 The cost of preparing shop drawings
3 The cost of shop handling, cutting, bending, etc.
4 The cost of selling
5 The cost of shop overhead and profit
6 The cost of transporting from the shop to the project
7 The cost of specialties, such as spacers, saddles, chairs, ties, etc.

It is customary to determine the weight of reinforcing steel based on the lengths and sizes of bars and the nominal weights given in Table 7-21, with no extra charge made for waste. Reinforcing bars are usually available in stock lengths of 40 and 60 ft.

Size extras for reinforcing steel If an estimator wishes to determine the approximate cost of reinforcing steel for a project, he should list each size bar separately and then determine the total weight by size. In addition to the base price cost of the reinforcing, an extra charge based on the sizes of the bars will be made. These extras are subject to change and should be verified before preparing an estimate.

Quantity extras for reinforcing steel This extra cost varies with the total quantity of steel purchased.

Cost for detailing and listing reinforcing steel Before fabricating reinforcing steel it is necessary for the fabricating shop to prepare drawings which show how the bars are to be fabricated. A charge is made for this service, based on the complexity of the drawings and the quantity of reinforcing.

Cost for fabricating reinforcing steel The cost for this operation varies with the sizes of the bars and the complexity of the operations.

Cost for reinforcing steel delivered to a project If an estimator does not wish to consider the several extra costs, an estimate of the cost can usually be obtained from a supplier, who will quote a total cost for the entire lot, fabricated and delivered to the project.

Labor placing reinforcing steel bars The rates at which workers will place reinforcing steel bars will vary with the following factors:

1 Sizes and lengths of bars
2 Shapes of the bars
3 Complexity of the structure
4 Distance and height the steel must be carried
5 Allowable tolerance in spacing bars
6 Extent of tieing required
7 Skill of workers

Less time is required to place a ton of steel when the bars consist of large sizes and long lengths than when they are small sizes and short lengths.

Straight bars may be placed more rapidly than bars with bends and end hooks.

If the bars must be placed in complicated structures such as stairs, the rate of placing will be less than for simple structures such as walls, floors, etc.

Steel bars should be stored as near the structure as possible, preferably not more than 50 to 100 ft away, in order to reduce the time required to carry it. If steel must be carried to upper floors or parts of a structure, additional time will be required.

Rigid tolerances on the spacing of bars will reduce the rate of placing steel somewhat.

TABLE 7-23   Rates of Placing Reinforcing-steel Bars, in Hours per Ton

Class of worker	Size of bars			
	5⁄8 in. and less		3⁄4 in. and over	
	Length of bars			
	Over 15 ft	Under 15 ft	Over 15 ft	Under 15 ft
Bars not tied in place				
Handymen	14–16	16–18	11–13	13–15
Steel setters	11–13	13–15	7–9	9–11
Bars tied in place				
Handymen	16–18	18–20	12–14	14–16
Steel setters	12–14	14–16	8–10	10–12

TABLE 7-24 Properties of Representative Styles of Welded-wire Fabric

Style	Weight per 100 sq ft, lb	Spacing of wire, in.		Gauge number		Sectional area of wires per ft, sq in.	
		Longi-tudinal	Trans-verse	Longi-tudinal	Trans-verse	Longi-tudinal	Trans-verse
44-1010	31	4	4	10	10	0.043	0.043
44-88	44	4	4	8	8	0.062	0.062
44-66	62	4	4	6	6	0.087	0.087
44-44	85	4	4	4	4	0.120	0.120
66-1010	21	6	6	10	10	0.029	0.029
66-88	30	6	6	8	8	0.041	0.041
66-66	42	6	6	6	6	0.058	0.058
66-44	58	6	6	4	4	0.080	0.080
66-22	78	6	6	2	2	0.108	0.108
48-1012	20	4	8	10	12	0.043	0.013
48-912	23	4	8	9	12	0.052	0.013
48-812	27	4	8	8	12	0.062	0.013
412-1012	19	4	12	10	12	0.043	0.009
412-812	25	4	12	8	12	0.062	0.009
412-610	36	4	12	6	10	0.087	0.014
412-49	49	4	12	4	9	0.120	0.017
412-48	51	4	12	4	8	0.120	0.021
416-812	25	4	16	8	12	0.062	0.007
416-610	35	4	16	6	10	0.087	0.011
416-49	48	4	16	4	9	0.120	0.013
416-28	64	4	16	2	8	0.162	0.015
612-66	32	6	12	6	6	0.058	0.029
612-44	44	6	12	4	4	0.080	0.040
612-22	59	6	12	2	2	0.108	0.054
612-14	61	6	12	1	4	0.126	0.040
612-03	72	6	12	0	3	0.148	0.047

More time is required to tie bars at every intersection than when little or no tieing is required.

Reinforcing steel may be placed by laborers, handymen, or skilled steel setters. The latter should place the steel at the fastest rate, with little or no supervision required.

Table 7-23 gives representative rates of placing reinforcing-steel bars. The rates are based on carrying the steel by hand not more than 100 ft from the stockpiles to the structure. The steel will be placed by either handymen or steel setters, but not by both.

Welded-wire fabric For certain types of concrete projects, such as sidewalks, pavements, floors, canal linings, etc., it may be more economical to use welded-wire fabric for reinforcing instead of steel bars. This fabric is made from cold-drawn steel wire, electrically welded at the intersections of longitudinal and transverse wires, to form rectangles or squares. It is available in flat sheets or rolls, the latter frequently being 60 in. wide and 150 ft long. It is usually priced by the square foot or roll, the price depending on the weight.

The quantity required will equal the total area to be reinforced, with 5 to 10 percent area added for side and end laps.

The fabric to be used is designated by specifying the style, such as 412-610. This style designates a rectangular fabric with longitudinal wires spaced 4 in. apart, transverse wires spaced 12 in. apart, using No. 6 gauge longitudinal wires, and No. 10 gauge transverse wires.

Table 7-24 gives the properties of representative styles of welded-wire fabric. Many other styles are available.

Based on a width of 60 in. and a length of 150 ft, the weight of a roll of fabric will vary from 75 to 1,000 lb or more.

Labor placing welded-wire fabric Fabric is placed by unrolling it over the area to be reinforced, cutting it to the required lengths, lapping the edges and the ends and tieing at frequent spacings. On large regular areas a man should place it at a rate of 0.25 hr per 100 sq ft, while for irregular areas, requiring cutting and fitting, the rate of placing may be 0.5 hr per 100 sq ft.

CONCRETE

Cost of concrete The cost of concrete in a structure includes the cost of aggregate, cement, water, equipment, and of labor mixing, transporting, and placing the concrete. When ready-mixed concrete is used, some of the costs are transferred from the job to the central mixing plant. The cost of the several items just listed will vary with the size of the job, the location, the quality of the concrete, the extent to which equipment is used instead of labor, and the distribution of concrete within the job.

Quantities of materials for concrete The estimator should determine the quantity of each class of concrete in the job. With this information he can determine the quantities and costs of aggregate, cement, and water for each class or for each structural element.

Concrete structures are designed for concretes having specified strengths, usually expressed in pounds per square inch in compression 28 days after it is placed in the structure. In order to produce a concrete with a specified strength it is common practice to employ a commercial laboratory to design the mix. Such a laboratry specifies the weight or volume of fine aggregate, coarse aggregate, cement, and water to produce a batch or a cubic yard of concrete having the required strength.

Estimators seldom have the laboratory design information when estimating the cost of a project. Tables giving the approximate quantities of aggregate, cement, and water for different qualities of concrete are available. The information given in these tables is sufficiently accurate for estimating purposes.

Table 7-25 gives the approximate quantities of cement, water, and coarse and fine aggregates required to produce 1 cu yd of concrete having the indicated 28-day compressive strengths. It should be noted that the

TABLE 7-25 Quantities of Cement, Water, and Aggregates Required for 1 Cu Yd of Concrete Having the Indicated 28-day Compressive Strength

Sk of cement	Gal of water	Weights of saturated surface-dry aggregate, lb		28-day compressive strength, psi
		Fine	Coarse	
1-in. coarse aggregate				
4.9	39.1	1,370	1,860	2,250
5.6	39.1	1,345	1,847	2,750
6.0	39.0	1,260	1,860	3,000
6.5	39.0	1,235	1,820	3,300
7.2	39.8	1,150	1,875	3,700
8.0	40.0	1,120	1,840	4,250
2-in. coarse aggregate				
4.5	36.0	1,350	1,980	2,250
5.1	35.6	1,275	1,980	2,750
5.5	35.7	1,265	1,980	3,000
6.0	36.0	1,200	1,980	3,300
6.7	36.8	1,140	2,010	3,700
7.4	36.8	1,110	2,000	4,250

TABLE 7-26   Approximate Output of Construction Types of Concrete Mixers Based on a 1-Min Mixing Time*

Size mixer	Batches per hr		Approx. output, cu yd per hr
	Range	Average	
3½S	25–33	30	4.0
6S	25–33	30	6.5
11S	24–32	28	11.0
14S	20–30	24	12.0
16S	20–30	24	14.0
28S	18–26	22	23.0

* The higher number of batches per hour may be used if the job is organized to eliminate lost time in charging and discharging the mixer. For a poorly organized job, the lower number of batches should be used.

aggregate is saturated surface dry when weighed. If the stockpiles of aggregates contain surface moisture, as they usually do, the quantity of water added should be decreased by the amount of water present on the surface of the aggregate. The weights of the aggregate should be increased to produce net weights that correspond to those given in the table.

Output of concrete mixers   The output of concrete mixers varies with the size of the batch, the method of charging, the method of discharging, and the time the batch must be mixed.

The nominal size of a concrete mixer is indicated by a number representing the quantity of concrete per batch, measured in cubic feet. The batch may not always equal the indicated size. For example, it might be more practical to use 2 sacks of cement for a batch less than the maximum possible size instead of using 2.2 sacks to produce a batch equal to the size of the mixer.

When the full batch is discharged immediately into a hopper or a bucket instead of into several buggies or wheelbarrows, it will be possible to reduce the time per batch, thus increasing the number of batches per hour.

Table 7-26 gives the approximate output of construction types of concrete mixers for various sizes, based on a 1-min mixing time.

Labor mixing and placing concrete   The labor required to mix and place concrete varies with the number of operations performed by labor, the location of the aggregate piles with respect to the mixer, the location of cement storage, the length of haul of concrete, the condition of runways, the hauling equipment, buggies or wheelbarrows, and the distribution of the placing area. Factors which decrease the amount of labor, and thus the cost of labor, are reducing the length of haul of aggregate, storing cement near the mixer, locating the mixer near the center of placing concrete, constructing runways wide enough for easy travel and if possible level or sloping slightly from the mixer, using buggies instead of wheelbarrows, and distributing the placing over an area sufficiently large to eliminate congestion and interference. Also, the use of laborsaving equipment will reduce the amount of labor required.

In estimating the man-hours of labor required for a given concrete pour, the estimator should allow for the time required for getting ready to start the pour, for cleaning out the mixer and the buggies, and for putting away tools and equipment after the pour is completed. This time, amounting to approximately 30 min, will be the same regardless of the length of pour. For this and other reasons, pours of less than 4 hr should be avoided when possible.

Example As an example illustrating the method of determining the amount of labor required to mix and place 1 cu yd of concrete, assume that 80 cu yd of concrete is to be mixed and placed with an 11S mixer. Aggregate will be handled with wheelbarrows. Cement will be stacked near the mixer. The concrete will be hauled about 40 ft in buggies and deposited for a beam-and-slab floor. Two sacks of cement will be used per batch, which will be 11 cu ft. A mixing time of 1 min is specified.

Assuming that the output of the mixer will be 11 cu yd per hr, the actual mixing time will be 7.3 hr. Add 0.5 hr for getting ready and cleaning up after the pour is completed. This gives a total time of 7.8 hr.

The weight of sand per batch will be approximately $\frac{11}{27} \times 1,350 = 550$ lb, and the weight of gravel approximately $\frac{11}{27} \times 1,850 = 750$ lb. While the volume of a concrete wheelbarrow is about 3 cu ft, the practical load for aggregate is about 2 cu ft, or 220 lb. Use three wheelbarrows for sand and four for gravel. At least one extra laborer will be needed on each stockpile. This requires five men handling gravel and four men handling sand. Concrete buggies of the type used on this job have a capacity of 5 to 6 cu ft, but the load should be limited to about 3 cu ft. It will require four buggies to haul a batch.

The labor-hours for the job will be as given below:

Hauling gravel, 5 men $\times$ 7.8 hr	=	39.0 hr
Hauling sand, 4 men $\times$ 7.8 hr	=	31.2 hr
Handling cement, 1 man	=	7.8 hr
Hauling concrete, 4 men $\times$ 7.8 hr	=	31.2 hr
Helping empty buggies, 1 man	=	7.8 hr
Spreading and leveling concrete, 3 men	=	23.4 hr
Moving runways and general utility, 2 men =		15.6 hr
Total laborers	=	156.0 hr
Mixer operator	=	7.8 hr
Foreman, 1 man	=	7.8 hr
Labor per cu yd, 156 ÷ 80	=	1.95 hr
Mixer[1] operator per cu yd	=	0.125 hr
Foreman[1] per cu yd	=	0.125 hr

[1] Additional time is allowed for the mixer operator to service the mixer and for the foreman, who must be at the job before the laborers arrive and after they leave.

In mixing concrete with a 16S or a 28S mixer, it may be desirable to use a batching plant for the aggregate. This will require a clamshell for handling the aggregate. Eliminate the labor hauling gravel and sand, and add a clamshell operator plus a man on the batching plant.

Example   A concrete bridge pier containing 210 cu yd of concrete is to be poured in one operation. A 28S concrete mixer will be used. Aggregate will be lifted to a two-compartment batching plant with a clamshell. The concrete will be transported directly from the mixer to the pier forms with a crane and a 1½-cu yd bucket. Cement will be stored at the job in sacks about 10 ft from the mixer.

Using 2-in. maximum size aggregate, a 2,500-lb concrete will require five sacks of cement per batch, assumed to be 1 cu yd.

The mixer should mix 22 cu yd of concrete per hour. The actual pour should require $210 \div 22 = 9.5$ hr. Add 0.5 hr to get ready and to clean up after the pour. This gives a total of 10 hr.

The labor by classification will be as given below:

Clamshell operator	10 hr
Man on batching plant	10 hr
Laborers handling cement, 3 men	30 hr
Mixer operator	10 hr
Crane operator	10 hr
Signalman on forms	10 hr
Laborers spreading concrete, 3 men	30 hr
Laborers on vibrator, 2 men	20 hr
Utility men, 2 men	20 hr
Carpenter checking forms	10 hr
Foreman	12 hr

The labor required per cubic yard of concrete will be as given below:

Clamshell operator, 10 hr @ 210 cu yd	= 0.05 hr
Mixer operator	= 0.05 hr
Crane operator	= 0.05 hr
Carpenter	= 0.05 hr
Foreman, 12 hr ÷ 210 cu yd	= 0.06 hr
Laborers,[1] 120 hr ÷ 210 cu yd	= 0.57 hr

Table 7-27 gives approximate labor-hours by classification required to mix and place concrete using various sizes of mixers and methods.

Ready-mixed concrete   If ready-mixed concrete is available, it is frequently more economical and satisfactory to purchase it than to mix concrete at the job. This is especially true where working space around a job is limited and when small quantities of concrete are needed at various times during construction.

When ready-mixed concrete is used, the costs at the job will be reduced

[1] For many locations it may be necessary to classify some of these men as other than laborers

TABLE 7-27 Labor-hours per Cubic Yard Required to Mix and Place Concrete*

Size mixer	Method of handling	Common labor	Foreman	Mixer operator	Hoisting engineer	Crane operator	Carpenter
None	Hand	3–3½	0.33				
6S	Wheelbarrow	2¼	0.17				
6S	Hoist spout	1¾	0.17		0.17		
11S	Wheelbarrow	2	0.12	....	....	....	0.12
11S	Hoist spout	1¾	0.12	0.12	0.12	....	0.12
14S	Wheelbarrow	2	0.10	0.10	....	....	0.10
14S	Hoist spout	1¾	0.10	0.10	0.10	....	0.10
16S	Wheelbarrow	2	0.10	0.10	....	....	0.10
16S	Buggies	1⅞	0.10	0.10	....	....	0.10
16S	Clamshell, hoist and buggies	1½	0.10	0.10	0.10	0.10	0.10
28S	Clamshell, hoist and buggies	1½	0.08	0.08	0.08	0.08	0.08
Mass concrete							
16S	Clamshell, crane and bucket	0.8	0.09	0.08	0.08	0.08	0.08
28S	Clamshell, crane and bucket	0.6	0.06	0.05	0.05	0.05	0.05

* The information given in the top portion of the table should be used for buildings and similar structures. The information given under Mass Concrete should be used for large foundations, piers, dams, etc.

TABLE 7-28   Labor-hours Required to Place 1 Cu Yd of Ready-Mixed Concrete

Type of structure	Method of handling	Common labor	Fore-man	Hoist or crane operator	Carpenter
Large foundation	Discharge directly from truck, using chute	0.5	0.07	. . . .	0.07
Bridge pier	Crane and bucket	0.5	0.07	0.07	0.07
Slab at ground level	Crane and bucket	0.7	0.07	0.07	0.07
Slab above ground level	Crane or hoist, bucket, hopper, and buggies	1.3	0.07	0.07	0.07
Foundation wall	Crane, bucket, and tremies	0.7	0.07	0.07	0.07
Foundation wall	Hand buggies	1.0	0.07	. . . .	0.07

to handling and placing the concrete. If concrete can be delivered to any part of a structure, the cost of handling should be less than for job-mixed concrete.

Labor placing ready-mixed concrete   The labor required to place ready-mixed concrete will vary with the rate of delivery, the type of structure, and the location of the structure.

If the trucks can be driven to large foundations, constructed at or below the level of the natural ground, it may be possible to discharge the concrete directly into the forms, using a chute. Not more than five or six men may be needed to spread and vibrate the concrete.

Concrete for a slab, constructed at or near ground level, may be discharged into a bucket, hoisted by a crane, and distributed over the slab area, with little handling necessary. A crew of six to nine men should be able to place up to 15 cu yd of concrete per hour.

Concrete for a floor slab, constructed above ground level, may be discharged into a bucket, hoisted with a crane or a tower unit, deposited into a floor hopper, then hauled to its destination using power-driven or hand-pushed buggies.

Example   Estimate the labor required to place the concrete for a floor slab 100 ft long, 60 ft wide, whose average thickness is 6 in. The floor will be 16 ft above ground level. The concrete will be delivered to a 1-cu yd bucket, hoisted with a crane, deposited into a 1-cu yd floor hopper, and hauled in hand-pushed buggies, deposited, and spread to the required thickness.

The volume of concrete will be $\dfrac{100 \times 60 \times 6}{12 \times 27} = 111$ cu yd. Concrete will be delivered
at the rate of 15 cu yd per hr.
The crew should be about as follows:

> 1 crane operator
> 1 man on the ground handling the bucket
> 1 man emptying the bucket into the hopper
> 1 man filling the buggies
> 5 men pushing buggies
> 2 men helping empty the buggies
> 5 men spreading and screeding the concrete
> 1 carpenter on runways
> 1 man helping the carpenter with runways
> 3 utility men
> 1 foreman

Depending on the location of the project, the men might be classified as follows:

> Laborers, 18 men
> Crane operators, 1 man
> Carpenter, 1 man
> Foreman, 1 man
> Time to place concrete, 111 ÷ 15   = 7.4 hr
> Add time to get ready and clean up = 0.6 hr
>
>   Total time   = 8.0 hr
> Total labor-hours, 8 × 18 = 144
> Labor-hours per cu yd, 144 ÷ 111 = 1.3
> Foreman-hours per cu yd, 8 ÷ 111 = 0.07

Table 7-28 gives the representative labor-hours required to place ready-mixed concrete for various types of structures, based on delivering concrete at a rate of approximately 15 cu yd per hr. For lower or higher rates of delivery, the labor-hours may be slightly higher or lower, respectively.

Lightweight concrete When the strength of concrete is not a primary factor, and when a reduction in weight is desirable, a lightweight aggregate, such as cinders, burned clay, vermiculite, pumice, or other materials may be used instead of sand and gravel or crushed stone. Concrete made with this aggregate and portland cement may weigh as little as 20 lb per cu ft. However, most weights will be higher than this, with values usually running from about 40 to 100 lb per cu ft, depending on the aggregate and the amount of cement used.

The insulating properties of lightweight concrete are better than for conventional concrete.

Perlite concrete aggregate While several lightweight aggregates are used, only Perlite will be discussed to illustrate the properties and costs of

TABLE 7-29   Representative Properties of Concrete Made with Perlite Aggregate*

Mix proportions by volume				Materials required for 1 cu yd				Compressive strength 28 days, psi	Weight when placed, lb per cu ft
Cement, sk	Perlite, cu ft	Water, gal	Air-entraining agent, pt	Cement, sk	Perlite, cu ft	Water, gal	Air-entraining agent, pt		
1	4	9	1.00	6.75	27	61.0	6.75	440	50.5
1	5	11	1.25	5.40	27	59.5	6.75	270	45.5
1	6	12	1.50	4.50	27	54.0	6.75	180	40.5
1	7	14	1.75	3.85	27	54.0	6.75	130	38.0
1	8	16	2.00	3.38	27	54.0	6.75	95	36.5

* From Perlite Institute file 6a/Pe.

lightweight concrete. This aggregate is a volcanic lava rock that has been expanded by heat to produce a material weighing approximately 8 lb per cu ft. Concrete made from this aggregate may be used for subfloors, fireproofing steel columns and beams, roof decks, concrete blocks, etc.

The materials used in making concrete include portland cement, aggregate, water, and an air-entraining agent. Mixing may be accomplished in a plaster or a drum-type concrete mixer, or transit-mixed concrete may be purchased in some localities. Perlite is generally available in bags containing 4 cu ft, which weigh about 32 lb.

Table 7-29 gives representative properties of concrete made with Perlite aggregate. The dry weights will be approximately 70 per cent of the wet weights.

The cost of Perlite concrete When lightweight concretes are used for floor fills or roof decks, they are usually placed in thinner layers than when conventional concrete is used. The reduction in volume, without a corresponding reduction in the number of laborers, will result in a higher labor cost per cubic yard of concrete. The cost of labor required to mix and place this concrete for a floor fill or roof deck should be increased at least 50 percent over that required for slabs constructed with conventional concrete.

Example Estimate the total cost and the cost per cubic yard for furnishing materials, mixing and placing 36 cu yd of 1.6 Perlite concrete for a floor fill 2½ in. thick placed at ground level.
The cost will be

Cement, 36 cu yd × 4.5 sk per cu yd = 162 sk @ $1.60	= $	259.20
Perlite, 36 cu yd × 27 = 972 cu ft @ $0.52	=	505.44
Water, 36 cu yd × 60 = 2.2 M gal @ $0.40	=	0.88
Air-entraining agent, 36 cu yd × 6.75 pints per cu yd = 243 pt @ $0.18	=	43.74
Concrete mixer, 36 cu yd × 0.15 hr per cu yd = 5.4 hr @ $1.55	=	8.37
Mixer operator, 5.4 hr @ $5.25	=	28.35
Laborers, 36 cu yd × 3 hr per cu yd = 108 hr @ $4.75	=	513.00
Cement finisher, 36 cu yd × 1 hr per cu yd = 36 hr @ $6.75	=	243.00
Foreman, 5.4 hr @ $8.50	=	45.90
Total cost	=	$1,647.88
Cost per cu yd, $1,647.88 ÷ 36 cu yd	=	45.50

TILT-UP CONCRETE WALLS

General description A method of building concrete walls that is proving economical and successful is to cast the walls on the concrete-floor slab, usually at ground level, and tilt them into position. For this method of construction, the walls are divided into several panels of convenient size. The concrete floor, on which the wall sections are to be cast, should be cleaned and coated with paraffin, a nonstaining oil, or a layer of heavy kraft paper.

Side forms, equal in height to the thickness of the wall, are placed around the panels to be poured. Stiffeners should be attached to the side forms to maintain true alignment. The reinforcing steel is placed in position, with steel dowels projecting through holes drilled in the side forms to permit the walls to be secured in position after they are tilted up. The kraft paper is wet and stretched smooth, and the concrete is placed in the forms. After the concrete has cured properly, the forms are removed and the wall panels are tilted to final position as units of a wall.

Frames for window and door openings should be installed before placing the concrete in the forms. Conduits can be placed in the walls easily.

A tight seal between the bottoms of the wall panels and the concrete floor or grade beam is obtained by spreading a layer of cement grout on the floor under the wall prior to raising the panel.

The width of wall panels is limited by the shape and size of the building and the lifting capacity of the equipment used to tilt the wall up. The panels are tied into a solid continuous structure by joining two adjacent panels with a reinforced-concrete column. The projecting reinforcing dowels are embedded in the column. Corners are secured in the same manner.

In order to reduce the danger of structural failure in the panels while they are being tilted into position, it is advisable to use strongbacks, steel beams temporarily bolted to the panels. For this purpose, bolts should be embedded in the concrete. After the walls are erected, the strongbacks are removed and the bolts can be cut off flush with or slightly inside the concrete surface.

The advantages of this method of construction include low cost of forms, low cost of placing reinforcing, and low cost of placing concrete, as high hoisting into vertical forms is not necessary.

Considerable care should be exercised in securing a satisfactory parting membrane under the panel to be cast or the wall may partly bond to the concrete floor. Lifting will be difficult or even impossible if this should occur. A heavy unit of lifting equipment, such as a crane, will be needed to tilt up the panels.

If proper care is exercised, a very satisfactory job can be obtained.

Example Estimate the total direct cost of furnishing materials, equipment, and labor to erect a concrete tiltup wall in place. The total length of the wall will be 118 ft 9 in. The height will be 12 ft 0 in., and the thickness will be 9 in.

The wall will be cast in eight panels, each 13 ft 9 in. wide. This width will permit the erection of 15- by 16-in. connector and support columns located along the walls and at the corners of the building.

The reinforcing steel for the wall will consist of ¾-in.-round horizontal bars, spaced 12 in. on centers, and ¾-in.-round vertical bars, spaced 8 in. on centers. The horizontal bars will project 10 in. outside the two vertical edges of the wall panels to tie the wall sections to the columns.

The concrete columns will be 15 by 16 in., with six 1-in.-round reinforcing bars per

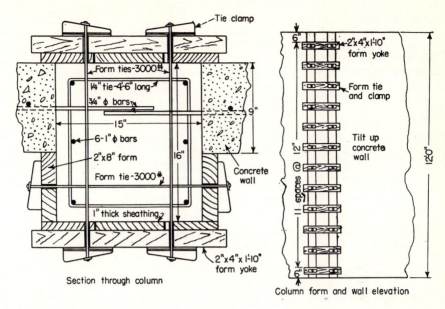

Fig. 7-19 Details of columns and forms for a tilt-up concrete wall.

column. Ties of ¼-in. steel bars will be spaced 12 in. apart around the column reinforcing bars. Figure 7-19 shows a section through a column.

Assume that lumber costs $220.00 per M fbm at the job.

The costs will be

Wall forms, 4 sets required, assume 2 uses		
Sides:		
8 pc, 2 in. × 10 in. × 14 ft 0 in. = 187 fbm @ $0.11	= $	20.57
8 pc, 2 in. × 10 in. × 12 ft 0 in. = 160 fbm @ $0.11	=	17.60
Side stiffeners:		
8 pc, 2 in. × 6 in. × 16 ft 0 in. = 128 fbm @ $0.11	=	14.08
8 pc, 2 in. × 6 in. × 14 ft 0 in. = 112 fbm @ $0.11	=	12.32
Nails, 8 lb @ $0.25	=	2.00
Kraft paper, 16 sq @ $1.60	=	25.60
Column forms, 7 required, assume 2 uses		
Sheathing:		
28 pc, 1 in. × 6 in. × 12 ft 0 in. = 168 fbm @ $0.11	=	18.48
14 pc, 1 in. × 8 in. × 12 ft 0 in. = 112 fbm @ $0.11	=	12.32
14 pc, 2 in. × 8 in. × 12 ft 0 in. = 224 fbm @ $0.11	=	24.64
Yokes, 168 pc, 2 in. × 4 in. × 1 ft 10 in. = 205 fbm @ $0.11	=	22.55
Form ties, 252 @ $0.24	=	60.48
Tie clamps, 504 @ $0.025	=	12.60
Labor building forms		
Wall forms:		
Carpenters, 8 forms × 2 hr each = 16 hr @ $7.67	=	122.72
Helpers, 8 forms × 1 hr each = 8 hr @ $5.79	=	46.32

Column forms:

Carpenters, 336 sq ft × 6 hr per 100 sq ft = 20 hr @ $7.67	=	153.40
Helpers, 336 sq ft × 4 hr per 100 sq ft = 13 hr @ $5.79	=	75.27
Foreman, 10 hr @ $8.50	=	85.00

Reinforcing steel

For walls:

96 pc, ¾ in. × 15 ft 4 in. = 2,210 lb @ $0.16	=	353.60
160 pc, ¾ in. × 11 ft 10 in. = 2,840 lb @ $0.16	=	454.40
Tie wire, 12 lb @ $0.20	=	2.40

For columns:

42 pc, 1 in. × 11 ft 10 in. = 1,320 lb @ $0.16	=	211.20
84 ¼-in. steel ties × 4 ft 6 in. long = 63 lb @ $0.20	=	12.60
Tie wire, 4 lb @ $0.20	=	0.80

Labor placing reinforcing steel

Steel setter, 3.2 tons × 12 hr per ton = 38 hr @ $7.25	=	275.50

Concrete, ready mixed

Walls, 8 × 13 ft 9 in. × 12 ft 0 in. × 9 in. = 36.4 cu yd @ $21.00	=	764.40
Columns, 7 × 16 in. × 15 in. × 12 ft 0 in. = 5.3 cu yd @ $21.00	=	111.30
Grout, 4 cu ft @ $1.60	=	6.40

Labor placing concrete

Laborers, 42 cu yd × 1.5 hr per cu yd = 63 hr @ $5.79	=	364.77
Foreman, 8 hr @ $8.50	=	68.00

Tilting up the panels

Crane, 8 panels × 1.5 hr each = 12 hr @ $14.57	=	174.84
Crane operator, 12 hr @ $7.45	=	89.40
Crane oiler, 12 hr @ $5.75	=	69.00
Laborers, 3 men × 12 hr = 36 hr @ $5.79	=	208.44
Foreman, 12 hr @ $8.50	=	102.00
Strongbacks, bolts, etc.	=	8.00

Total cost	=	$4,003.00
Cost per cu yd of concrete, $4,003.00 ÷ 41.7 cu yd	=	96.00
Cost per sq ft of wall surface $4,003.00 ÷ 1,426 sq ft	=	2.81

CONCRETE BRIDGE PIERS

Piers for highway and railway bridges are frequently built of concrete. If a solid-rock foundation is available at the site of the pier, it is customary to excavate into or down to the rock and construct the pier on the rock. However, if the rock is not present or if it is so deep that excavating to it is too expensive, it will usually be necessary to drive piles to support the pier. The piles may be steel or concrete. They should extend into the bottom of the pier far enough to transmit the load safely from the pier to the piles.

The cost of the pier includes only the cost of forms, reinforcing steel, and concrete. If irregular surface shapes, such as projections or depressions, are required, adequate provisions for the costs of these irregularities must be included in the estimate. Likewise, provision should be made for the cost of special surface finishes that may be required.

Forms for piers   Forms for piers are made of wood or steel. In selecting the material for forms, consideration should be given to the number of times the forms will be used and to the kind of surface finish required for the pier. Additional costs for smooth-form linings may be less than the extra cost of labor necessary to produce the required surface finish after using rough forms of strip lumber.

Steel forms   Steel forms may be economical if they can be used enough times to justify the higher initial cost. The salvage value of steel forms which are fabricated for special piers, will usually be low. Forms are fabricated from steel plates and shapes, such as angles and I beams, into sections which will conform to the shape of the pier when they are assembled. The sections are lifted into position by cranes or other suitable hoisting equipment and fastened together to form a rigid structure. The cost of the form sections should be obtained from a fabricator. The cost of erecting will include the costs of labor and equipment at the job.

Wood forms   If wood forms are used, they will include sheathing, studs, wales, and possibly braces. The opposite sides of the forms are held in position by form ties which pass completely through the forms and bear against the wales.

The sheathing may be *S4S* or *D* and *M* lumber, plywood, or a combination of planking with thin plywood lining. All elements of the forms should be designed to resist safely the pressure of the concrete for the temperature and the rate of pour.

For piers not more than 12 ft 0 in. high, it is satisfactory to use 1-in.-thick sheathing, 2- by 4-in. studs, and double 2- by 4-in wales.. For piers higher than 12 ft 0 in., it may be necessary to use 2- by 6-in. or larger studs and double 2- by 6-in. or larger wales. As this height is approximate, the actual height for which an increase in the size of studs and wales is necessary should be determined for the particular job.

Example   Figure 7-20 shows the dimensions and details of a reinforced-concrete pier for a highway bridge. Four identical piers are required for the project. Because a horizontal construction joint is permitted midway between the top and bottom of the pier, it is possible to use the side forms two times on each pier, or eight times on the project. The end forms will be used one time on each pier, for a total of four times on the project.

Forms   Concrete will be placed in the forms at a rate of 3 ft per hr at a temperature of 70°F. Table 7-1 indicates a maximum pressure of about 536 lb per sq ft. Table 7-3 will be used to design the forms, based on a pressure of 550 lb per sq ft.
The lumber and spacings of members will be

Sheathing, 1 × 6 in. *S4S*, net width, 5½ in.
Studs, 2 × 4 in. *S4S*, spaced 21 in. on centers.
Wales, double 2- × 4-in. *S4S*, spaced 26 in. on centers
Form ties, spaced 33 in. on centers

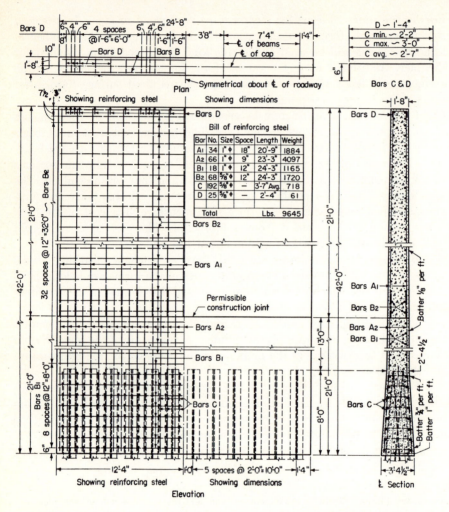

Bill of reinforcing steel

Bar	No.	Size	Space	Length	Weight
A₁	34	1" φ	18"	20'-9"	1884
A₂	66	1" φ	9"	23'-3"	4097
B₁	18	1" φ	12"	24'-3"	1165
B₂	68	⅝" φ	12"	24'-3"	1720
C	192	⅝" φ	—	3'-7"Avg.	718
D	25	⅝" φ	—	2'-4"	61
Total				Lbs.	9645

Fig. 7-20 Details of a concrete bridge pier. (*Texas Highway Department.*)

The area served by a form tie will be 26 × 33 in. ÷ 144 = 5.96 sq ft. The stress on a tie will be 5.96 × 536 = 3,200 lb. If 3,000-lb ties are used the maximum spacing should be reduced to $\dfrac{3,000}{3,300} \times 33 = 31$ in. Use a spacing of 30 in. on centers for the ties.

If 2- by 6-in. studs and double 2- by 6-in. wales are used with 1-in. sheathing, the quantity of lumber for studs and wales will be increased by about 17 percent. Unless the possible extra rigidity of the form panels provided by using the larger studs and wales is desired it is more economical to use 2- by 4-in studs and wales, which will be used for this estimate.

The insides of the forms will be lined with ¼-in.-thick water-proof plywood.

Figure 7-21 shows the details of the form design.

As the piers will have construction joints at the midelevations, the forms will be erected for the lower half of the piers and filled with concrete. After the concrete has set sufficiently,

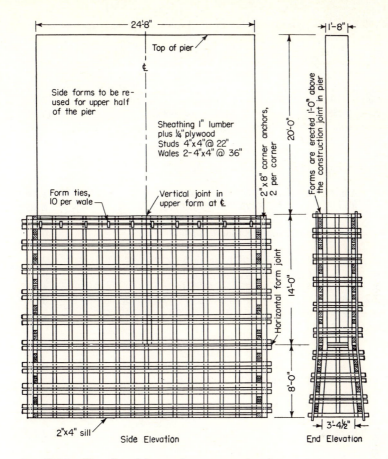

Fig. 7-21   Details of forms for a concrete pier.

the forms will be removed and raised to the higher elevation. Because of the batter in the piers, it will not be possible to reuse the lower end forms at the higher elevation.

The Cost of Forms   The cost of form lumber per use will be based on the number of times the lumber can be used. Assume that the lumber for the side forms can be used an average of eight times, and that the lumber for the ends can be used an average of four times.

The cost per use should be about as follows:

Side lumber,

$$\frac{\$220.00 \text{ per M fbm}}{8 \times 1,000} = \$0.0275 \text{ per fbm}$$

End lumber,

$$\frac{\$220.00 \text{ per M fbm}}{4 \times 1,000} = \$0.055 \text{ per fbm}$$

Plywood for sides, \$0.32 per sq ft ÷ 8 = \$0.04 per use

Plywood for ends, \$0.32 per sq ft ÷ 4 = \$0.08 per use

The forms will be held in position by four guy wires attached near the tops of each side.

The costs will be

Materials:

Side sheathing, 2 × 22 × 24.67 + 20% waste = 1,300 fbm @ $0.0275	= $	35.75
End sheathing, 110 sq ft + 20% waste = 132 fbm @ $0.055	=	7.26
Side plywood, 1,038 + 10% waste = 1,145 sq ft @ $0.04	=	45.80
End plywood, 108 + 10% waste = 119 sq ft @ $0.08	=	9.57

Studs:

36 pc, 2 in. × 4 in. × 8 ft 0 in. = 192 fbm @ $0.0275	=	5.28
38 pc, 2 in. × 4 in. × 14 ft 0 in. = 355 fbm @ $0.0275	=	9.76

Sills:

8 pc, 2 in. × 4 in. × 12 ft 0 in. = 64 fbm @ $0.0275	=	1.76
8 pc, 2 in. × 4 in. × 14 ft 0 in. = 75 fbm @ $0.0275	=	2.06
1 pc, 2 in. × 4 in. × 10 ft 0 in. = 7 fbm @ $0.055	=	0.39
1 pc, 2 in. × 4 in. × 12 ft 0 in. = 8 fbm @ $0.055	=	0.44

Wales:

88 pc, 2 in. × 4 in. × 14 ft 0 in. = 821 fbm @ $0.0275	=	22.60
16 pc, 2 in. × 4 in. × 6 ft 0 in. = 64 fbm @ $0.055	=	3.52
28 pc, 2 in. × 4 in. × 4 ft 0 in. = 75 fbm @ $0.055	=	4.12
Wale supports, 176 pc, 1 in. × 4 in. × 1 ft 0 in. = 59 fbm @ $0.055	=	3.24
Scabs for wales, 44 pc, 2 in. × 4 in. × 3 ft 0 in. = 88 fbm @ $0.0275	=	2.42

Corner anchors:

8 pc, 2 in. × 8 in. × 14 ft 0 in. = 149 fbm @ $0.0275	=	4.10
8 pc, 2 in. × 8 in. × 8 ft 0 in. = 86 fbm @ $0.0275	=	2.36
Form oil, 1,146 sq ft ÷ 600 sq ft per gal = 2 gal @ $1.20	=	2.40
Form ties, 11 wales × 10 ties per wale = 110 @ $0.21	=	23.10
Form tie clamps, 220 @ $0.025	=	5.50
Nails, 3,475 fbm × 13 lb per M fbm = 46 lb @ $0.25	=	11.50
Bolts, 50 lb @ $0.05 per use	=	2.50
Guy wire, 50 lb @ $0.05 per use	=	2.50
		———
Total cost of form materials	= $	207.93

Labor:

Building forms:

Carpenters, 1,146 sq ft × 4 hr per 100 sq ft = 46 hr @ $7.67	= $	352.82
Helper, 1,146 sq ft × 1.5 hr per 100 sq ft = 17 hr @ $5.79	=	98.43

Erecting and removing forms:

Carpenters, 1,146 sq ft × 4.5 hr per 100 sq ft = 52 hr @ $7.67	=	398.84
Helper, 1,146 sq ft × 1.5 hr per 100 sq ft = 17 hr @ $5.79	=	98.43
Crane operator, 8 hr @ $6.78	=	54.24
Crane oiler, 8 hr @ $4.68	=	37.44
Foreman, based on using 4 carpenters = 24 hr @ $8.50	=	204.00
		———
Total labor cost		$1,244.20

Equipment:

Power saws, 3,475 fbm ÷ 400 fbm per hr = 8.7 hr @ $0.46	= $	4.00
Crane, 8-ton truck-mounted diesel, 8 hr @ $14.73	=	117.84
		———
Total cost of equipment		$ 121.84

Total cost of forms for first use:

Materials	=	$ 207.93
Labor	=	1,244.20
Equipment	=	121.84

Total cost	=	$1,573.97
Cost per sq ft, $1,573.97 ÷ 1,146	=	1.38

The Cost of Reinforcing Steel

Steel reinforcing, 9,645 lb @ $0.18	=	$1,736.10
Tie wire, 24 lb @ $0.24	=	5.76

Labor placing reinforcing steel:

Steel setters, 4.8 tons × 12 hr per ton = 57.6 hr @ $7.25	=	417.60
Helpers, 4.8 tons × 4 hr per ton = 19.2 hr @ $4.75	=	91.20
Crane operator, 8 hr @ $6.78	=	54.24
Crane oiler, 8 hr @ $4.68	=	37.44
Crane, 8-ton truck-mounted diesel, 8 hr @ $14.73	=	117.84

Total cost of reinforcing steel	=	$2,460.18
Cost per lb, $2,460.18 ÷ 9,645 lb	=	0.255

The Cost of Concrete

The concrete will be handled with a crane and bucket.

The volume of concrete will be	=	84.3 cu yd
Add for overrun and waste	=	2.7 cu yd
Total volume	=	87.0 cu yd

Materials:

Cement, 87 cu yd × 4.5 sk per cu yd + 2% waste = 400 sk @ $1.62	=	648.00
Sand, 87 cu yd × 1,320 lb per cu yd + 10% waste = 63 tons @ $2.45	=	154.35
Gravel, 87 cu yd × 2,060 lb per cu yd + 5% waste = 94 tons @ $2.95	=	277.30
Water, 87 cu yd × 108 gal per cu yd = 9 M gal @ $2.40	=	21.60

Equipment:

Mixer, 87 cu yd ÷ 14 cu yd per hr = 6 hr @ $2.99	=	17.94
Crane and bucket, 6 hr @ $14.73	=	88.38
Internal vibrator, 6 hr @ $0.45	=	2.70
Sundry equipment, 6 hr @ $1.75	=	10.50

Labor mixing and placing concrete:
Assume 7 labor-hr for the placing.

Mixer operator, 7 hr @ $4.85	=	33.95
Crane operator, 7 hr @ $6.78	=	47.46
Crane oiler, 7 hr @ $4.68	=	32.76
Carpenter, 7 hr @ $7.67	=	53.69
Laborers handling cement, sand and gravel etc., 10 men × 7 hr = 70 hr @ $3.25	=	227.70
Foreman, 7 hr @ $8.50	–	50.50
Utility laborers, 4 men, 28 hr @ $3.25	=	91.00

Total cost of concrete	=	$1,766.83

The total direct cost of the pier will be

Forms	=	$1,573.97
Reinforcing steel	=	2,460.18
Concrete	=	1,766.83
Total cost	=	$5,800.98
Cost per cu yd, $5,800.98 ÷ 84.3 cu yd	=	69.00

PROBLEMS

7-1 Estimate the direct cost of materials, equipment, and labor for wood forms for a concrete foundation wall for a building whose outside dimensions are 46 ft 6 in. wide, 88 ft 0 in. long, 9 ft 8 in. high, and 9 in. thick. The concrete will be placed in the forms at a rate of 3 ft per hr at a temperature of 70°F. The concrete will be vibrated as it is placed.

Use 1- by 6-in. D and M lumber for sheathing and 2- by 4-in. lumber for studs and wales. Assume that the lumber can be used five times.

Use 3,000-lb form ties. Check or limit the allowable spacing of the ties for load safety.

Use the average national wage rates for carpenters and helpers, as given in this book.

The cost of materials will be

Lumber, per M fbm, $220.00

Nails, per lb, $0.25

Form ties, see book for costs

Form tie clamps, see book for costs

7-2 Estimate the total direct cost and the cost per square foot for the materials, equipment, and labor for wood forms for a retaining wall 126 ft 0 in. long, 14 ft 6 in. high, and 12 in. thick.

Use 1- by 6-in. D and M sheathing and 2- by 4-in. lumber for studs and wales. Assume that the lumber will be used four times.

The forms will be filled at the rate of 4 ft per hr at a temperature of 80°F.

Use the average national wage rates for carpenters and helpers, as given in this book.

The cost of materials will be

Lumber, $220.00 per M fbm

Nails, $0.25 per lb

See this book for the cost of form ties and clamps.

7-3 In designing the forms for use in building a concrete wall, it is desirable to select lumber sizes that will require the smallest quantity of lumber. Two plans are under study, namely plan A and plan B. Determine which plan requires the least number of fbm per square foot of wall surface.

Item	Plan A	Plan B
Sheathing	1 in. D and M	1 in. D and M
Studs	2- by 4-in. $S4S$	2- by 6-in. $S4S$
Wales	2- by 4-in. $S4S$	2- by 6-in. $S4S$

Concrete will be placed in the forms at the rate of 6 ft per hr at a temperature of 70°F.

7-4 Estimate the total direct cost and the cost per square foot of surface for materials, equipment, and labor for making, erecting, and removing wood forms for 32 concrete columns. The columns will be 18 in. by 18 in. by 11 ft 8 in. high. Adjustable steel clamps will be used with the forms.

Assume that the lumber will be used four times.

Determine the cost for the first use and for subsequent uses which do not require remaking the forms.

A 12-in.-blade table saw will be used for sawing and ripping the lumber.

Use the average national wage rates for carpenters and helpers.

The cost of materials will be

Lumber, $220.00 per M fbm

Nails, $0.25 per lb

Column clamps, $0.15 per use each

7-5 When designing concrete columns for a multistory building it is determined that loads from the floors require 16- by 16-in. columns to support the second and third floors, and 14- by 14-in. columns to support the fourth and fifth floors. All columns will be 9 ft 8 in. clear height. If these two column sizes are used, the area of the longitudinal reinforcing steel in the columns will be equal to 4 percent of the gross area of the columns.

If the columns that will support the fourth and fifth floors retain the 16- by 16-in. size, the area of the reinforcing steel for these columns can be reduced to $1\frac{1}{2}$ percent of the area of the columns.

If the larger columns are specified for the fourth and fifth floors, the forms from the two lower floors can be reused, without modification or remaking. Only the cost of erecting and removing will be necessary. The amount and cost of concrete will be increased, whereas the quantity and cost of reinforcing steel will be reduced.

If the choice of size is based solely on the cost of labor making and removing forms, the cost of reinforcing steel, and the cost of concrete, should the larger size be retained for the higher floors?

The following costs will apply:

Carpenter and helper, use average wage rates in book

Reinforcing steel in place, $0.32 per lb

Concrete in place, $36.00 per cu yd

7-6 Estimate the total direct cost and the cost per square foot of surface area for furnishing, erecting, and removing forms for interior concrete beams. The beams will be 14 in. wide and 16 in. deep and have a total length of 196 ft. Wood shores, size 4 by 4 in. S4S will be used to support the beam forms. The distance from the lower floor to the bottom of the beam will be 10 ft 4 in.

Assume that the lumber will be used four times.

Use the national average wage rates for carpenters and helpers.

The cost of materials will be

Lumber, $220.00 per M fbm

Nails, $0.25 per lb

7-7 If the forms for the beams of Prob. 7-6 are reused for other beams of the same sizes, estimate the total direct cost and the cost per square foot of surface area.

7-8   Estimate the costs for Prob. 7-6 using adjustable shores instead of wood shores.

7-9   Estimate the total direct cost and the cost per square foot for furnishing, erecting, and removing forms for eight bays of concrete slab only for beam-and-slab-type concrete floors. Supporting stringers and shores will not be required for the forms.

Each bay will be 7 ft 0 in. wide and 22 ft 4 in. long, clear dimensions. The floor slab will be 6 in. thick.

Assume that the lumber will be used four times.

Use the national average wage rates for carpenters and helpers.

Other costs will be

Lumber, $220.00 per M fbm

Nails, $0.25 per lb

Electric saw, with blade, $0.20 per hr

7-10   Estimate the total direct cost and the cost per square foot of floor area for furnishing, erecting, and removing wood forms for 1,642 sq ft of slab only for beam-and-slab-type concrete floors, using adjustable shores. The ceiling height will be 11 ft 8 in. The slab will be 6 in. thick. Supporting stringers will be required.

Assume that the lumber will be used four times.

Material and labor costs will be the same as for Prob. 7-9.

7-11   Estimate the total direct cost and the cost per square foot for the forms for Prob. 7-10 using wood shores instead of adjustable shores.

7-12   Estimate the total direct cost and the cost per pound for furnishing and placing the reinforcing steel for the foundation wall of Prob. 7-1. The vertical reinforcing will consist of No. 6 bars 9 ft 6 in. long, spaced 12 in. apart around the wall, located at the center of the wall. The horizontal reinforcing will consist of No. 6 bars spaced not over 12 in. apart, with the bottom bar placed 6 in. above the bottom of the wall. The horizontal bars are available in lengths up to 40 ft. All splices must lap not less than 30 diameters of the bars. At each corner one bar from each row of horizontal reinforcing must be bent and extend not less than 30 diameters of the bar beyond the bend for bond purposes. The reinforcing is to be securely wired in place by steelsetters.

Material costs and labor costs per hour will be

Reinforcing steel, $0.16 per lb

Tie wire, $0.25 per lb

Steel setters, $7.25 per hr

7-13   Estimate the total direct cost and the cost per pound of reinforcing steel for a concrete floor slab resting on the ground. The floor will be 66 ft 8 in. long, 52 ft 4 in. wide, and 6 in. thick. The reinforcing will be No. 4 bars spaced not more than 12 in. apart each way in the slab. The bars laid next to and parallel to the edges of the slab will be placed 3 in. from the edges of the slab. The ends of the bars will extend to within 2 in. from the edges of the slab.

The bars are available in lengths not to exceed 40 ft 0 in. All splices must lap not less than 30 diameters of the bars. The reinforcing will be tied at each intersection by steel setters.

Material and labor costs per hour will be

Reinforcing steel, $0.16 per lb

Tie wire, $0.25 per lb

Steel setters, $7.25

7-14   Estimate the total direct cost and the cost per 100 sq ft for furnishing and placing 66-44 welded-wire fabric for a concrete slab 88 ft 0 in. wide and 98 ft 6 in. long located on the surface of the ground.

Material and labor costs per hour will be

Welded-wire fabric, $9.40 per 100 sq ft

Tie wire, $0.25 per lb

Steel setters, $7.25 per hr

7-15   Estimate the total direct cost and the cost per cubic yard of concrete for furnishing and placing a beam-and-slab-type concrete floor to be constructed above the ground. The slab will be 6 in. thick and will have a total area of 3,460 sq ft. The exterior beams extending below the slab will have a cross section 14 in. wide and 24 in. from the top of the slab to the bottom of the beam. The total length of this beam will be 246 ft 0 in. The interior beams extending below the slab will have a cross section 14 in. wide and 16 in. deep below the bottom of the slab. The total length of these beams will be 216 ft 0 in.

Ready-mixed concrete will be delivered directly to the job and discharged into a tower bucket which will hoist it to a concrete floor hopper set at the elevation of the slab under construction. The rate of delivery will be 12 cu yd per hr.

Power-driven buggies will haul the concrete from the floor hopper to the area of placement.

The hoisting and placing equipment will include

1 heavy single-tube steel tower, 50 ft high

1 tower hoist bucket, 27-cu-ft capacity

1 single-gate 27-cu-ft concrete floor hopper

1 double-drum 20-hp gasoline-engine-operated hoisting unit

3 9-cu-ft power-driven concrete buggies

The top of the slab will be hand-screeded to the specified thickness, but will not be finished.

Material and labor costs will be

Concrete delivered to the job, $19.75 per cu yd

1 carpenter, $6.67 per hr

1 carpenter helper, $5.79 per hr

3 buggy operators, $5.75 per hr each

2 screed men, $5.79 per hr each

1 hoisting engineer, $7.48 per hr

6 laborers, $5.79 per hr each

1 foreman, $8.50 per hr

REFERENCES

1   "Formwork for Concrete, Publication SP-4," American Concrete Institute, Detroit, Mich., 1963.
2   R. L. PEURIFOY, "Formwork for Concrete Structures," McGraw-Hill Book Company, New York, 1964.

8
Floor Finishes

General Many types of finishes are applied to floors. Several of the more popular types will be discussed in this chapter.

CONCRETE-FLOOR FINISHES

In Chap. 7 the methods of estimating the cost of materials, equipment, and labor for concrete subfloors were discussed. The costs included all operations through the screeding of the rough concrete floors but did not include finishing the floors. These costs will be discussed in this chapter.

For most concrete floors it is necessary to add other materials to the top of the subfloors in order to produce finished floors of the desired types. The types of finishes most commonly used are monolithic topping and separate topping, to either of which may be added coloring pigments or hardening compounds.

Monolithic topping One of the best and most economical finishes for concrete floors is monolithic topping. After the concrete for the floor has been screeded to the desired level and has set sufficiently, it is floated with a wood float or a power-floating machine until all visible water disappears. Then the surface of the concrete is dusted with dry cement or a mixture of cement and sand, in approximately equal quantities. After the concrete

TABLE 8-1   Quantities of Materials for 100 Sq Ft
of Monolithic Topping

Thickness, in.	Proportion		Cement, sk	Sand, cu ft
	Cement	Sand		
$\frac{1}{64}$	1	0	0.125	0
$\frac{1}{32}$	1	0	0.25	0
$\frac{1}{16}$	1	1	0.25	0.25
$\frac{1}{8}$	1	1	0.50	0.50

or mixture is applied to the floor, it is lightly floated with a wood float, then finished with a steel trowel.

Materials required for a monolithic topping   The thickness of the monolithic topping may vary from $\frac{1}{64}$ to $\frac{1}{8}$ in. For finishes less than $\frac{1}{16}$ in. thick, cement only should be used, while for finishes thicker than $\frac{1}{16}$ in., fine sand may be added to the cement. Table 8-1 gives the quantities of materials required for finishing 100 sq ft of concrete floor using various proportions and thicknesses.

Labor finishing concrete floors using monolithic topping   In estimating the amount of labor required to finish a concrete floor, the estimator must consider several factors which may affect the total time required. After the concrete is placed and screeded, the finishers must wait until the concrete is ready for finishing. The length of wait may vary from as little as 1 hr under favorable conditions to as much as 6 hr under unfavorable conditions. The factors which affect the time are the amount of excess water in the concrete, the temperature of the concrete, and the temperature of the atmosphere. If the concrete is free of excess water and the weather is warm, finishing may be started soon after the concrete is placed. If the concrete contains excess water and the weather is cold and damp, finishing must be delayed several hours.

If concrete-placing operations are stopped at the normal end of a day, the concrete finishers may have to work well into the night in order to finish the job. This requires the payment of overtime wages at the prevailing rates.

Table 8-2 gives the labor-hours required to finish 100 sq ft of monolithic concrete floor under various conditions.

Separate concrete topping   A separate concrete topping may be applied immediately after the subfloor is placed, or it may be applied several days or weeks later. The latter method is not considered so satisfactory as the first method. The surface of the subfloor should be left rough to assure a

TABLE 8-2   Labor-hours Required to Finish 100 Sq Ft
of Monolithic Concrete Floor

| | Conditions | | |
Classification	Favorable	Average	Unfavorable
Cement finisher	0.75	1.25	1.75
Helper	0.75	1.25	1.75

bond with the topping. The topping may vary from ¾ to 1 in. thick. Concrete for the topping is usually mixed in the proportions 1 part cement, 1 part fine aggregate, and 1½ to 2 parts coarse aggregate, up to ⅜ in. maximum size, with about 5 gal of water per sack of cement.

After the concrete topping has set sufficiently, it is floated with a wood float and finished with a steel trowel. It may be desirable to dust the surface with cement during the finishing operation.

Material required for a separate topping   The quantities of materials required for a separate topping will vary with the thickness of the topping and the proportions of the mix.

Table 8-3 gives quantities of materials required for 100 sq ft of floor using various thicknesses and proportions.

Labor mixing, placing, and finishing separate topping   The concrete for a separate topping should be mixed with a small concrete mixer, such as a 3½S or a 6S, if one is available, as 0.31 cu yd will cover 100 sq ft of area 1 in. thick.

TABLE 8-3   Quantities of Materials for 100 Sq Ft of Separate
Topping

Thickness, in.	Proportions by volume			Cement, sk	Sand, cu yd	Gravel, cu yd
	Cement	Sand	Gravel			
½	1	1	1½	1.75	0.07	0.10
	1	1	2	1.62	0.06	0.12
	1	1½	3	1.18	0.07	0.13
¾	1	1	1½	2.62	0.10	0.15
	1	1	2	2.42	0.09	0.18
	1	1½	3	1.70	0.10	0.18
1	1	1	1½	3.50	0.13	0.20
	1	1	2	3.24	0.12	0.23
	1	1½	3	2.27	0.13	0.25

TABLE 8-4   Labor-hours Required to Mix, Place, and Finish 100 Sq Ft of Separate Topping

Thickness, in.	Finishers per crew	Labor-hours*			
		Finisher	Helper	Laborer	Foreman
½	1	1.25	1.25	1.25	0
	2	1.25	1.25	1.25	0.63
	3	1.25	1.25	0.80	0.42
¾	1	1.25	1.25	1.25	0
	2	1.25	1.25	1.25	0.63
	3	1.25	1.25	0.80	0.42
1	1	1.25	1.25	1.25	0
	2	1.25	1.25	1.25	0.63
	3	1.25	1.25	1.25	0.42
	4	1.25	1.25	1.00	0.31

* Add time for a hoisting engineer if one is required.

A finisher and a helper finishing 80 sq ft per hr will require 0.25 cu yd of topping per hr for a 1-in.-thick layer. A crew of three finishers and three helpers should finish 240 sq ft per hr under average conditions. This will require 0.75 cu yd of 1-in.-thick topping per hr. A 3½S mixer with three laborers can mix and place the topping. If the topping is placed above the ground floor, a hoisting engineer will be needed on a part-time basis.

Table 8-4 gives approximate labor-hours required to mix, place, and finish 100 sq ft of topping under average conditions.

Labor finishing concrete floors with a power machine Gasoline-engine- or electric-motor-driven power machines may be used to float and finish concrete floors. One manufacturer, the Whiteman Manufacturing Company, makes two sizes, 46-in. and 35-in. ring diameter. Each machine is supplied with three trowels, one set for floating and one for finishing the surface. The trowels are easily interchanged. The machines rotate at 75 to 100 rpm.

On an average job the 46-in.-diameter machine should cover 3,000 to 4,000 sq ft per hr, one time over, including time for delays. The 35-in.-diameter machine should cover 2,400 to 3,000 sq ft per hr. For most jobs it is necessary to go over the surface three to four times during the floating and the same during the finishing operation.

In operating a power machine a crew of two to three men will be required, one operator and one or two finishers, to hand-finish inaccessible areas.

Table 8-5 gives approximate labor-hours required to finish 100 sq ft of

Fig. 8-1 Power-operated trowel finishing concrete slab. (*Whiteman Manufacturing Company.*)

TABLE 8-5 Approximate Labor-hours Required to Finish 100 Sq Ft of Surface Area, Using Power Finishers

Operation	Machine operator	Finisher
46-in.-diameter machine, 750 sq ft per hr		
Floating	0.133	0.266
Finishing	0.133	0.266
Total time	0.266	0.532
35-in.-diameter machine, 600 sq ft per hr		
Floating	0.167	0.333
Finishing	0.167	0.333
Total time	0.334	0.666

surface area, based on going over the surface four times during each operation.

Example Estimate the cost of materials and labor required to hand-finish 5,000 sq ft of floor area using a $\frac{1}{32}$-in.-thick monolithic topping. Assume that the weather will be clear, with an average temperature of 70 deg, which is an average condition. The placing of the concrete will start at 8:00 A.M. and will be completed at 4:00 P.M. One cement finisher will report at 8:00 A.M. to direct the screeding of the slab, and the rest of the finishers will report at 9:00 A.M. and will remain on the job until the finishing is completed. Assume that the finishing will be completed around 7:00 P.M. It will be necessary to pay overtime wages to the finishers and helpers, at $1\frac{1}{2}$ times the regular wage rates, for all time in excess of 8 hr.

The number of finishers may be determined as follows:

$$\text{No. of man-hr required, } \frac{5,000}{100} \times 1.25 = 62.5$$

Time required to finish the job, 9:00 A.M. to 7:00 P.M. = 10 hr
No. of finishers required, 62.5 hr ÷ 10 hr = 6.25
Use 6 finishers, which may cause a delay in the time of completion until about 7:30 P.M.

The cost will be

Cement, 5,000 sq ft × 0.25 sk per 100 sq ft = 12.5 sk @ $1.70	= $ 21.25
Cement finishers:	
Regular time, 6 men × 8 hr = 48 hr @ $6.15	= 295.20
Overtime, 62.5 − 48 = 14.5 hr @ $9.23	= 133.84
Helpers:	
Regular time, 6 men × 8 hr = 48 hr @ $4.25	= 204.00
Overtime, 62.5 − 48 = 14.5 hr @ $6.38	= 92.51
Add the cost of a foreman after 4:00 P.M. only, at overtime wage rates =	
3.5 hr @ $12.75	= 44.62
Total cost	= $791.42
Cost per sq ft, $791.42, ÷ 5,000 sq ft	= 0.158

TERRAZZO FLOORS

General A terrazzo floor is obtained by applying a mixture of marble chips or granules, portland cement, and water laid on an existing concrete or wood floor. White cement is frequently used. The thickness of the terrazzo topping usually varies from $\frac{1}{2}$ to $\frac{3}{4}$ in. Several methods are used to install the topping on an existing floor.

Terrazzo topping bonded to a concrete floor If terrazzo is to be placed on and bonded to a concrete floor, the concrete surface should be cleaned and moistened. Then a layer of underbed not less than $1\frac{1}{4}$ in. thick, made by mixing 1 part portland cement, 4 parts sand, by volume, and enough water to produce a stiff mortar, is spread uniformly over the concrete surface.

While the underbed is still plastic, brass or other metal strips are installed in the mortar to produce squares having the specified dimensions.

TABLE 8-6 Labor-hours Required to Place Terrazzo Floors

Operation	Labor-hours
Cleaning concrete floor, per 100 sq ft	0.75–1.25
Placing roofing felt on wood floors, per 100 sq ft	0.3–0.5
Placing netting on wood floor, per 100 sq ft	0.5–0.7
Mixing and placing 1¼-in.-thick underbed, per 100 sq ft:	
Mechanic	1.0–1.5
Helper	2.5–3.5
Mixing and placing ¾-in.-thick terrazzo topping and metal strips, per 100 sq ft*:	
Mechanic	2.0–3.0
Helper	2.5–3.0
Finishing terrazzo topping, per 100 sq ft	7.0–8.0
Mixing and placing 100 lin ft of 6-in.-terrazzo cove base:	
Mechanic	14.0–18.0
Helper	14.0–18.0
Finishing 100 lin ft of 6-in.-terrazzo cove base	8.0–10.0

* These rates are for metal strips placed to form 5-ft 0-in. squares. For other size squares, apply the following factors:
4 ft 0 in., 1.10
3 ft 0 in., 1.20
2 ft 0 in., 1.30

The terrazzo topping is made by dry-mixing about 200 lb of granulated marble and 100 lb cement, or in other proportions as specified. Water is added to produce a reasonably plastic mix. This mix is placed on the underbed inside the metal strips, after which it is rolled to increase the density, then hand-troweled to bring the top surface flush with the tops of the metal strips.

After the surface has hardened sufficiently, it is rubbed with a machine-powered coarse carborundum stone. The surface is then covered with a thin layer of grout made with white cement. At the time of final cleaning, this coat of grout may be removed with a machine-powered fine-grain carborundum hone.

Terrazzo placed on a wood floor When terrazzo is placed on a wood floor, it is common practice to cover the floor with roofing felt and galvanized wire netting, such as No. 14 gauge, 2-in. mesh, which is nailed to the floor. A concrete underbed not less than 1¼ in. thick, as described in the previous article, is installed to a uniform depth over the floor.

Metal strips and terrazzo topping are installed and finished, as described in the previous article.

Labor required to place terrazzo floors The labor rates required to place terrazzo floors will vary considerably with the sizes of the areas placed. The rates for large areas will be less than for small areas. If terrazzo bases are required, additional labor should be allowed for placing and finishing the bases.

Table 8-6 gives representative labor-hours for the several operations required in placing terrazzo floors. Use the lower rates for large areas and the higher rates for small areas.

Example Estimate the cost of 100 sq ft of terrazzo floor placed on a concrete slab, with a 1¼-in. concrete underbed, ¾-in.-thick terrazzo topping, and brass strips to form 4-ft 0-in. squares. Use white cement for the topping.

The quantities will be

Underbed, 100 × 1.25/12 = 10.5 cu ft
Topping, 100 × 0.75/12 = 6.25 cu ft

The cost will be

Sand, including waste, 12 cu ft @ $0.15	= $ 1.80
Gray cement, 3 sk @ $1.70	= 5.10
Marble, 625 lb @ $0.05	= 31.25
White cement, including grout, 3.5 sk @ $4.20	= 14.70
Brass strips, including waste, 60 lin ft @ $0.25	= 15.00
Labor cleaning concrete floor, 1 hr @ $4.75	= 4.75
Mechanic placing underbed, 1.25 hr @ $7.25	= 9.06
Helper placing underbed, 3.0 hr @ $4.80	= 14.40
Mechanic placing topping and strips, 2.75 hr @ $7.25	= 19.94
Helper placing topping and strips, 3.0 hr @ $4.80	= 14.40
Helper finishing terrazzo topping, 7 hr @ $4.80	= 33.60
Finisher and sundry equipment, 7 hr @ $3.50	= 24.50
Total cost	= $188.50
Cost per sq ft, $188.50 ÷ 100	= 1.885

ASPHALT TILE

General Asphalt tiles are available in sizes 3 by 3 in., 6 by 6 in., 9 by 9 in., 12 by 12 in., 6 by 12 in., 6 by 24 in., 9 by 18 in., 12 by 24 in., and 18 by 24 in., and in thicknesses of ⅛ in. and 3⁄16 in. The cost of tile will vary with the thickness, color, and quality.

When estimating the quantity of tile required for a given floor area, an allowance should be included for waste due to cutting at the edges of the floor and to irregularities in the locations of the walls around the floor. Wastage may amount to 2 to 15 percent of the net area of a floor.

Laying asphalt tile on a concrete floor Before laying tile, the floor should be cleaned, after which a primer may be applied at a rate of 200 to

TABLE 8-7 Labor-hours Required to Lay 100 Sq
Ft of Asphalt Tile

Type of floor	Size tile, in.	Labor-hours
Concrete	12 × 12	1.5–2.0
	9 × 9	2.0–2.5
	6 × 6	2.8–3.3
Wood	12 × 12	1.8–2.2
	9 × 9	2.3–2.8
	6 × 6	2.0–3.5

300 sq ft per gal. An asphalt cement is then applied at a rate of approximately 200 sq ft per gal; then the tile are laid and rolled with a smooth-wheel roller.

Laying asphalt tile on a wood floor Prior to laying tile, the floor should be finished with a sanding machine or by some other approved method. A layer of felt is bonded to the floor with a linoleum paste, applied at a rate of approximately 150 sq ft per gal; and rolled thoroughly. Then an asphalt cement or emulsion is applied at a rate of approximately 150 sq ft per gal. The tile are then laid and rolled.

Labor laying asphalt tile The labor required to lay asphalt tile will vary considerably with the size and shape of the floor covered. If specified color patterns are required, the amount of labor will be greater than for single colors.

Table 8-7 gives representative labor-hours for the operations required in laying asphalt tile.

Example Estimate the cost of materials and labor required to lay 100 sq ft of asphalt tile $\frac{3}{16}$ in. thick, using 12- by 12-in. squares, on a concrete floor.

The cost will be

Asphalt primer, 0.5 gal @ $1.45	= $ 0.73
Asphalt cement, 0.5 gal @ $1.50	= 0.75
Tile, including waste, 105 sq ft @ $0.22	= 23.10
Tile setter, 1.8 hr. @ $7.25	= 13.05
Total cost	= $37.63
Cost per sq ft, $37.63 ÷ 100	= 0.376

PROBLEMS

8-1 A concrete slab for the second-story floor of a building will be 58 ft 6 in. wide, 136 ft 0 in. long, and 6 in. thick. The concrete will be delivered at a rate of 15 cu yd per hr, starting at 7:00 A.M., and continued until all concrete has been placed.

The slab will be finished with a monolithic topping $\frac{1}{32}$ in. thick, using hand finishers.

The finishers will report at 9:00 A.M. and will remain until the job is finished, estimated to be 3 hr after the last concrete is placed. Overtime wages must be paid at $1\frac{1}{2}$ times the regular wage rates for all time in excess of 8 hr. The cost of a foreman will be charged to finishing for only the time after the last concrete is placed in the slab.

Assume average working conditions.

Estimate the total cost and the cost per square foot for the materials and the labor for the topping.

Assume that the unit costs for materials and labor will be the same as those used in the example on page 203.

8-2 Estimate the cost of finishing the floor in Prob. 8-1 using one or more, as needed, 35-in.-diameter gasoline-engine-powered floor trowels. Use the costs from Prob. 8-1 and from Appendix B in preparing your estimate.

8-3 Estimate the total direct cost and the cost per square foot for furnishing the materials and labor required for a 4,480-sq-ft terrazzo floor placed on a concrete slab, with a $1\frac{1}{4}$-in.-thick underbed, $\frac{3}{4}$-in.-thick terrazzo topping, and brass strips to form 4-ft 0-in. squares. Use gray cement and marble chips for the topping.

The floor consists of large rooms and corridors at approximately ground level.

Use the unit costs for materials and labor from the example on page 205.

8-4 Estimate the total direct cost and the cost per square foot for furnishing and laying $\frac{3}{16}$-in.-thick asphalt tile, size 9- by 9-in. squares, on a concrete floor whose area is 2,360 sq ft.

Use the unit costs for materials and labor from the example on page 206.

9

Floor Systems

General This chapter describes several types of floor systems which may be used as substitutes for concrete-beam-and-slab, concrete slab only, or pan-and-joist concrete floors. An examination of the cost developed for each of the floor systems and a comparison with the cost of an all-concrete floor will reveal that reductions in the cost frequently may be effected by the choice of the floor system.

While these analyses are made for floor systems only, the results should demonstrate that similar analyses for other parts of structures may also permit the selection of methods and materials that will produce reductions in the costs.

STEEL-JOIST SYSTEM

General description Open-web steel joists are shop fabricated into the form of a Warren truss by an arc-welding process. The chord members consist of T sections, angles, or bars. The web is made of a single round bar, of uniform diameter, which is welded to the top and bottom chord members. A steel bearing plate is welded to each end of a joist to provide proper bearing area. The bearing plates are designed to permit the ends of the joist to rest on masonry walls or structural-steel beams.

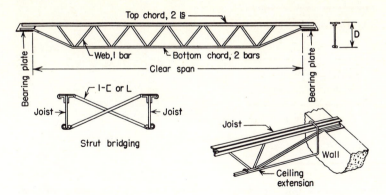

Fig. 9-1 Steel joist.

Floor and ceiling If fireproof construction is required, a concrete slab, usually 2 to 3 in. thick, may be installed on ribbed metal lath, which rests on the tops of the joists, and a plastered ceiling on ribbed metal lath supported by the bottom chords of the joists. If wood floors are to be installed, the joists may be purchased with wood nailer strips attached to the top chords.

Bridging Required bridging for steel joists includes struts of the proper length made from small channels or angles. The struts are fabricated to permit the ends to be clamped to the top chord of one joist and the bottom chord of an adjacent joist. Two struts are installed for each line of bridging to give a cross-bracing effect.

Cross bridging should be installed in accordance with the specifications of the designer or the manufacturer of the joists which will be approximately as follows:

Span, ft	Lines of bridging
Up to 14	1 row near center
14–21	2 rows at $\frac{1}{3}$ points
21–32	3 rows at $\frac{1}{4}$ points

Metal lath To support the concrete floor, a continuous layer of ribbed metal lath is installed over the joists, with the ribs perpendicular to the joists. The ribs are attached to the top chords of the joists with wire or lath clips, spaced not over 12 in. apart along the joists. Rib lath is designated by the size of the rib and the combined weight of the rib and the lath, expressed in pounds per square yard. Metal lath should conform to the information given in Table 9-1.

TABLE 9-1   Data for the Use of Metal Lath

Joist spacing, in.	Size of rib, in.	Weight, lb per sq yd
Up to 19	$\frac{3}{8}$	3.4
19–24	$\frac{3}{8}$	4.0
24–30	$\frac{3}{4}$	5.4

Floor lath should be lapped at least $1\frac{1}{2}$ in. beyond the center of supporting joists at the end of sheets and should be securely wired to the next sheet.

Ceiling extensions   Ceiling extensions are installed at the ends of joists at the same elevation as the bottom chord when a plastered ceiling is to be attached. These extensions support the metal lath along the bearing wall or supporting beam.

Joist wall anchors   A joist wall anchor is installed at the ends of every second or third joist bearing on a masonry wall in order more effectively to anchor the joists to the wall. It consists of a short length of small round bar, bent to a modified V shape, installed through a hole at the end of the joist. The anchor is embedded in mortar when the wall is extended above the joists.

Joist beam anchors   Every joist resting on structural steel, if not bolted or welded, should be anchored by means of a hairpin anchor bent around the flange of the supporting member. The anchor is installed through a hole at the end of the joist.

Bolted end connections   The ends of a joist may be connected to the flanges of the structural-steel supporting members by bolts which are installed through holes in the bearing plates of the joist and in the flanges of the supporting members. Two bolts should be installed at each end of a joist.

Welded end connections   The ends of a joist are frequently field-welded to the structural-steel supporting members. A 1-in.-long weld on each side of the bearing plate is satisfactory.

Size and dimensions of steel joist   Steel joists are available in a great many sizes and lengths for varying loads and spans. Standard open-web joists, which are designed in accordance with the Steel Joist Institute standard specifications, are available for spans varying from 3 to 42 ft or more, while special joists are available for spans in excess of 60 ft. Lengths generally vary in 6-in. steps. Joists may be installed with any desired spacing from 12 to 72 in. or more, provided that the maximum safe load is not exceeded.

Cost of steel joists The cost of steel joists varies so much with the number and sizes of the joists, the accessories required, and the location of the job that a table of costs is of little value to an estimator. Before preparing an estimate for a particular job, the estimator should submit the plans and specifications to a representative of a manufacturer for a quotation. Care should be exercised to be sure that the quotation includes all necessary accessories such as bridging, ceiling extension rods, lath clips, anchors, and end connections. The quotation should specify whether the prices are f.o.b. the shop or the job. If the prices are f.o.b. the shop, the estimator must add the cost of transporting the materials to the job.

Labor erecting steel joists The labor cost of erecting steel joists and accessories is usually estimated by the ton. In order to arrive at a reasonable unit price per ton, it is necessary to determine the probable rate at which the joists will be erected. The rate will vary with the size and length of the joists, the method of supporting them, the type of end connections, the type of bridging, the height that they must be lifted, and the complexity of the floor area.

In some locations, union regulations require that all labor erecting steel joists must be performed by union mechanics, while in other locations helpers are permitted to assist in the erection.

TABLE 9-2   Dimensions of Ceco* Joists

Type	Depth, in.	Length, ft	Top chord 2 angles	Bottom chord 2 ϕ bars	Web 1 ϕ bar	Weight, lb per lin ft
81	8	4–16	$1 \times 1 \times \frac{1}{8}$	$\frac{7}{16}$	$\frac{3}{8}$	3.9
82	8	8–16	$1 \times 1 \times \frac{1}{8}$	$\frac{1}{2}$	$\frac{3}{8}$	4.1
102	10	10–20	$1 \times 1 \times \frac{1}{8}$	$\frac{1}{2}$	$\frac{5}{16}$	4.2
103	10	10–20	$1\frac{1}{4} \times 1\frac{1}{4} \times \frac{1}{8}$	$\frac{9}{16}$	$\frac{7}{16}$	4.9
104	10	10–20	$1\frac{1}{2} \times 1\frac{1}{2} \times \frac{1}{8}$	$\frac{5}{8}$	$\frac{1}{2}$	5.9
123	12	12–24	$1\frac{1}{4} \times 1\frac{1}{4} \times \frac{1}{8}$	$\frac{9}{16}$	$\frac{1}{2}$	5.2
124	12	12–24	$1\frac{1}{2} \times 1\frac{1}{2} \times \frac{1}{8}$	$\frac{5}{8}$	$\frac{1}{2}$	6.2
125	12	12–24	$1\frac{3}{4} \times 1\frac{3}{4} \times \frac{1}{8}$	$\frac{21}{32}$	$\frac{9}{16}$	7.2
126	12	12–24	$1\frac{1}{2} \times 1\frac{1}{2} \times \frac{3}{16}$	$\frac{3}{4}$	$\frac{9}{16}$	8.1
145	14	14–28	$1\frac{3}{4} \times 1\frac{3}{4} \times \frac{1}{8}$	$\frac{21}{32}$	$\frac{9}{16}$	7.3
146	14	14–28	$1\frac{1}{2} \times 1\frac{1}{2} \times \frac{3}{16}$	$\frac{3}{4}$	$\frac{19}{32}$	8.5
147	14	14–28	$1\frac{3}{4} \times 1\frac{3}{4} \times \frac{3}{16}$	$\frac{13}{16}$	$\frac{19}{32}$	9.8
166	16	16–32	$1\frac{1}{2} \times 1\frac{1}{2} \times \frac{3}{16}$	$\frac{3}{4}$	$\frac{19}{32}$	8.9
167	16	16–32	$1\frac{3}{4} \times 1\frac{3}{4} \times \frac{3}{16}$	$\frac{13}{16}$	$\frac{5}{8}$	10.2

* Ceco Steel Products Corporation.

TABLE 9-3 Labor-hours Required to Erect a Ton of Steel Joists

Length of span	Labor-hr per ton	
	Ironworker	Helper*
Irregular construction, small areas		
6–10	6.5	3.25
10–16	6.0	3.0
16–24	5.5	2.75
24–30	5.0	2.5
Regular construction, large areas		
16–24	4.5	2.25
24–32	4.0	2.00

* For each floor above the first floor, add 1.5 helpers-hours if the joists are carried up by hand.

Table 9-3 gives the approximate labor-hours required to erect a ton of steel joists, including the installation of bridging and accessories. If helpers are not permitted to assist the ironworkers, the time shown for helpers should be added to that shown for the ironworkers.

Labor installing metal lath on top of steel joists Ribbed metal lath is frequently installed on the top of steel joists to support the concrete slab. The lath is fastened to the joists by passing hairpin wire clips over the ribs, under the top chords of the joists, and over the ribs on the opposite sides of the joists. The free ends of the clips are then twisted together. The clips should be spaced not over 12 in. apart along a joist. The sides and ends of the lath should be securely wired to the joist. The sides and ends of the lath should be securely wired to the adjacent sheets.

Two skilled lathers working together should install 20 to 30 sq yd of lath per hr, depending on the complexity of the floor area. This is equivalent to 6.5 to 10 labor-hours per 100 sq yd.

Labor placing welded-wire fabric Welded-wire fabric is frequently used to reinforce the concrete slab placed on steel joists. Table 7-27 gives the properties of representative samples of welded-wire fabric. Many sizes and weights are manufactured. An experienced worker should place approximately 400 sq ft of welded-wire fabric per hr on straight-run jobs. If cutting and fitting are necessary, the rate will be reduced. As the fabric is furnished in rolls containing approximately 750 sq ft, it may be necessary to use mechanical equipment to hoist it to floors above the ground level.

Concrete for slabs   The subject of mixing and placing concrete for the rough floor has been discussed in Chap. 7. As the information given in Chap. 7 was developed primarily for floors having a greater thickness than is generally used with floors supported by steel joists, the amounts of labor required to haul, spread, and screed the concrete for joist-supported floors should be increased to provide for the additional time needed. An increase of approximately 25 percent for the operations affected should be adequate.

Example   Estimate the total direct cost of steel joists, metal lath, and a $2\frac{1}{2}$-in.-thick light-weight concrete slab for a floor area 48 ft 0 in. wide by 72 ft 0 in. long. The floor will be divided into four bays, each 18 ft 0 in. wide by 48 ft 0 in. long. The floor is one story above the ground level.

The steel joist will be type 125, 18 ft 0 in. long, spaced 24 in. apart, with two rows of bridging per joist. The joist will be welded to the supporting steel members, using two 1-in.-long welds at each end of each joist.

The joists will be covered with $\frac{3}{8}$-in.-rib metal lath, weighing 3.4 lb per sq yd. The lath will be fastened to the joist with lath clips spaced not over 12 in. apart.

The concrete will be reinforced with style 66-1010 welded-wire fabric. The top surface of the concrete will be screeded to the required thickness but will not be finished under this estimate.

The area of the floor will be 48 ft 0 in. by 72 ft 0 in. = 3,456 sq ft.

The volume of concrete will be

$3,456 \times 2.5/36$	= 240 cu yd
Add for possible overrun	= 10 cu yd
Total volume	= 250 cu yd

The cost will be

Joists:

Joists, 100 each @ $32.55	=	$3,255.00
Bridging, 24 in. long, 384 struts @ $0.36	=	138.24
Ceiling extensions, 2 × 100 = 200 @ $0.61	=	122.00
Welding electrodes, 5 lb @ $0.38	=	1.90
Freight on joists to the job, 6.5 tons @ $5.25	=	34.13
Labor unloading, 6.5 tons × 1 hr per ton = 6.5 hr @ $5.79	=	37.63
Ironworkers erecting, 6.5 tons × 4.5 hr per ton = 29.2 hr @ $7.25	=	211.70
Helpers erecting, 6.5 tons × 2.25 hr per ton = 14.6 hr @ $5.79	=	84.53
Foreman, 8 hr @ $8.50	=	68.00
Welding equipment, 8 hr @ $2.25	=	18.00
Sundry equipment	=	16.00
Total cost of joists	=	$3,987.13

Metal lath:

Metal lath, including 2% waste, 392 sq yds @ $1.05	= $	411.60
Lath clips, 1,800 @ $0.035	=	63.00
Tie wire, 4 lb @ $0.24	=	0.96
Lather installing, 392 sq yd × 8 hr per 100 sq yd = 31.4 hr @ $7.24 =		227.34
Scaffolds	=	9.00
Total cost of metal lath	= $	711.90

Reinforcing steel:
 Welded-wire fabric, 3,456 sq ft plus 5% for waste = 3,629 sq ft @
  $0.041   = $ 148.79
 Tie wire, 4 lb @ $0.24   =   0.96
 Steel setter placing fabric, 3,629 sq ft × 0.3 hr per 100 sq ft = 10.9 hr
  @ $7.25   =   78.83

      Total cost of reinforcing   = $ 228.58
Concrete slab:
 Concrete, ready-mixed, 250 cu yd @ $21.00   = $5,250.00
 Hoisting engineer, 250 cu yd ÷ 10 cu yd per hr = 25 hr @ $6.63   =   165.80
 Buggy men, 5 men × 25 hr = 125 hr @ $5.79   =   723.80
 Emptying buggies, 2 men = 50 hr @ $5.79   =   289.50
 Spreading concrete, 4 men = 100 hr @ $5.79   =   579.00
 Moving runways, 2 men = 50 hr @ $5.79   =   289.50
 Foreman, 25 hr @ $8.50   =   212.50
 Hoisting equipment, 25 hr @ $4.20   =   105.00
 Runways   =   90.00
 Sundry equipment   =   60.00

      Total cost of concrete   = $7,765.10
Summary of costs:
 Joists   = $3,987.13
 Metal lath   =   711.90
 Reinforcing steel   =   228.58
 Concrete   =   7,765.10

      Total cost   = $12,692.71
      Cost per sq ft, $12,692.71 ÷ 3,456   =   3.67

COMBINED CORRUGATED-STEEL FORMS AND REINFORCEMENT FOR FLOOR SYSTEM

Description A complete floor system which is suitable for floors supporting light to heavy loads is constructed with a deep corrugated-steel combined form and reinforcing unit and concrete. The steel sheets which are used in constructing this system are sold under the trade name Cofar and are manufactured from extremely high-strength steel, varying in thickness from 20 to 24-gauge, in lengths up to 14 ft 4 in., with cover widths as shown in Table 9-4. The deep corrugated-steel sheets serve as longitudinal reinforcing for positive moment. T wires (transverse wires), spaced not over 6 in. apart, and welded across the corrugations in manufacture, provide temperature reinforcement in the slab as well as mechanical anchorage in transferring positive shear from the concrete to the steel. Conventional reinforced-concrete design procedures are followed for simple and continuous spans.

Table 9-5 gives the quantities of concrete in cubic yards per square

TABLE 9-4   Properties of Corrugated Sheets and Reinforcing*

Gauge No.	Approximate cover width, in.	Handling weight including T wires, psf	Reinforcing area, sq in. per ft of width	Per ft of width	
				Moment of inertia, in.^{4}	Section modulus, in.^{3}
24	29¼	1.90	0.35	0.075	0.12
22	31¼	2.20	0.45	0.094	0.15
20	30¼	2.60	0.54	0.106	0.17

* Courtesy of Granco Steel Products Company Catalogue No. BC 571.

foot of floor area for various thicknesses of floor slab. The thickness is measured from the top of the slab to the bottom of the corrugation.

Installing corrugated sheets   The sheets are equally suited to concrete or steel-frame construction. They should be fastened to the supports by welds, screws, bolts, or other means. Conventional negative reinforcing steel is installed over the supports. Conduit for electric wires is laid on the sheets with openings made through them. Provisions for large openings, such as for stairs, should be prefabricated by the manufacturer.

To avoid excessive form stresses and deflection while supporting the wet concrete between permanent supporting members, one or two lines of temporary supports should be installed until the concrete has gained

TABLE 9-5   Cubic Yards of Concrete per Square Foot of Floor Area for 1½-in.-deep Corrugated Forms

Depth, in.	Cu yd per sq ft
3.5	0.00887
4.0	0.01042
4.5	0.01196
5.0	0.01350
5.5	0.01505
6.0	0.01659

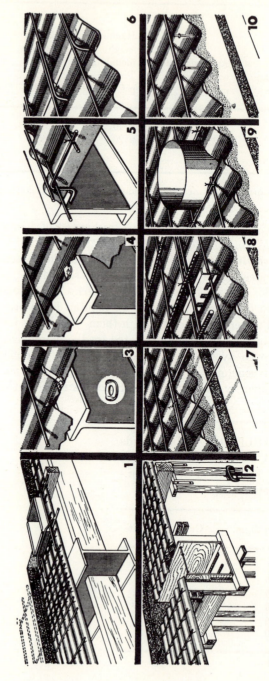

Fig. 9-2 Installation of corrugated-steel forms. (*Granco Steel Products Company.*)

strength. The spacing of the temporary supports should conform to the recommendations of the manufacturer.

Labor installing corrugated sheets The labor required to install corrugated sheets will vary somewhat with the sizes and weights of the sheets, the types of supports, and the complexity of the floor. Based on a floor with 18-ft 0-in. by 24-ft 0-in. bays, with supporting steel beams spaced 8 ft 0 in. apart, requiring 22-gauge 1.25-in.-deep corrugations, the rates of placing should be about as follows:

Operation	Labor-hour per 100 sq ft
Hoisting, placing, and welding	1.1
Installing accessories	0.8
Installing negative reinforcing bar	0.8
Installing and removing temporary supports	1.1
Total labor-hour per 100 sq ft	3.8

Example Estimate the total direct cost and the cost per square foot for materials and labor for a bay 18 ft 0 in. wide by 24 ft 0 in. long using corrugated forms and reinforcement and a concrete slab.

The sheets will be 24 gauge with 1.25-in.-deep corrugations, with No. 3 gauge wire transverse reinforcing, spaced 6 in. on centers. The permanent supporting members will be steel beams, spaced 8 ft 0 in. on centers, parallel to the 18-ft 0-in. side of the bay.

The concrete will have a maximum depth of 3.5 in. Negative reinforcing, installed over each permanent beam, will consist of No. 4 bars, 4 ft 0 in. long, spaced 6 in. on centers.

Temporary intermediate supports for the corrugated sheets will consist of one 2- by 8-in. wood stringer and adjustable shores, spaced 4 ft 6 in. apart. The stringer will be installed parallel with the 18 ft 0 in. sides of the bay.

The area of the bay will be 18 ft × 24 ft = 432 sq ft.

The cost will be

Materials:
Corrugated sheets, f.o.b. the factory, 432 sq ft @ $0.64 = $276.48
Freight on sheets, 432 sq ft @ $0.05 = 21.60
Negative reinforcing steel, 108 pc × 4 ft 0 in. long × 0.668 = 288 lb @
$0.18 = 51.84
High chairs, tie wire, etc., 432 sq ft @ $0.03 = 12.96
Concrete, ready-mixed, 432 sq ft × 0.00887 cu yd per sq ft = 3.9 cu yd
@ $21.00 = $ 81.90
Equipment:
Stringers, 5 uses, 112 fbm @ $0.048 = 5.38
Shores, 15 each @ $0.45 = 6.75
Hoisting equipment, 3 hr @ $4.20 = 12.60
Welding equipment, 4 hr @ $2.25 = 9.00
Sundry equipment = 4.50

Labor:

Installing sheets, 432 sq ft × 3.8 hr per 100 sq ft = 16.5 hr @ $7.25	=	119.63
Foreman, based on 4 ironworkers, 4.1 hr @ $8.50	=	34.85
Time to place concrete, 3.9 cu yd ÷ 10 cu yd per hr = 0.4 hr		
Laborers placing concrete, 12 men × 0.4 hr = 4.8 hr @ $5.79	=	27.79
Carpenter setting stringers, etc., 4.5 hr @ $7.67	=	34.51
Foreman, 0.4 hr @ $8.50	=	3.40
Hoisting engineer, for all materials 4.1 hr @ $6.63	=	27.18
Total cost	=	$730.37
Cost per sq ft, $730.37 ÷ 432	=	1.69

10
Masonry

Masonry units Masonry units which are commonly used for construction include brick, tile, concrete blocks, and stone, natural or artificial. They are bonded by a suitable mortar, with metal ties frequently added to increase the strength of the bond. Most of the units are available in several sizes, grades, and textures. Plans and specifications for a structure designate the kind of unit; the size, grade, texture, kind of mortar, and thickness of joint; and the quality of workmanship required.

Estimating the cost of masonry In estimating the cost of a structure to be constructed entirely or partly of masonry units, the estimator should determine separately the quantity and cost of each kind of unit required. An appropriate allowance should be made for waste, resulting primarily from breakage. Determine the quantities and cost of materials for the mortar, including an allowance for waste. Estimate the cost of labor laying the masonry units. If construction equipment is used in mixing the mortar, hoisting the masonry units, or for other purposes, the cost of such equipment should be added to the other costs.

Mortar Most mortar for masonry units is made by mixing portland cement, lime, and sand or by mixing a commercial masonry cement with sand. The quantity of each ingredient may vary to produce a mortar suitable for the particular job. Coloring is sometimes added. If a fine sand is used, the workability of the mortar will be much better than if coarse sand is used.

TABLE 10-1   Quantity of Lime
per Cubic Foot of Putty

Kind of lime	Lb per cu ft of putty
Hydrated lime	45
Pulverized quicklime	25
Lump quicklime	23

Lime for mortar   The lime used in making masonry mortar is available as lump quicklime, pulverized quicklime, or hydrated lime. Quicklime or hydrated lime may be purchased in paper bags, containing 50 lb, or it may be purchased in bulk. Prices are frequently quoted by the ton.

Quicklime is approximately pure calcium oxide, while hydrated lime is calcium hydroxide. Before quicklime can be used in making mortar, it must by hydrated, or slaked, by mixing it with water and allowing it to season for several days. The putty will remain usable for several weeks after it is slaked.

Table 10-2 gives the approximate quantities of materials required for 1 cu yd of mortar for various mixing proportions.

BRICK MASONRY

Bricks may be classified by the material from which they are made, the method of molding them, the purpose for which they will be used, the size, etc. The costs, which vary considerably, are usually based on 1,000 units, either at the factory or delivered to the destination.

Sizes of bricks   Bricks are manufactured in a great many sizes. The Common Brick Manufacturers Association adopted a standard size, with nominal

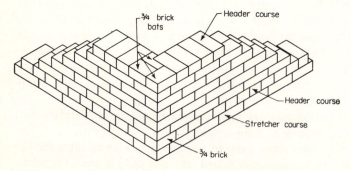

Fig. 10-1   Brick wall laid in common bond.

TABLE 10-2 Quantities of Materials Required for 1 Cu Yd of Mortar

Masonry cement and sand

Proportions by volume		Quantity	
Cement	Sand	Cement, sk	Sand, cu yd
1	2	13.5	1
1	2.5	11.1	1
1	3	9.0	1
1	3.5	7.7	1
1	4	6.8	1

Pulverized quicklime, portland cement, and sand

Proportions by volume			Quantity		
Lime putty	Cement	Sand	Lime, lb	Cement, sk	Sand, cu yd
1	0	3	225	0	1
2	1	9	150	3	1
1	1	6	112.5	4.5	1
0.5	1	4.5	75	6	1
0.2	1	3	45	9	1
0.1	1	3	22.5	9	1

Hydrated lime, portland cement, and sand

Proportions by volume			Quantity		
Dry lime*	Cement	Sand	Lime, lb	Cement, sk	Sand, cu yd
1	0	3	450	0	1
2	1	9	300	3	1
1	1	6	225	4.5	1
0.5	1	4.5	150	6	1
0.2	1	3	90	9	1
0.1	1	3	45	9	1

* Hydrated lime is usually sold in paper bags weighing 50 lb and containing 1 cu ft.

dimensions of $2\frac{1}{4}$ by $3\frac{3}{4}$ by 8 in., corresponding to the thickness, width, and length, respectively. However, unequal degrees of burning will frequently cause variations in these dimensions for the finished products. All manufacturers have not installed molds to produce standard sizes. While face bricks are made in sizes corresponding to the standard-size common bricks, other sizes are also available. It is good practice to specify the size of bricks desired, in estimating or purchasing them, instead of referring to them as "standard bricks."

In this book, the $2\frac{1}{4}$- by $3\frac{3}{4}$- by 8-in. brick is used for all examples and estimates.

Thickness of brick walls   Brick walls are usually designated as 4, 8, 12, 16, and 20 in. thick, although these dimensions are not entirely correct in all instances. The actual thickness will vary with the number of bricks of thickness and the thickness of the vertical mortar joints between rows of bricks. A three-brick-thick wall is referred to as a 12-in. wall, but if $\frac{1}{2}$-in.-thick mortar joints are used, the actual thickness will be $12\frac{1}{4}$ in. In a similar manner, a four-brick-thick wall will actually be $16\frac{1}{2}$ in. thick.

Estimating the quantity of bricks   In estimating the cost of bricks for a structure, the estimator should determine the quantity of bricks for each type required. At least two methods of estimating the quantities of bricks are used. One method is to determine the total volume of the walls, usually in cubic feet, then to multiply this volume by the probable number

TABLE 10-3   Number of Standard Bricks per 100 Sq Ft of Wall Area

Nominal thickness of wall, in.	Thickness of horizontal mortar joint, in.				
	$\frac{1}{4}$	$\frac{3}{8}$	$\frac{1}{2}$	$\frac{5}{8}$	$\frac{3}{4}$
End joints $\frac{1}{4}$ in. thick					
4	698	665	635	608	582
8	1,396	1,330	1,270	1,216	1,164
12	2,095	1,995	1,905	1,824	1,746
16	2,792	2,660	2,540	2,432	2,328
End joints $\frac{1}{2}$ in. thick					
4	677	645	615	588	564
8	1,354	1,290	1,230	1,176	1,128
12	2,031	1,935	1,845	1,764	1,692
16	2,708	2,580	2,460	2,352	2,256
20	3,385	3,225	3,075	2,940	2,820

TABLE 10-4 Cubic Yards of Mortar Required per 1,000
Standard-size Bricks for Full Joints, Using Common Bond
(No Allowance Included for Waste)

Nominal thickness of wall, in.	Thickness of horizontal mortar joint, in.				
	¼	⅜	½	⅝	¾
Vertical joints ¼ in. thick					
4	0.211	0.291	0.376	0.458	0.542
8	0.301	0.384	0.474	0.562	0.649
12	0.346	0.432	0.523	0.614	0.703
16	0.429	0.519	0.616	0.709	0.805
Vertical joints ½ in. thick					
4	0.262	0.348	0.433	0.518	0.627
8	0.365	0.456	0.546	0.637	0.751
12	0.414	0.506	0.600	0.693	0.809
16	0.433	0.520	0.621	0.715	0.832

of bricks per cubic foot. Another method is to determine the area of the wall, with a separate area for each different thickness, then multiply the respective area by the number of bricks per square foot of area, considering the thickness of the wall. The latter method should give more accurate results than the former. For example, wall thicknesses may be 12 or 13 in. without any variation in the number of bricks required provided that there is no change in the thickness of the bed and end mortar joints. The area method of estimating quantities will be correct, whereas the volume will give different quantities.

Regardless of the method used, the estimator should adjust the volume or area for openings in the wall. Failure to make this adjustment introduces an error which is too large for accurate estimating.

The net exposed area of a standard-size brick laid flat is 8 by 2¼ in., which equals 18 sq in. The effective area is increased by the vertical and horizontal mortar joints. If each joint is ½ in. thick, the effective area will be 8½ by 2¾ in., which equals 23⅜ sq in. The number of bricks required per square foot of wall area for one brick thickness will be 6.17. For an 8-in.-thick wall, the number will be 12.34.

Table 10-3 gives the number of bricks per 100 sq ft of wall area for various wall and mortar-joint thicknesses, using 2¼- by 3¾- by 8-in. bricks. No allowance is made for waste.

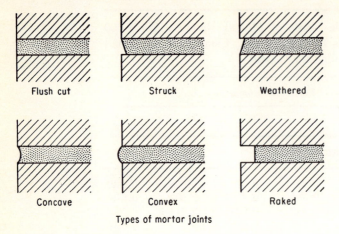

Types of mortar joints

Fig. 10-2   Mortar joints for masonry.

Quantity of mortar   The quantity of mortar required for brick masonry will vary with the thickness of the mortar joints and the extent to which all joints are filled with mortar. The inside vertical joints are not always filled with mortar, especially on secondary backup walls. It is difficult accurately to estimate the quantity of mortar that will be wasted in laying bricks; however, this will usually amount to 10 to 25 percent of the quantity required in laying the bricks.

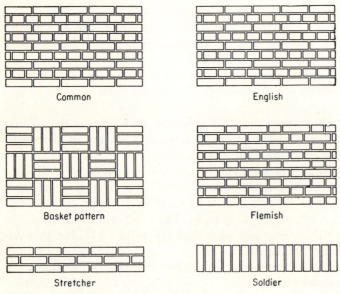

Fig. 10-3   Bonds for brick masonry.

Table 10-4 gives the quantities of mortar required for 1,000 standard-size bricks for various joint thicknesses.

Types of joints for brick masonry Several types of mortar joints are specified for brick masonry which will affect the rate of laying bricks. The more common types of joints are shown in Fig. 10-2.

Flush-cut joints are made by passing the trowel across the surface of the bricks and removing any excess mortar. This operation requires very little time.

Struck and weathered joints are made with a trowel after the mortar has gained some stiffness. This operation requires more time than cut joints.

Concave joints are made by tooling the mortar with a wood or non-staining metal rod before the mortar gains final set.

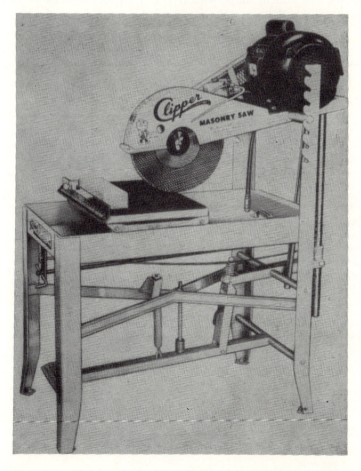

Fig. 10-4 Masonry saw. (*Whiteman Manufacturing Company.*)

Raked joints are made by removing the mortar to a depth of ¼ to ⅜ in., using a special tool.

Bonds Figure 10-3 illustrates the more commonly used bonds for brick walls. The amount of labor required to lay bricks will vary with the type of bond used.

Labor laying bricks The labor required to lay bricks varies with a number of factors such as the quality of work, the type of bricks, the kind of mortar used, the shape of the walls, the kind of bond used, the thickness of the wall, weather conditions, and the inclination of the bricklayers.

If the walls must be plumbed accurately with straight courses and uniformly thick joints, the labor required will be greater than for a less rigid quality. If the joints must be tooled, more labor will be required than if the joints are simply cut flush with a trowel.

If the walls are irregular in shape with frequent openings, pilasters, or other changes in shape, the labor requirements will be greater than for long straight walls.

TABLE 10-5 Labor-hours Required to Lay 1,000 Bricks Using Lime-Cement Mortar and Common Bond*

Type of structure and quality of work	Type of joint	Hr per 1,000 bricks	
		Mason	Helper
Common brick			
Commercial buildings, warehouses, stores, shops,	Cut	7.5	7.5
walls 12 in. or more thick	Struck	8.0	8.0
Public buildings, requiring first-grade workmanship,	Cut	9.0	8.0
walls 12 in. or more thick, 2 or 3 stories high	Struck	9.5	8.5
Backing-up face brick or stone partition, walls	Cut	7.5	7.5
8 in. or more thick	Struck	8.0	8.0
Face brick			
Commercial buildings, stores, office buildings with	Cut	15.0	10
4-in. face brick, ordinary workmanship	Struck	15.5	10
	Concave	16.0	10
Public buildings, schools, libraries, churches, etc.,	Cut	18.0	11
with 4-in. face brick, first-class workmanship	Struck	18.5	11
	Concave	19.0	11

* For other than common bond, the labor-hours for brickmasons should be increased approximately 10 to 15 per cent.

If the walls contain various patterns requiring changes in the bond, the labor requirements will be greater than for a uniform bond.

Less labor is required to lay bricks in a thick wall than in a thin one, as less time is spent per brick for plumbing, leveling, and moving scaffolds.

The labor required to lay bricks in cold or wet weather will be greater than when the weather is mild and dry.

Table 10-5 gives the approximate labor-hours required to lay 1,000 bricks for various types of structures under different conditions. The indicated time includes mixing the mortar, hodding bricks and mortar, and laying the bricks.

Example Estimate the total direct cost and the cost per 1,000 bricks for constructing the walls of a rectangular building 116 ft 0 in. long, 68 ft 0 in. wide, with brick walls 16 ft 0 in. high. The walls will be 12 in. thick, with the bricks laid in common bond. The total area of openings for doors and windows will be 484 sq ft. Common brick will be used for the wall. The mortar will be 1:1:6 hydrated lime, portland cement, and sand, respectively. Ordinary workmanship is specified, with $\frac{1}{2}$-in.-thick struck joints.

The quantities are determined as follows:

The gross outside area of the wall will be $2(116 + 68) \times 16$	= 5,888 sq ft
Deduct for four corners, 4×16	= −64 sq ft
Deduct area of openings	= −484 sq ft
Net area	= 5,340 sq ft

The cost will be

Bricks, 5,340 sq ft $\times$ 1,845 per 100 sq ft =	98,523	
Add for waste, 2%	= 1,970	
Total quantity	= 100,493	
Bricks, 100,493 @ $55.00 per M		= $5,527.11
Quantity of mortar, 98,523 bricks $\times$ 0.6 cu yd per M brick	= 59.1 cu yd	
Add 10% for waste	= 5.9 cu yd	
Total quantity	= 65.0 cu yd	
Lime, 65 cu yd $\times$ 225 lb per cu yd = 7.3 tons @ $48.00		= 350.40
Cement, 65 cu yd $\times$ 4.5 sk per cu yd = 293 sk @ $1.60		= 468.80
Sand, including waste, 69 cu yd @ $4.20		= 289.80
Mortar mixer, 32 hr @ $1.75		= 56.00
Scaffolds and sundry equipment		= 85.00
Bricklayers, 98,523 bricks $\times$ 8 hr per M = 788 hr @ $8.23		= 6,485.24
Helpers, 98,523 bricks $\times$ 8 hr per M = 788 hr @ $5.60		= 4,412.80
Foreman, based on 6 bricklayers, 131 hr @ $8.50		= 1,113.50
Mixer operator, 131 hr @ $5.80		= 759.80
Total cost		= $19,548.45
Cost per M bricks, $19,548.45 ÷ 98,523		= 198.40

TABLE 10-6   Sizes, Weights, and Quantity of Mortar for Concrete Blocks, Using ⅜-in. Joints*

Actual size: thickness, height, length, in.	Approx weight per block, lb		Quantities per 100 sq ft of wall area	
	Standard	Light-weight	No. blocks	Mortar, cu yd
3⅝ × 4⅞ × 11⅝	11–13	8–10	240	0.15
5⅝ × 4⅞ × 11⅝	17–19	12–14	240	0.16
7⅝ × 4⅞ × 11⅝	22–24	14–16	240	0.17
3⅝ × 7⅝ × 11⅝	17–19	12–14	150	0.11
5⅝ × 7⅝ × 11⅝	26–28	17–19	150	0.12
7⅝ × 7⅝ × 11⅝	33–35	21–23	150	0.13
5⅝ × 3⅝ × 15⅝	17–19	11–13	225	0.18
7⅝ × 3⅝ × 15⅝	22–24	14–16	225	0.19
3⅝ × 7⅝ × 15⅝	23–25	16–18	113	0.10
5⅝ × 7⅝ × 15⅝	35–37	24–27	113	0.11
7⅝ × 7⅝ × 15⅝	45–47	28–32	113	0.12

* These quantities do not include any allowance for waste. Add 2 to 5 per cent for blocks and 20 to 50 per cent for mortar.

Face brick with common-brick backing Buildings are frequently erected with a 4-in.-thick veneer of face bricks, while the balance of the wall thickness is obtained with common bricks. If a common bond is used, it is customary to lay face bricks as headers every sixth course for bond purposes. As a header course will require twice as many face bricks as a stretcher course, the total number of face bricks required must be increased over the number that would be required for a uniform 4-in. thickness. This is equivalent to one extra stretcher course in six courses, amounting to an increase of 16⅔ percent in the number of face bricks required. The number of common bricks may be reduced in an amount equal to the increase in the number of face bricks.

CONCRETE BLOCKS

Concrete blocks are made in various sizes, using portland cement, sand and gravel, or cement and lightweight aggregate. Table 10-6 lists the popular sizes. As the blocks are frequently made in local or nearby plants, with limited sizes available, it may not be possible to obtain all sizes and types in a given community. While blocks of any size and type may be obtained by purchasing them from a distant plant, the increased cost of freight may make their cost unreasonably high. The designer should check the source of supply before specifying given sizes for a project.

Quantities of materials required for concrete blocks Table 10-6 gives the nominal and actual sizes, approximate weights, and quantities of mortar required for joints for the more popular sizes of blocks.

The actual sizes of blocks usually will be ⅜ in. less than the nominal sizes. When ⅜-in.-thick mortar joints are used, the dimensions occupied by blocks will equal the nominal sizes.

TABLE 10-7   Labor-hours Required to Handle and Lay 1,000 Concrete Blocks*

Nominal size block: thickness, height, length, in.	Labor-hr per 1,000 blocks	
	Mason	Helper
Standard blocks		
4 × 5 × 12	33–38	33–38
6 × 5 × 12	38–44	38–44
8 × 5 × 12	44–50	44–50
4 × 8 × 12	38–44	38–44
6 × 8 × 12	44–50	44–50
8 × 8 × 12	50–55	50–55
6 × 4 × 16	38–44	38–44
8 × 4 × 16	44–50	44–50
4 × 8 × 16	38–44	38–44
6 × 8 × 16	44–50	44–50
8 × 8 × 16	52–57	52–57
Lightweight blocks		
4 × 5 × 12	30–35	30–35
6 × 5 × 12	35–40	35–40
8 × 5 × 12	40–45	40–45
4 × 8 × 12	35–40	35–40
6 × 8 × 12	40–45	40–45
8 × 8 × 12	45–50	45–50
6 × 4 × 16	35–40	35–40
8 × 4 × 16	40–45	40–45
4 × 8 × 16	35–40	35–40
6 × 8 × 16	40–45	40–45
8 × 8 × 16	47–52	47–52

* Add the cost of a hoisting engineer if required.

The quantities given in Table 10-6 are based on 100 sq ft of wall area, with no allowance for waste or breakage of the blocks. The quantities of mortar given do not include any allowance for waste, which frequently will amount to 20 to 50 percent of the net amount required.

Labor laying concrete blocks   Concrete blocks are laid by masons. Joints are made by spreading mortar along the inside and outside horizontal and vertical edges. Joints may be cut smooth with a steel trowel, or they may be tooled as for brick. The joints are more resistant to the infiltration of moisture when they are tooled, as the tooling increases the density of the mortar.

Table 10-7 gives the labor-hours required to lay 1,000 concrete blocks of various sizes. The time given for masons includes laying the blocks and tooling the joints, if required. The rates provide for different classes of work. The time given for laborers includes supplying mortar and blocks for the masons. Add the cost of a hoisting engineer if required.

Example   Estimate the total direct cost and the cost per 1,000 for furnishing and laying standard-weight concrete blocks, whose nominal sizes are 8 by 8 by 12 in., to back up the brick facing for a wall for a rectangular building 120 ft 0 in. long, 80 ft 0 in. wide, and 18 ft 0 in. high. The wall will be 12 in. thick, consisting of a 4-in.-thick face of brick and 8-in.-thick concrete blocks. The total area of the openings will be 516 sq ft. Mortar joints will be ⅜ in. thick, using 1:1:6 hydrated lime, portland cement, and sand, respectively.

The quantities are determined as follows:

Gross perimiter of building, 400 lin ft
Net perimeter of concrete block portion of building, = 400 ft 0 in. −
    8 × 4 in. = 397 ft 4 in.
Gross area of concrete block portion of building = 397 ft 4 in. × 18
    ft 0 in.   = 7,152 sq ft
Deduct area of openings   = −516 sq ft
Deduct area of 4 corners, 4 × 8 in. × 18 ft   = −48 sq ft
                                                                             ─────────
    Net area of concrete portion   = 6,588 sq ft

The cost will be

Materials:
    Concrete blocks, 6,588 sq ft × 1.5 per sq ft = 9,882 @ $0.32   = $ 3,162.24
    Add 2% for waste, 198 @ $0.32   =   63.36
    Mortar required, 65.88 sq × 0.13 cu yd per sq   = 8.6 cu yd
    Add 40% for waste, 0.4 × 8.6   = 3.5 cu yd
                                                                   ──────────
        Total quantity   = 12.1 cu yd
    Lime, 12.1 cu yd × 225 lb per cu yd = 1.4 tons @ $48.00   =   67.20
    Cement, 12.1 cu yd × 4.5 sk per cu yd = 55 sk @ $1.60   =   88.00
    Sand, including waste, 13 cu yd $4.20   =   54.60
Equipment:
    Mortar mixer, 12.1 cu yd ÷ 1½ cu yd per hr = 8 hr @ $1.75   =   14.00
    Scaffolds and sundry equipment   =   48.00

Labor:
Masons, 9,882 blocks × 52 hr per M = 515 hr @ $8.23 = 4,238.45
Helpers, 9,882 blocks × 52 hr per M = 515 hr @ $5.60 = 2,884.00
Foreman, based on 6 masons, 515 ÷ 6 = 86 hr @ $8.50 = 731.00
Mixer operator, 86 hr @ $5.80 = 498.80

Total cost = $11,849.65
Cost per M blocks, $11,849.65 ÷ 9,882 = 1,200.00

LOAD-BEARING CLAY TILE

Load-bearing clay tile are used for many purposes, such as for exterior and interior walls and for backing up face brick. They are constructed with webs sufficiently strong to support considerable loads. They are available in various sizes and types including half blocks, corner blocks, and jamb blocks for window and door openings. Surfaces are ordinarily scored to assist plaster and stucco in adhering; however, they are also available with surfaces sufficiently attractive on one side to permit them to be used as a finished wall without further treatment.

Quantities of materials required for clay-tile walls Table 10-8 gives the sizes, weights, and quantities of tile and mortar required for the more popular sizes. The quantities of mortar are based on tile of full size and a ⅜-in.-thick full mortar joint for bedding and buttered ends.

Labor laying clay tile The labor required to lay load-bearing clay tile will vary with the size of the tile, the number of openings and offsets in the wall, and the quality of workmanship. Openings and changes in wall shapes should be spaced to permit the use of full tile lengths without cutting.

Table 10-9 gives the approximate labor-hours required to lay 1,000 load-bearing tile, using ordinary workmanship, for walls with typical open-

TABLE 10-8 Sizes, Weights, and Quantities of Mortar for Load-bearing Clay Tile

Size: thickness, height, length, in.	Weight per tile, lb	Tile per 100 sq ft of wall	Cu yd mortar per 1,000 tile, ⅜-in. joint*
4 × 5 × 12	10	219	0.52
8 × 5 × 12	16	219	1.03
4 × 12 × 12	22	95	0.58
6 × 12 × 12	30	95	0.85
8 × 12 × 12	36	95	1.07
12 × 12 × 12	48	95	1.52

* Includes 20 percent allowance for waste.

TABLE 10-9 Labor-hours Required to Handle and Lay 1,000 Load-bearing Clay Tile

Size tile, in.	Labor-hr per 1,000 tile	
	Mason*	Helper*
4 × 5 × 12	24–30	24–30
8 × 5 × 12	30–36	30–36
4 × 12 × 12	36–45	36–45
6 × 12 × 12	45–53	45–53
8 × 12 × 12	53–64	53–64
12 × 12 × 12	64–80	64–80

* Use the lower values for long, straight walls, requiring ordinary workmanship, and the higher values for walls with irregular shapes, requiring higher-class workmanship.

ings. The table is based on using one helper for each mason. If the materials must be hoisted, add the labor for a hoisting engineer for the job.

STONE MASONRY

Several kinds of stone, both natural and artificial, are used in structures such as buildings, walls, piers, etc. Natural stones used for construction include sandstone, limestone, dolomite, slate, granite, marble, etc. Artificial limestone is available in many areas.

Each kind of stone and work should be estimated separately. The cost of stone in place may be estimated by the cubic yard, ton, cubic foot, square foot, or linear foot. Because of the various methods of pricing stonework, an estimator should be very careful to see that he uses the correct method in preparing his estimate.

Bonds for stone masonry Figure 10-5 illustrates the more common bonds for stone masonry.

Rubble masonry is formed of stones of irregular shapes which are laid in regular courses or at random with mortar joints.

Ashlar masonry is formed of stones cut with rectangular faces. The stones may be laid in courses or at random with mortar joints.

Mortar for stone masonry The mortar used for setting stones may be similar to that used for brick masonry. Sometimes special nonstaining white or stone-set cement may be specified instead of gray portland cement.

MASONRY

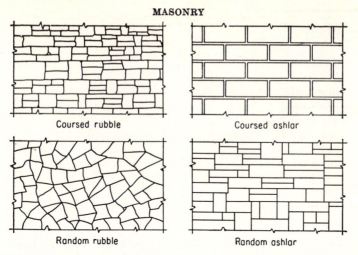

Fig. 10-5   Bonds for stone masonry.

Hydrated lime is usually added to improve the working properties of the mortar.

The quantity of mortar required for joints will vary considerably with the type of bond, the thickness of the joints, and the size of stones used.

Table 10-10 gives representative quantities of mortar required per cubic yard of stone.

Weights of stone   Table 10-11 gives the ranges in weights of stones commonly used for masonry. The given weights are expressed in pounds per net cubic foot of volume.

TABLE 10-10   Quantities of Mortar Required per Cubic Yard of Stone Masonry

Type of bond	Quantity of mortar, cu ft
Coursed rubble	6.5–8.5
Random rubble	7.5–9.5
Cobblestone	6.5–9.5
Coursed ashlar, ¼-in. joints	1.5–2.0
Random ashlar, ¼-in. joints	2.0–2.5
Coursed ashlar, ½-in. joints	3.0–4.0
Random ashlar, ½-in. joints	4.0–5.0

TABLE 10-11   Weights of
Building Stones

Stone	Weight, lb per cu ft
Dolomite	155–175
Granite	165–175
Limestone	150–175
Marble	165–175
Sandstone	140–160
Slate	160–180

Cost of stone   The cost of stone varies so much with the kind of stone, the extent of cutting done at the quarry, and the location where it will be used that no estimate which requires accurate pricing should be made without obtaining current prices for the particular stone. The cost of freight to the destination must be added to the cost at the source in order to determine the cost at the job.

Stones suitable for rubble masonry may be priced by the ton for the specified kind of stone and sizes of pieces.

Stones may be purchased at a quarry in large rough-cut blocks, hauled to the job, and then cut to the desired sizes and shapes. The cost of such blocks may be based on the volume, with the largest dimensions used in determining the volume, or the cost may be based on the weight of the stone.

Stones used for ashlar masonry may be priced by the ton, cubic foot, or square foot of wall area for a specified thickness.

A cordovan limestone frequently used for random ashlar construction in Texas is available in such sizes as 4 in. thick, 2 to 10 in. or more in height, in steps of 2 in., and in random lengths. The top and bottom beds and the back sides will be sawed at the quarry, while the end faces may be sawed at the job, if desired. The exposed face is usually left rough. The cost of this stone will vary from $27.00 to $40.00 or more per ton delivered to a job within the delivery area.

This stone weighs about 140 lb per cu ft. Thus a ton will produce about $2,000 \div 140 = 14.3$ cu ft. If the stone is 4 in. thick, the area of wall covered per ton will be about $3 \times 14.3 = 43$ sq ft. With some loss from cutting and breakage, the net area per ton should be about 40 sq ft.

Consider a stone whose weight is 150 lb per cu ft and whose cost is $40.00 per ton. If this stone is furnished with the top and bottom beds, and the back face sawed for a wall thickness of 4 in., in random lengths, the cost per cubic foot and per square foot of wall area should be about as given in the accompanying table. The costs include an allowance of 10 percent for waste resulting from breakage and sawing end joints.

Item	Cost per ton	Cost, including 10% waste	
		Per cu ft	Per sq ft
Stone f.o.b. job	$40.00	$3.30	$1.10
Cutting end joints	......	1.20	0.40
Total cost	$40.00	$4.50	$1.50

Labor setting stone masonry   Table 10-12 gives representative values of the labor-hours required to handle and set stone masonry. The lower rates should be used for large straight walls, with plain surfaces, and the higher rates for walls with irregular surfaces, pilasters, closely spaced openings, etc. The rates for hoisting will vary with the size of stones, heights hoisted, and equipment used.

Example   Estimate the cost per 100 sq ft for furnishing and setting stone in random ashlar bond for a building. The stone will be sawed on the top and bottom beds and on the front and back faces. It will be furnished in random lengths to be sawed to the desired lengths at the job. The thickness will be 4 in. The stone will be set with ½-in. mortar joints. The stone will be hoisted with a hand derrick.

The stone, which will weigh 150 lb per cu ft, will be delivered to the job at a cost of $40.00 per ton. Allow 10 percent for waste.

The quantity of stone required for 100 sq ft will be

$$\frac{100 \times 4 \times 1.1}{12} = 36.7 \text{ cu ft.}$$

The weight of the stone will be $\dfrac{36.7 \times 150}{2,000} = 2.76$ tons.

The cost will be

Stone, f.o.b. the job, 2.76 tons @ $40.00	=	$110.40
Cutting end joints, 36.7 cu ft @ $1.20	=	44.04
Mortar, 36.7 ÷ 27 = 1.36 cu yd of stone × 4.5 cu ft per cu yd = 6.15 cu ft @ $1.10	=	6.76
Metal wall ties, 100 sq ft ÷ 2 sq ft per tie = 50 @ $14.00 per M	=	0.70
Stone setter, 36.7 cu ft × 0.12 hr per cu ft = 4.4 hr @ $8.23	=	36.21
Helpers, 36.7 cu ft × 0.6 hr per cu ft = 22 hr @ $5.60	=	123.20
Stone setter pointing stone, 36.7 cu ft × 0.05 hr per cu ft = 1.8 hr @ $8.23	=	14.81
Helper pointing stone, 36.7 cu ft × 0.03 hr per cu ft = 1.1 hr @ $5.60	=	6.16
Stone setter cleaning stone, 100 sq ft × 1.5 hr = 1.5 hr @ $8.23	=	12.35

TABLE 10-12 Labor-hours Required to Handle and Set Stone Masonry

| | Labor-hours required | | | |
| | Per cu yd | | Per cu ft | |
Operation	Skilled	Helper	Skilled	Helper
Unloading stones from truck	0.2– 0.4	0.8– 1.6	0.01–0.02	0.03–0.06
Rough squaring	4.0–10.0	....	0.15–0.37	
Smoothing beds	6.0–12.0	....	0.22–0.45	
Setting stone by hand:				
Rubble	2.0– 4.0	3.0– 6.0	0.08–0.15	0.11–0.22
Rough squared	3.0– 6.0	4.0– 9.0	0.11–0.22	0.15–0.33
Ashlar, 4- to 6-in. veneer	10.0–15.0	10.0–15.0	0.37–0.55	0.37–0.55
Setting stones using a hand derrick:				
Heavy cut stone, ashlar, cornices, copings, etc.	2.0– 3.0	12.0–18.0	0.08–0.11	0.45–0.67
Light cut stone, ashlar, sills, lintels, cornices, etc.	2.5– 4.0	12.5–20.0	0.09–0.15	0.46–0.75
Setting stones using a power crane:				
Heavy cut stone:				
Crane operator	1.0– 2.0	....	0.04–0.08	
Stone setter	1.0– 2.0	....	0.04–0.08	
Helpers	....	7.0–14.0	....	0.26–0.52
Medium cut stone:				
Crane operator	1.5– 3.0	....	0.06–0.11	
Stone setter	1.5– 3.0	....	0.06–0.11	
Helpers	....	9.0–18.0	....	0.33–0.67
Pointing cut stone:				
Heavy stone	0.5– 0.8	0.2– 0.4	0.02–0.03	0.01–0.02
Veneer, 4 to 6 in. thick	1.0– 1.5	0.2– 0.8	0.04–0.06	0.02–0.03
Cleaning stone:				
Stone setter	1.5 hr per 100 sq ft			
Helper	0.75 hr per 100 sq ft			

Helper cleaning stone, 100 sq ft × 0.75 hr = 0.75 hr @ $5.60	=	4.20
Foreman, based on 2 stone setters, 3.9 hr @ $8.50	=	33.15
Equipment, saws, derrick, scaffolds, etc.	=	18.00
Total cost	=	$409.98
Cost per sq ft, $409.98 ÷ 100	=	4.10

PROBLEMS

10-1 Estimate the total direct cost and the cost per 1,000 bricks for constructing a brick wall for a rectangular building. The outside dimensions will be 56 ft 6 in. wide and 108 ft

0 in. long, with walls 12 in. thick and 15 ft 6 in. high. The total area of the openings will be 324 sq ft. Common brick, laid in common bond, will be used, with ½-in.-thick cut mortar joints.

The mortar, which will be made of hydrated lime, portland cement, and sand, in the proportions 1:1:6, respectively, by volume, will be mixed in a 3½-cu-ft mortar mixer.

Six bricklayers will be used on the job.

Use the national average wage rates for the bricklayers and helpers (laborers) as given in this book, with the foreman to be paid $0.25 per hr more than the bricklayers.

The material costs will be as follows:

Bricks, f.o.b. the job, per M	$48.00
Lime, per ton	62.50
Cement, per sack	1.60
Sand, per cu yd	4.20
Scaffolds for the job	35.00

10-2 Estimate the total cost and the cost per 1,000 bricks for Prob. 10-1 for bid purposes, using current local prices for materials and labor. Include the cost of a performance bond.

10-3 Estimate the total direct cost of constructing the wall of Prob. 10-1, using 8- by 8- by 12-in. nominal-size standard-weight concrete blocks on the inside and a 4-in. face-brick veneer on the outside. Use the same mortar as for Prob. 10-1. The blocks will be laid with ⅜-in. joints and the brick with ½-in. concave joints. The bricks are to be laid with first-class workmanship, in a common bond. Use one corrugated-metal wall tie for each 4 sq ft of wall area.

Six bricklayers will be used on the wall.

Use the same wage rates as for Prob. 10-1.

Estimate the costs of the blocks and bricks separately; then combine the costs for the total.

Material costs will be as follows:

Bricks, $72.00 per M

Concrete blocks, $0.34 each

Lime, $42.00 per ton

Cement, $1.60 per sack

Sand, $4.20 per cu yd

Wall ties, $14.60 per M

Scaffolds, $35.00 for the job

10-4 Estimate the total cost of the wall for Prob. 10-3 for bid purposes, using current local prices for materials and labor. Include the cost of a performance bond.

10-5 Estimate the total direct cost and the cost per square foot for furnishing and laying cordovan limestone 4 in. thick in random ashlar for the walls for a building whose net surface area will be 7,460 sq ft. The stones will be hoisted with a hand derrick.

Use the same prices for materials and labor as in the example on page 235 of this book.

Use ¼-in. tooled mortar joints, with the mortar to be the same as for Prob. 10-1. Use one metal wall tie for each 4 sq ft of wall area.

10-6 Estimate the total cost of the stone for Prob. 10-5 for bid purposes, including the cost of a bid bond.

Use the same prices for materials and labor as for Prob. 10-5.

Carpentry

General information   This chapter will deal with the furnishing of materials, work, and equipment to construct houses, warehouses, shop buildings, roof trusses, etc., which are constructed primarily with lumber and timber.

The production rates given are based on using power equipment to fabricate the lumber when the use of such equipment is practical. Many of the tables for production give two rates, one for work requiring ordinary workmanship and the other for work requiring first-grade workmanship.

Lumber sizes   The sizes of most lumber, with the exception of certain specialties such as moldings, are designated by the dimensions of the cross sections, using the nominal dimensions, which are the dimensions prior to finishing the lumber. After lumber is sawed lengthwise, it may be passed through one or more finishing operations, which will reduce its actual size to less than the identifying dimensions. Thus a 2- by 4-in. plank will actually be $1\frac{1}{2}$ in. thick and $3\frac{1}{2}$ in. wide after it is surfaced on four sides. This lumber is designated as 2- by 4-in. *S4S*.

If a 1- by 6-in. plank is dressed and matched, with a tongue and groove, the actual dimensions will be $\frac{3}{4}$ in. thick by $5\frac{1}{8}$ in. net width. This lumber is designated as 1- by 6-in. *D* and *M*, and is often identified by the name center match.

Because the finishing and dressing operations will result in a reduction in the actual widths of the lumber, a quantity estimate must include an allowance for this side wastage or shrinkage. For a plank nominally 6 in. wide, the net width will be 5½ in. Thus the side shrinkage will be ½ in.

The side shrinkage for quantity purposes will be $\dfrac{0.5 \times 100}{5.5} = 9.1$ percent.

In order to provide enough lumber for a required area, the quantity must be increased 9.1 percent to replace side shrinkage only.

Lumber is available in lengths which are multiples of 2 ft, with the actual lengths not less than the designated lengths. If other than standard lengths are needed, they must be cut from standard lengths. This operation may result in end waste. For example, if a 1- by 6-in. D and M plank 10 ft 6 in. long is needed, it must be cut from a plank that is 12 ft 0 in. long with a 1-ft 6-in. end waste, unless the piece cut off can be used elsewhere. This will

represent an end waste of $\dfrac{1.5 \times 100}{10.5} = 14.3$ percent.

The combined side and end wastes for the cited examples will be

Side waste	= 9.1 percent
End waste	= 14.3 percent
Total waste	= 23.4 percent

Note that the side and end wastes are determined on the basis of the dimensions of lumber required instead of the nominal dimensions.

Table 11-1 lists the commonly used sizes of lumber, with the nominal and actual cross section dimensions. Lumber that is less than 1 in. thick is considered to be 1 in. thick for quantity purposes. In the table the first dimension is the thickness and the second dimension is the width.

Grades of lumber   Lumber is usually specified by the type, such as pine, fir, oak, etc., size, and grade. Pine and fir are most commonly used for structural members, while oak, maple, redwood, etc., are used for floors, siding, and special purposes.

Pine and fir are graded according to quality as follows:

Select grades, A, B, C, and D
Common grades, Nos. 1, 2, 3, and 4

Select grades are used for trim, facings, moldings, etc., which may be finished in natural color or painted. Grade A is the best quality, and grade D is the poorest.

Common grades are used for structural members such as sills, joist, studs, rafters, planks, etc. Grade No. 1 is the best quality, and grade No. 4 is the poorest.

The lumber used for timber structures is generally designated as stress grade.

TABLE 11-1   Dimensions and Properties of Lumber

Nominal size, in.	Standard dressed size, in.	Area of section, sq in.	Moment of inertia, I	Section modulus, S
Boards				
1 × 3	¾ × 2½	1.875	0.977	0.781
1 × 4	¾ × 3½	2.625	2.680	1.531
1 × 6	¾ × 5½	4.125	10.398	3.781
1 × 8	¾ × 7¼	5.438	23.817	6.570
1 × 10	¾ × 9¼	6.938	49.446	10.695
1 × 12	¾ × 11¼	8.438	88.989	15.820
Dimension				
2 × 3	1½ × 2½	3.750	1.953	1.563
2 × 4	1½ × 3½	5.250	5.359	3.063
2 × 6	1½ × 5½	8.250	20.797	7.563
2 × 8	1½ × 7¼	10.875	47.635	13.141
2 × 10	1½ × 9¼	13.875	98.932	21.391
2 × 12	1½ × 11¼	16.875	177.979	31.641
2 × 14	1½ × 13¼	19.875	290.775	43.891
3 × 4	2½ × 3½	8.750	8.932	5.104
3 × 6	2½ × 5½	13.750	34.661	12.604
3 × 8	2½ × 7¼	18.125	79.391	21.901
3 × 10	2½ × 9¼	23.125	164.886	35.651
3 × 12	2½ × 11¼	28.125	296.631	52.734
3 × 14	2½ × 13¼	33.125	484.625	73.151
3 × 16	2½ × 15¼	38.125	738.870	96.901
4 × 4	3½ × 3½	12.250	12.505	7.146
4 × 6	3½ × 5½	19.250	48.526	17.646
4 × 8	3½ × 7¼	25.375	111.148	30.661
4 × 10	3½ × 9¼	32.375	230.840	49.911
4 × 12	3½ × 11¼	39.375	415.283	73.828
4 × 14	3½ × 13¼	46.375	678.475	102.411
4 × 16	3½ × 15¼	53.375	1034.418	135.660
6 × 6	5½ × 5½	30.250	76.255	27.729
6 × 8	5½ × 7½	41.250	193.359	51.563
6 × 10	5½ × 9½	52.250	392.963	82.729
6 × 12	5½ × 11½	63.250	697.068	121.229
6 × 14	5½ × 13½	74.250	1127.672	167.063
6 × 16	5½ × 15½	85.250	1706.776	220.229
6 × 18	5½ × 17½	96.250	2456.380	280.729

TABLE 11-1   (Continued)

Dressed and matched*

S2S and CM	
1 × 4	¾ × 3⅛
1 × 6	¾ × 5⅛
1 × 8	¾ × 6⅞
1 × 10	¾ × 8⅞
1 × 12	¾ × 10⅞

Flooring	
1 × 2	¾ × 1⅛
1 × 3	¾ × 2⅛
1 × 4	¾ × 3⅛
1 × 6	¾ × 5⅛

Shiplap	
1 × 4	¾ × 3⅛
1 × 6	¾ × 5⅛
1 × 8	¾ × 6⅞
1 × 10	¾ × 8⅞
1 × 12	¾ × 10⅞

* This lumber is also available in other thicknesses.

Cost of lumber   Lumber is usually priced by the thousand feet board measure, M fbm, with the exception of certain types used for trim, moldings, etc., which may be priced by the linear foot. A board foot is a piece of lumber whose nominal dimensions are 1 in. thick, 12 in. wide, and 1 ft long, or it may be 2 in. thick, 6 in. wide, and 1 ft long. The number of feet board measure in a plank 2 in. by 8 in. by 18 ft 0 in. long is $(2 \times 8 \times 18)/12 = 24$ fbm. This same operation should be applied to determine the number of feet board measure in any piece of lumber.

The cost of lumber varies with many factors, including the following:

1   Kind of lumber, pine, fir, oak, etc.
2   Grade of lumber, select or common
3   Size of pieces, thickness, width, and length
4   Extent of milling required, rough, surfaced, dressed and matched
5   Whether dried or green
6   Quantity purchased
7   Freight cost from mill to destination

TABLE 11-2   Names, Sizes, and Numbers of Nails and Spikes per Pound

Size	Length, in.	Trade name					Roofing			
		Common	Spikes	Casing	Finishing	Shingle	Barbed	No. 8	No. 9	No. 10
	3/4	...	...	...	.....	...	714	205	252	290
	7/8	876	...	...	.....	...	469	179	219	253
2d	1	876	...	1,010	1,351	...	411	158	193	224
3d	1¼	568	...	635	807	429	251	128	156	183
4d	1½	316	...	473	584	274	176	108	131	154
5d	1¾	271	...	406	500	235	151	93	113	133
6d	2	181	...	236	309	204	103			
7d	2¼	161	...	210	238					
8d	2½	106	...	145	189					
9d	2¾	96	...	132	172					
10d	3	69	41	94	121					
12d	3¼	63	38	87	113					
16d	3½	49	30	71	90					
20d	4	31	23	52	62					
30d	4½	24	17	46						
40d	5	18	13	35						
50d	5½	14	10							
60d	6	11	9							
	7	...	7							
	8	...	4							
	9	...	3½							
	10	...	3							
	12	...	2½							

The better grades of lumber cost more than the poorer grades. Select grades are more expensive than common grades. In the select grades, the costs decrease in the order A, B, C, and D, while in the common grades, the costs decrease in order Nos. 1, 2, 3, and 4.

The cost of lumber is higher for thick and wide planks than for thin and narrow ones. Lengths in excess of about 20 to 24 ft may require special orders. Prices quoted may be for lengths through 20 ft, with increasingly higher prices charged for greater lengths.

Rough lumber is the product of the sawing operation, with no further surfacing operation. After the lumber is sawed, it may be given a smooth surface on one or more sides and edges in which case it is designated surfaced one edge, *S1E*, surfaced two sides, *S2S*, or surfaced on all edges and sides, *S4S*. It may be surfaced on both sides with the edges matched with tongues and grooves, designated as dressed and matched, *D and M*. These operations reduce the thickness and width and increase the cost.

Freshly cut lumber contains considerable moisture which should be removed before the lumber is used in order to prevent excessive shrinkage after it is placed in the structure. It usually is placed in a heated kiln to expel the excess moisture.

Nails and spikes　The trade names, sizes, and weights of the more popular nails and spikes are given in Table 11-2.

Bolts　The bolts commonly used as connectors for timber structures are available in sizes varying from $\frac{1}{4}$ in. to larger than 1 in. and in lengths varying from 1 in. to any desired length. The length does not include the head. Bolts with square heads and either square or hexagonal nuts are used. Two washers should be used with every bolt, one for the head and one for the nut. If the head and the nut bear against steel plates, it is satisfactory to use stamped-steel washers, but if there are no steel plates under the heads and nuts, cast-iron ogee washers should be used to increase the bearing area of the wood.

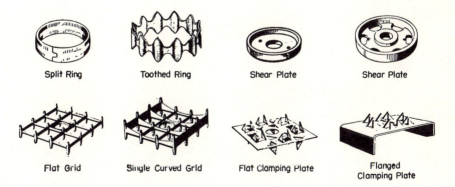

Split Ring　　Toothed Ring　　Shear Plate　　Shear Plate

Flat Grid　　Single Curved Grid　　Flat Clamping Plate　　Flanged Clamping Plate

Fig. 11-1　Timber connectors. (*Timber Engineering Company.*)

TABLE 11-3 Approximate Pounds of Nails Required for Carpentry (Includes 10 Per Cent Waste)

Size and type lumber	Number of nails per support	Size and kind of nails	Pounds per M fbm of lumber for nail spacings*					
			12 in.	16 in.	20 in.	24 in.	36 in.	48 in.
1 × 4	2	8d common	63	47	38	31	21	16
1 × 5	2	8d common	49	37	30	25	17	12
1 × 6	2	8d common	41	31	25	21	14	10
1 × 8	2	8d common	32	24	19	16	11	8
1 × 10	3	8d common	37	28	22	19	12	10
1 × 12	3	8d common	31	23	18	15	10	8
2 × 4	2	20d common	110	83	66	55	37	28
2 × 6	2	20d common	72	54	42	36	24	18
2 × 8	2	20d common	56	42	34	28	19	14
2 × 10	3	20d common	65	49	40	33	22	17
2 × 12	3	20d common	56	42	34	28	19	14
3 × 4	2	60d common	204	153	122	102	68	51
3 × 6	2	60d common	136	102	82	68	45	34
3 × 8	2	60d common	101	76	61	51	34	26
3 × 10	3	60d common	124	93	74	62	41	31
3 × 12	3	60d common	104	78	62	52	35	26
Framing, studs, etc.:								
2 × 4	...	16d common	15	11	9	7	5	4
2 × 6	...	16d common	11	8	6	5	4	3
2 × 8	...	16d common	11	8	7	6	4	3
Siding:								
Bevel, ½ × 4	1	6d finish	12	9				
Bevel, ½ × 6	1	6d finish	8	6				
Bevel, ½ × 8	1	6d finish	6	4½				
Drop, 1 × 4	2	8d casing	51	38				
Drop, 1 × 6	2	8d casing	33	25				
Drop, 1 × 8	2	8d casing	27	20				

			Lb per M fbm	
Flooring, *D* and *M* softwood:				
1 × 3	1	8d brads	51	38
1 × 4	1	8d brads	37	28
1 × 6	1	8d brads	25	19
Joists:				
2 × 6	...	16d common	17	
2 × 8	...	16d common	10	
2 × 10	...	16d common	8	
2 × 12	...	16d common	7	
Bridging, 1 × 4	4 ea	8d common	83	
Rafters:				
2 × 4	...	16d common	18	
2 × 6	...	16d common	18	
2 × 8	...	16d common	15	
2 × 10	...	16d common	12	

			Lb per 100 sq ft
Flooring, *D* and *M* hardwood:			
$\frac{3}{8}$ × 2	...	3d finish	1.1
$\frac{3}{16}$ × 2¼	...	3d finish	0.9
$\frac{13}{16}$ × 2	...	8d cut	6.0
$\frac{13}{16}$ × 2¼	...	8d cut	5.5
$\frac{13}{16}$ × 3¼	...	8d cut	4.0
Plasterboard	...	3d barbed	2.5
Wood shingles:			
4½ in. exposed	2	3d shingle	4.5
5 in. exposed	2	3d shingle	4.1
5½ in. exposed	2	3d shingle	3.8
Asphalt shingles	2 ea	⅞ in. barbed	4.2

* If different sizes of common nails are used make the following changes: 10d for 8d, increase weight 53 per cent. 20d for 16d, increase weight 58 per cent.

TABLE 11-4   Weights of Bolts with Square Heads and Hexagonal Nuts in Pounds per 100

Length under head, in.	Diameter of bolt, in.										
	¼	⁵⁄₁₆	⅜	⁷⁄₁₆	½	⅝	¾	⅞	1	1⅛	1¼
1	2.7	5.0	7.2	11.2	14.9	28	43				
1¼	3.1	5.5	8.0	12.2	16.3	30	46	68			
1½	3.4	6.1	8.8	13.3	17.7	32	49	73	103	144	190
1¾	3.8	6.6	9.6	14.4	19.0	35	52	77	109	151	199
2	4.1	7.2	10.4	15.4	20.4	37	55	81	115	158	208
2¼	4.5	7.7	11.1	16.5	21.8	39	58	85	120	165	216
2½	4.8	8.2	11.9	17.5	23.2	41	61	90	126	172	225
2¾	5.2	8.8	12.7	18.6	24.6	43	64	94	131	179	234
3	5.5	9.3	13.5	19.7	26.0	45	68	98	137	187	242
3¼	5.9	9.9	14.3	20.7	27.4	48	71	102	142	194	251
3½	6.2	10.4	15.1	21.8	28.8	50	74	107	148	201	260
3¾	6.6	11.0	15.8	22.9	30.2	52	77	111	153	208	268
4	6.9	11.5	16.6	23.9	31.6	54	80	115	159	215	277
4¼	7.3	12.0	17.4	25.0	33.0	56	83	119	165	222	286
4½	7.6	12.6	18.2	26.1	34.4	58	86	124	170	229	294
4¾	8.0	13.1	19.0	27.1	35.7	61	89	128	176	236	303
5	8.3	13.7	19.8	28.2	37.1	63	93	132	181	243	312
5¼	8.6	14.2	20.5	29.3	38.5	65	96	136	187	250	321
5½	9.0	14.8	21.3	30.3	39.9	67	99	141	192	257	329
5¾	9.3	15.3	22.1	31.4	41.3	69	102	145	198	264	338
6	9.7	15.9	22.9	32.4	42.7	71	105	149	204	271	347

	1.4	2.2	3.1	4.3	5.6	8.7	12.5	17.0	22.3	28.2	34.8
6¼	10.0	16.4	23.7	33.5	44.1	74	108	153	209	278	355
6½	10.4	16.9	24.5	34.6	45.5	76	111	158	215	285	364
6¾	10.7	17.5	25.2	35.6	46.9	78	114	162	220	292	373
7	11.1	18.0	26.0	36.7	48.3	80	118	166	226	299	381
7¼	11.4	18.6	26.8	37.8	49.7	82	121	170	231	306	390
7½	11.8	19.1	27.6	38.8	51.1	84	124	175	237	313	399
7¾	12.1	19.7	28.4	39.9	52.4	87	127	179	242	320	407
8	12.5	20.2	29.2	41.0	53.8	89	130	183	248	327	416
8½	....	21.3	30.7	43.1	56.6	93	136	192	259	341	434
9	....	22.4	32.3	45.2	59.4	98	143	200	270	356	451
9½	....	23.5	33.9	47.4	62.2	102	149	209	281	370	468
10	....	24.6	35.4	49.5	65.0	106	155	217	293	384	486
10½	....	....	37.0	51.6	67.8	111	161	226	304	398	503
11	....	....	38.6	53.7	70.5	115	168	234	315	412	520
11½	....	....	40.1	55.9	73.3	119	174	243	326	426	538
12	....	....	41.7	58.0	76.1	124	180	251	337	440	555
12½	....	....	....	60.1	78.9	128	186	260	348	454	573
13	....	....	....	62.3	81.7	132	193	268	359	468	590
13½	....	....	....	64.4	84.5	137	199	277	370	482	607
14	....	....	....	66.5	87.2	141	205	285	382	496	625
14½	....	....	....	....	90.0	145	211	294	393	510	642
15	....	....	....	....	92.8	150	218	302	404	525	660
15½	....	....	....	....	95.6	154	224	311	415	539	677
16	....	....	....	....	98.4	158	230	320	426	553	694
Per inch additional	1.4	2.2	3.1	4.3	5.6	8.7	12.5	17.0	22.3	28.2	34.8

TABLE 11-5 Data for Cast-iron Ogee Washers

Diam of bolt, in.	Diam of washer, in.	Thickness of washer, in.	Weight per 100, lb
$\frac{3}{8}$	$2\frac{3}{16}$	$\frac{1}{2}$	31.1
$\frac{1}{2}$	$2\frac{3}{8}$	$\frac{1}{2}$	37.5
$\frac{5}{8}$	$2\frac{3}{4}$	$\frac{5}{8}$	56.0
$\frac{3}{4}$	3	$1\frac{1}{16}$	75.0
$\frac{7}{8}$	$3\frac{1}{2}$	$\frac{3}{4}$	106.0
1	$3\frac{3}{4}$	$\frac{7}{8}$	150.0
$1\frac{1}{4}$	$5\frac{3}{4}$	$1\frac{3}{16}$	450.0
$1\frac{1}{2}$	$6\frac{1}{8}$	$1\frac{1}{2}$	662.0

Timber connectors A factor which has contributed materially to the design and construction of timber structures was the introduction of modern timber connectors. These connectors are available through the Timber Engineering Company of Washington, D.C. They are classified under a general term as Teco connectors. They include split rings, toothed rings, shear plates, spike grids, and clamping plates. These connectors are illustrated in Fig. 11-1.

Split rings Split rings, in sizes 2½- and 4-in. diameters, are for use with wood-to-wood connections for medium and heavy loads. It is necessary to drill a hole for the bolt and a groove for the ring. A complete connector unit includes a split ring, bolt, and two steel washers.

Toothed rings Toothed rings, in sizes 2-, 2⅝-, 3⅜-, and 4-in. diameters, are for use with wood-to-wood connections in thin members with light loads. A complete unit includes a toothed ring, bolt, and two steel washers.

Shear plates Flanged shear plates of 2⅝-in.-diameter pressed steel and 4-in.-diameter malleable iron are for wood-to-wood or wood-to-steel connections. They are placed in precut grooves, or daps, in the timber members. A complete unit includes one or two plates, a bolt, and two steel washers.

Spike grids Spike grids, of malleable cast iron, are designed for use with piles, poles, and flat timber members. The flat grid joins two flat timber members, the single curve joins a curved pile and a flat surface. A complete unit includes a grid, a bolt, and two washers.

Clamping plates Clamping plates are used as railroad-tie spacers by placing them between the ties and the guard timbers.

Installing toothed rings and spike grids Toothed rings and spike grids are installed by using a high-strength threaded rod assembly in the bolthole. Pressure exerted by the assembly will force the connectors to penetrate the

TABLE 11-6 Weights and Dimensions of Timber Connectors

Timber connector	Size, in.	Dimensions of metal, in.		Shipping weight, lb per 100
		Depth	Thickness	
Split rings	2½	0.75	0.163	28
	4	1.00	0.193	70
Toothed rings	2	0.94	0.061	9
	2⅝	0.94	0.061	12
	3⅜	0.94	0.061	15
	4	0.94	0.061	18
Shear plates:				
Pressed steel	2⅝	0.375	0.169	35
Malleable iron	4	0.62	0.20	90
Spike grids:				
Flat	4⅛ × 4⅛	1.00	. . . . .	48
Single curve	4⅛ × 4⅛	1.38	. . . . .	70
Circular	3⅛ diam	. . . .	. . . . .	26
Clamping plates:				
Plain	5¼ × 5¼	. . . .	0.077	59
Flanged	5 × 8½	2.00	0.122	190

wood members. After the wood members are compressed sufficiently, the assembly is removed and replaced with a machine bolt of the proper size.

Table 11-6 gives the weights and dimensions of timber connectors.

Fabricating lumber Fabricating lumber includes such operations as sawing, ripping, chamfering edges, boring holes, etc. The rates at which these

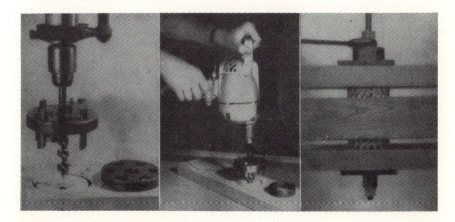

Fig. 11-2 Tools used to prepare lumber for Teco connectors. (*Timber Engineering Company.*)

operations can be performed will vary greatly with several factors, including the following:

1 Extent to which power equipment is used
2 Amount of fabricating required
3 Length of finished pieces
4 Size of pieces
5 Number of similar pieces fabricated with one machine setting
6 Care in sorting and storing lumber at the job

Lumber can be sawed five to ten times as fast with a power saw as with a handsaw, provided that the operations are not too complicated.

It requires considerably more time to fabricate rafters and stair stringers than it does to fabricate studs and joists.

Since a certain amount of time is required to handle and measure lumber, more time is required to fabricate a given quantity of lumber consisting of short lengths and small sections than when it consists of long lengths and large sections.

Time is required to set a machine for a given operation. If only a few pieces are fabricated, after which the machine must be reset, the total time per piece will be higher than when a great many similar pieces are fabricated. This may be illustrated by analyzing several operations.

Example Assume that you are to cut to length, with square corners, 60 pieces of 2-in. by 12-in. by 19-ft 6-in.-long lumber for floor joists, using a stationary cutoff saw with a long table. The saw blade will be set to the correct position, and a stop will be set at one end of the table to eliminate measurements. One helper will supply stock lumber, a mechanic will saw the lumber, and another helper will remove and stack the fabricated pieces.

The time required should be about as follows:

Operation	Units	Unit time, min	Total time, min
Set saw and stop	1	10	10
Saw lumber	60	0.75	45
Total time			55

$$\text{Quantity of lumber,} \quad \frac{60 \times 2 \times 12 \times 19.5}{12} = 2{,}340 \text{ fbm}$$

$$\text{Time per M fbm,} \quad \frac{55}{60 \times 2.34} = 0.39 \text{ hr}$$

If the same saw is used to fabricate 120 studs 2 in. by 4 in. by 8 ft 0 in. long, requiring a square cut at each end, the time should be about as follows:

Operation	Units	Unit time, min	Total time, min
Set saw and stop	1	10	10
Saw lumber	120	0.33	40
Total time			50

$$\text{Quantity of lumber, } \frac{120 \times 2 \times 4 \times 8}{12} = 640 \text{ fbm}$$

$$\text{Time per M fbm, } \frac{50}{60 \times 0.64} = 1.31 \text{ hr}$$

If this saw is used to fabricate 100 pieces of cross bridging 1 in. by 4 in. by 18 in. long, requiring a bevel cut at each end, the time should be about as follows:

Operation	Units	Unit time, min	Total time, min
Set saw and stop	1	10	10
Saw lumber	100	0.2	20
Total time			30

$$\text{Quantity of lumber, } \frac{100 \times 1 \times 4 \times 1.5}{12} = 50 \text{ fbm}$$

$$\text{Time per M fbm, } \frac{30}{60 \times 0.05} = 10 \text{ hr}$$

These examples illustrate the necessity of analyzing each operation to determine the probable time required. Simply assuming that work will be done at some given rate per unit of lumber, regardless of the size and length of lumber and the kind of operation, is not sufficiently accurate for estimating purposes.

ROUGH CARPENTRY HOUSES

House framing   Houses may be constructed using a balloon, braced, or platform frame. Figure 11-3 shows typical wall and partition sections for balloon and platform frames. For the balloon frame, the studs for the outside

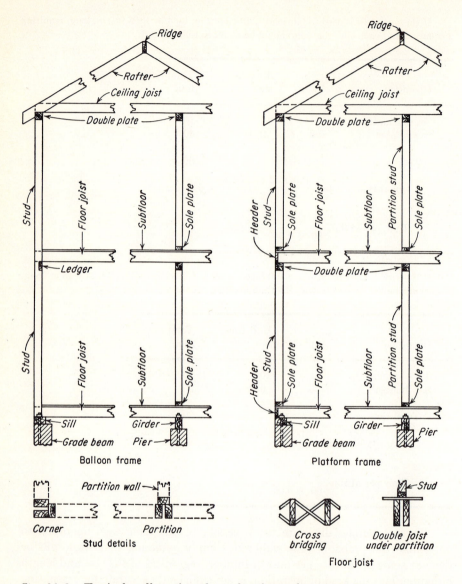

Fig. 11-3 Typical wall section through a frame house.

walls of a two-story building extend from the sill to the top plate on which the rafters rest. For the platform frame, the studs for the external walls are limited to one story in length. The lengths of studs for interior or partition walls for both types of frames are limited to one story.

Sills Sills are attached to masonry-grade beams, usually with anchor bolts set in the masonry at 6- to 8-ft intervals. The sills, which may be 2- by 4-in.,

2- by 6-in., 2- by 8-in., 4- by 6-in., or 6- by 8-in. *S4S* lumber, are bored to permit the bolts to pass through them. Unless the top of a grade beam is smooth and level, a layer of stiff mortar should be placed under each sill to assure full bearing and uniform elevation. At the corners and at end joints, the sills should be connected by half-lapped joints.

Table 11-8 gives the labor-hours required to fabricate and install sills, including boring the holes and tightening the nuts.

Floor girders Floor girders are installed under the first-floor joists to provide intermediate support for the joists. The girders may consist of single pieces such as 4- by 6-in., 6- by 8-in., etc., lumber, or they may be made by securely nailing together several pieces of lumber 2 in. thick. They are supported by piers usually spaced 8 to 10 ft apart. End joints should be constructed with half-laps, located above piers.

Floor and ceiling joists Floor and ceiling joists are generally spaced 12, 16, or 24 in. on centers, using 2- by 6-in., 2- by 8-in., 2- by 10-in., or 2- by 12-in. lumber, rough or *S4S*. When the unsupported lengths exceed 8 ft, one row of cross bridging should be installed at spacings not greater than 8 ft. The bridging may be made from 1- by 3-in. or 1- by 4-in. lumber, with two 8d nails at each end of each strut.

Studs Studs for one- and two-story houses usually are 2- by 4-in. lumber. Studs for walls, which must provide passages for plumbing pipes, should be 2- by 6-in. lumber. The studs may be spaced 12, 16, 20, or 24 in. on centers, with the 16-in. spacing used for houses whose walls will be plastered, since plaster laths are furnished in 48-in. lengths.

At all corners and where partition walls frame into other walls, it is common practice to install three studs, as illustrated in Fig. 11-3.

At all openings for windows and doors, two studs should be used, with one stud on each side of the opening cut to support the header over the opening.

TABLE 11-7 Labor-hours Required per M FBM of Sills

Size sill, in.	Labor-hours	
	Carpenter	Helper
2 × 4	24–30	8–10
2 × 6	18–22	6–8
2 × 8	14–18	5–6
4 × 6	15–20	5–7
4 × 8	14–19	5–7
6 × 8	13–18	4–6

TABLE 11-8 Labor-hours Required per M FBM of Floor Girders

Size girder, in.	Labor-hours	
	Carpenter	Helper
4 × 6	15–20	5–7
6 × 8	13–18	4–6
8 × 10	11–16	4–6
3–2 × 8	14–18	5–6
3–2 × 10	12–15	4–5

TABLE 11-9 Labor-hours Required per M FBM of Joists

Size joists, in.	Labor-hours	
	Carpenter	Helper
2 × 6	11–13	4–5
2 × 8	9–12	4–5
2 × 10	8–11	4–5
2 × 12	7–10	4–5
Cross bridging, per 100 sets	7–9	1–2

TABLE 11-10 Labor-hours Required per M FBM for Studs with One Bottom and Two Top Plates

Size studs, in.	Stud spacing, in.	Labor-hours	
		Carpenter	Helper
2 × 4	12	20–25	5–6
	16	22–26	5–6
	24	24–28	5–6
2 × 6	12	19–24	5–6
	16	21–25	5–6
	24	23–27	5–6

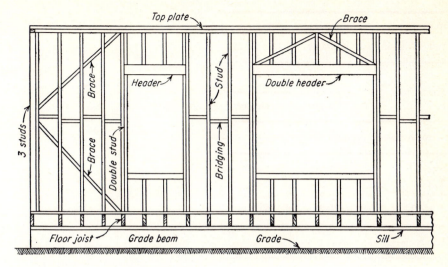

Fig. 11-4 Typical framing details for studs.

All the studs for a wall of a room are cut to length, properly spaced, and nailed to a bottom and top plate, on the subfloor if it is installed, then tilted into position. A second top plate may then be installed to join the sections or to tie partition walls to main walls.

Lateral rigidity for walls may be provided by diagonal sheathing or by cut-in braces installed at each corner of the building.

Rafters The labor required to frame and erect rafters will vary considerably with the type of roof. Double-pitch or gable roofs, with no dormers or gables, will require the least amount of labor, while hip roofs, with dormers or gables framing into the main roof, will require the greatest amount of labor.

Subfloors Subfloors may be laid straight or diagonally, using 1- by 6-in. or 1- by 8-in. planks. Two 8d common wire nails should be used at each joist. More labor will be required to lay diagonal than straight flooring. Also, the lumber wastage will be greater for diagonal flooring.

Roof decking The lumber used for roof decking may be 1- by 6-in. or 1- by 8-in. S4S, 1- by 6-in. or 1- by 8-in. center match, or 1- by 6-in. or 1- by 8-in. shiplap, fastened with at least two 8d common wire nails per rafter. The labor will be slightly higher for a two-story than for a one-story building. Also, the labor will be higher for a steep roof than for a flat roof.

Wall sheathing Sheathing for exterior or interior walls may consist of 1- by 6-in. or 1- by 8-in. S4S planks, 1- by 6-in. or 1- by 8-in. shiplap, or 1- by 6-in. or 1- by 8-in. center match. The planks may be installed horizontally or diagonally. Diagonal sheathing will require more labor and will increase the wastage, but it will give the building much greater rigidity than

TABLE 11-11   Labor-hours Required per M FBM for Rafters

Size rafters, in.	Labor-hours	
	Carpenter	Helper
Gable roofs—no dormers		
2 × 4	27–31	7–8
2 × 6	25–29	7–8
2 × 8	24–28	6–7
2 × 10	22–26	6–7
Hip roofs—no dormers		
2 × 4	30–34	8–9
2 × 6	29–33	8–9
2 × 8	28–32	7–8
2 × 10	26–30	7–8

TABLE 11-12   Labor-hours Required per M FBM for Subfloors

Size flooring, in.	Joist spacing, in.	Labor-hours	
		Carpenter	Helper
Straight flooring			
1 × 6	12	11–12	4–5
	16	10–11	4–5
	24	9–10	4–5
1 × 8	12	10–11	4–5
	16	9–10	4–5
	24	8–9	4–5
Diagonal flooring			
1 × 6	12	13–14	5–6
	16	12–13	5–6
	24	11–12	5–6
1 × 8	12	12–13	5–6
	16	11–12	5–6
	24	10–11	5–6

TABLE 11-13 Labor-hours Required per M FBM
for Roof Decking

Size decking, in.	Rafter spacing, in.	Labor-hours	
		Carpenter	Helper
1 × 6	12	12–15	5–7
	16	11–14	5–7
	20	10–13	5–7
1 × 8	12	11–14	5–7
	16	10–13	5–7
	20	9–12	5–7

horizontal sheathing. At least two 8d common wire nails should be used at each stud.

Board and batten siding Vertical boards, usually 1 by 10 in. or 1 by 12 in., and batts sometimes are used for external siding on houses. The batts are installed to cover the joints between the boards.

TABLE 11-14 Labor-hours Required per
M FBM for Wall Sheathing

Type and size sheathing, in.	Labor-hours	
	Carpenter	Helper
Horizontal sheathing		
1 × 6 *S4S*	11–13	4–5
1 × 8 *S4S*	10–12	4–5
1 × 6 shiplap	12–14	4–5
1 × 8 shiplap	11–13	4–5
1 × 6 center match	12–14	4–5
1 × 8 center match	11–13	4–5
Diagonal sheathing		
1 × 6 *S4S*	16–18	4–5
1 × 8 *S4S*	15–17	4–5
1 × 6 shiplap	17–19	4–5
1 × 8 shiplap	16–18	4–5
1 × 6 center match	17–19	4–5
1 × 8 center match	16–18	4–5

TABLE 11-15 Labor-hours Required per M FBM of Boards for Board and Batten Siding

Size boards, in.	Labor-hours	
	Carpenter	Helper
1 × 10 *S4S*	16–18	4–5
1 × 12 *S4S*	15–17	4–5

Drop siding Rustic and drop siding, which is available in dressed-and-matched or shiplap patterns, in the nominal and face widths given in Table 11-16, is available in nominal thicknesses of ⅝ and 1 in. Thicknesses less than 1 in. are considered to be 1 in. in determining the number of feet board measure of lumber.

TABLE 11-16   Labor-hours Required per M FBM for Drop Siding

Pattern	Width, in.		Labor-hours	
	Nominal	Face	Carpenter	Helper
Ordinary workmanship				
Dressed and matched	4	3¼	22–24	5–6
	5	4¼	21–23	5–6
	6	5³⁄₁₆	20–22	4–5
	8	7⅛	18–20	4–5
Shiplap	4	3⅛	22–24	5–6
	5	4⅛	21–23	5–6
	6	5¹⁄₁₆	20–22	4–5
	8	6⅞	18–20	4–5
	10	8⅞	16–18	4–5
First-class workmanship				
Dressed and matched	4	3¼	27–29	6–7
	5	4¼	26–28	6–7
	6	5³⁄₁₆	25–27	5–6
	8	7⅛	23–25	5–6
Shiplap	4	3⅛	27–29	6–7
	5	4⅛	26–28	6–7
	6	5¹⁄₁₆	25–27	5–6
	8	6⅞	23–25	5–6
	10	8⅞	21–23	5–6

TABLE 11-17 Labor-hours Required per M
FBM for Bevel Siding

Width, in.		Labor-hours	
Nominal	Exposed to weather	Carpenter	Helper
Ordinary workmanship			
4	2¾	28–30	5–6
5	3¾	24–26	5–6
6	4¾	21–23	4–5
8	6¾	18–20	4–5
First-class workmanship			
4	2¾	34–36	5–6
5	3¾	30–32	5–6
6	4¾	26–28	4–5
8	6¾	23–25	4–5

Bevel siding Bevel siding is available in ½ and ⅝ in. nominal thicknesses
and in the widths given in Table 11-17. The number of feet board measure
for thicknesses less than 1 in. is the product of the nominal width in feet
times the length in feet. The quantity of bevel siding required to cover a
given wall area will vary with the actual width and the width exposed to
weather.

Ceiling and partition Beaded and V-pattern ceiling and partition are
available in the nominal and face widths given in Table 11-18 for all thicknesses.

TABLE 11-18 Labor-hours Required per
M FBM for Ceiling and Partition

Width, in.		Labor-hours	
Nominal	Face	Carpenter	Helper
3	2⅜	26–28	5–6
4	3¼	24–26	5–6
5	4¼	22–24	4–5
6	5³⁄₁₆	20–22	4–5

TABLE 11-19   Labor-hours Required to Lay 100 Sq Ft of Surface Area for Wood Shingles

Kind of work	Labor classification	Length exposed to weather, in.					
		4	4½	5	5½	6	7½
Plain hip and gable roofs	Carpenter	3.5	3.2	3.0	2.8	2.5	2.0
	Helper	0.9	0.8	0.8	0.7	0.6	0.5
Irregular hip and gable roofs	Carpenter	5.0	4.6	4.3	4.0	3.7	3.2
	Helper	1.0	0.9	0.8	0.8	0.7	0.6
Plain walls	Carpenter	5.0	4.6	4.3	4.0	3.7	3.2
	Helper	1.0	0.9	0.8	0.8	0.7	0.6

Wood shingles   Wood shingles are available in three lengths, 16, 18, and 24 in., and in random widths. For use on roofs having a pitch not less than ¼, which is a rise-to-run ratio of 1:2, the recommended exposure to weather is as follows:

Total length, in.	Length exposed to weather, in.
16	5
18	5½
24	7½

Shingles are sold by the square but packed in bundles which will cover ¼ square each when the standard exposure is used. About 10 percent should be added for waste. Prices will vary with the quality and length.

Shingles are nailed to shingle laths, sizes 1 by 2 in. to 1 by 4 in., which are installed perpendicular to the roof rafters. The spacing of the lath should correspond to the length of shingles exposed to the weather. Each shingle should be fastened with two nails.

Fascia, frieze, and corner boards   The quantity of lumber for these boards should be estimated by the M fbm of lumber. The labor required to place the boards will be greater for short than for long boards.

Wood furring and grounds   The labor required to place wood furring strips and grounds is usually estimated by the 100 lin ft.

Door bucks   Door bucks may be made from 2- by 4-in. or 2- by 6-in. stock lumber or from 2- by 6-in. lumber having the back grooved out by the mill, depending on the thickness of the wall in which they will be used.

TABLE 11-20 Labor-hours Required to Place 100 Lin Ft of Fascia, Frieze, and Corner Boards

Item	Labor-hours	
	Carpenter	Helper
Fascia and frieze	4–5	0.8–1.0
Corner boards	3–4	0.6–0.8

The total time required per opening will include the time to make, set, plumb, and brace a buck, which will be slightly higher for large than for small openings.

Finished wood floors The lumber used for finished wood floors includes pine, maple, oak, beech, and birch. All these varieties are not generally available in all locations.

Table 11-23 lists some, but not necessarily all, of the sizes of wood flooring produced. All the sizes listed may not be available in all locations. This is especially true with respect to hardwood flooring.

The percentages of side waste listed in Table 11-23 are determined by dividing the reduction in width by the net width. For example, if the nominal width is 4 in. and the face width is $3\frac{1}{4}$ in., the loss in width is 0.75 in. For these dimensions the table lists the percent to be added for side waste to be $(0.75 \times 100)/3.25 = 23.1$. The reason for using this procedure is that when the quantity of lumber required is determined, the area of the floor is known. If the quantity of side waste listed in Table 11-23 is added to the area of the floor for the particular nominal width of flooring selected, this allowance will be sufficient for the side waste. An additional amount may be required for end waste, resulting from sawing to shorter lengths.

TABLE 11-21 Labor-hours Required to Place 100 Lin Ft of Wood Furring Strips and Grounds

Kind of work	Labor-hours	
	Carpenter	Helper
Furring strips on masonry walls	2.0–4.0	0.4–0.6
Furring strips on wood floors	2.5–4.5	0.4–0.6
Grounds on masonry walls	2.5–4.0	0.5–0.8
Grounds on frame walls	1.5–3.0	0.4–0.6

TABLE 11-22 Labor-hours Required to Make and Install Wood Door Bucks

| Size opening | Labor-hours per opening | |
	Carpenter	Helper
Up to 3 ft 0 in. by 7 ft 0 in.	2.0–2.5	0.5–0.6
Over 3 ft 0 in. by 7 ft 0 in.	2.5–3.0	0.6–0.7

TIMBER STRUCTURES

Fabricating timber for structures Fabricating timber members for structures involves starting with commercial sizes of lumber and sawing them to the correct lengths and shapes, boring holes, dapping, chamfering, etc. Fabricating may be done with hand tools or with power-driven equipment.

The time required to fabricate timbers is illustrated by the following examples.

TABLE 11-23 Dimensions of Wood Flooring

| Thickness, in.* | | Width, in. | | Add for side waste, percent |
Nominal	Worked	Nominal	Face	
Softwood				
$\frac{1}{2}$	$\frac{7}{16}$	3	$2\frac{1}{8}$	41.2
$\frac{5}{8}$	$\frac{9}{16}$	4	$3\frac{1}{8}$	28.0
1	$\frac{3}{4}$	5	$4\frac{1}{8}$	21.2
$1\frac{1}{4}$	1	6	$5\frac{1}{8}$	17.1
$1\frac{1}{2}$	$1\frac{1}{4}$	6	$5\frac{1}{8}$	17.1
Hardwood				
1	$2\frac{5}{32}$	$2\frac{1}{4}$	$1\frac{1}{2}$	50.00
1	$2\frac{5}{32}$	$2\frac{3}{4}$	2	37.5
1	$2\frac{5}{32}$	3	$2\frac{1}{4}$	33.3
1	$2\frac{5}{32}$	4	$3\frac{1}{4}$	23.0

* Generally all widths of softwood flooring are available in each listed thickness. In some locations each listed nominal width of hardwood may not be available. Also hardwood may be available in other thicknesses.

Example Determine the probable time required to cut a 2-in. by 8-in. by 12-ft 0-in.-long plank to a specified length, requiring two square cuts with a handsaw.

The approximate time should be as given in the accompanying table.

Operation	No. of units	Time per unit, min	Total time, min
Pick up and place on sawhorses	1	0.3	0.3
Measure for length	1	0.2	0.2
Mark for cuts	2	0.1	0.2
Saw the plank	2	1.0	2.0
Total time	...	...	2.7

Total time assuming a 45-min hour, $2.7 \times {}^{60}\!/_{45} = 3.6$ min
Average time per cut, $3.6 \div 2 = 1.8$ min

If the plank is cut by an electric handsaw, the only reduction in time will be in the two sawing operations. The total time should be about as follows:

Operation	Time, min
Pick up and place on sawhorses	0.3
Measure for length	0.2
Mark for cuts	0.2
Saw the plank, 2 cuts × 0.25 min	0.5
Total time	1.2

Total time assuming a 45-min hour, $1.2 \times {}^{60}\!/_{45} = 1.6$ min
Average time per cut, $1.6 \div 2 = 0.8$ min

The time to bore two 1-in.-diameter holes in a 2-in.-thick plank should be about as follows:

Operation	Time, min
Pick up and place on sawhorses	0.3
Mark location of holes	0.3
Bore, 2 holes × 0.75 min	1.5
Total time	2.1

Total time assuming a 45-min hour, $2.1 \times {}^{60}\!/_{45} = 2.8$ min
Average time per hole, $2.8 \div 2 = 1.4$ min

TABLE 11-24   Approximate Time in Minutes Required to Perform Various Operations in Fabricating Timbers

Nominal size of timber, in.	Operation				
	Sawing		Ripping, power, per lin ft	Boring 1 hole	
	Hand	Power		Hand	Power
2 × 4	1.3	0.6	0.2	1.4	0.8
2 × 6	1.5	0.7			
2 × 8	1.8	0.8			
2 × 12	2.1	0.9			
3 × 6	1.8	0.8	0.3	2.0	1.0
3 × 8	2.1	0.9			
3 × 12	2.5	1.1			
4 × 4	1.8	0.8	0.4	2.5	1.2
4 × 6	2.2	1.0			
4 × 8	2.6	1.2			
4 × 12	3.0	1.5			
6 × 6	2.5	1.1	0.6	3.0	1.5
6 × 8	3.0	1.5			
6 × 12	4.0	2.1			
6 × 18	5.5	3.0			
8 × 8	4.0	2.0	0.8	3.5	1.7
8 × 12	6.0	3.0			
8 × 18	9.0	4.5			
8 × 24	12.0	6.0			
10 × 10	6.0	3.0	1.0	4.0	2.0
10 × 12	8.0	4.0			
10 × 18	11.0	5.5			
12 × 12	10.0	5.0	1.2	4.5	2.5
12 × 18	14.0	7.0			
12 × 24	18.0	9.0			

Table 11-24 gives the approximate time required to fabricate timbers. The rates given include a reasonable amount of time for handling the timbers in addition to the actual operation. The time for operations will vary with the number of timbers fabricated to a given pattern, the extent to which jigs are used for measurements, and the distance from the stockpile to the working area.

Example   Use the information in Tables 11-24 and 11-25 to determine the probable time required to fabricate and assemble a joint for three 3- by 8-in. members, using two 4-in. split rings and one ¾-in. bolt per assembly.

    Boring and grooving, 4 × 1 = 4 min
    Installing 2 rings, 2 × 1   = 2 min
    Installing 1 bolt, 1 × 3   = 3 min
                                  ————
        Total time   = 9 min

TABLE 11-25 Approximate Time in Minutes Required to Install Connectors and Bolts

Connector	Size, in.	Power grooving, per groove	Installing	
			Connector	Bolt
Split rings	All	1	1	3
Shear plates	...	1	1	3
Toothed rings	2	...	5	3
	2⅝	...	5	3
	3⅝	...	6	3
	4	...	6	3
Spike grids	3⅛	...	5	3
	4⅛ × 4⅛	...	7	3
Clamping plates	All	...	4	3

MILL BUILDINGS

General descriptions   Wooden structures of the mill-building type are frequently used for textile mills, factories, and warehouses. The timber members include columns, girders, beams, trusses, flooring, and roofing. The construction is generally open, with no thin sections, sharp projections, or concealed spaces. While such a structure is not fireproof, it is classified as slow-burning. By treating the timbers with fire-resistant chemicals, it is possible to reduce the rate of burning to less than would be experienced with untreated timbers. Connections between members are made with nails, lag screws, bolts, timber connectors, and metal or wood column caps and bases.

Columns   Columns for mill buildings are usually square in sizes from 8 to

ESTIMATING CONSTRUCTION COSTS

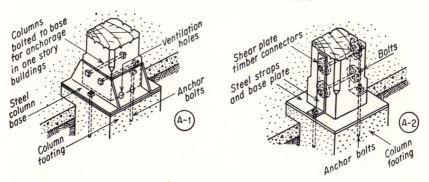

Fig. 11-5   Typical details of bases for wood columns. (*National Lumber Manufacturing Co.*)

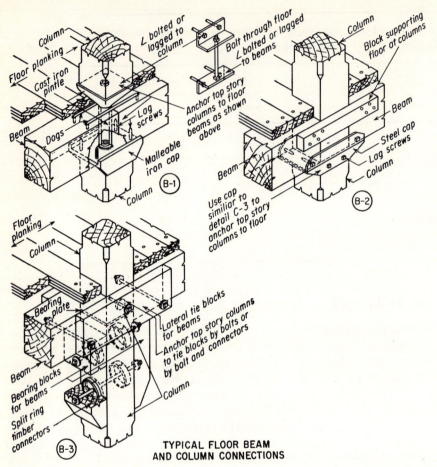

Fig. 11-6 Typical details for floor-beam and column connections. (*National Lumber Manufacturing Co.*)

12 in. The sides are chamfered or rounded to reduce the fire hazard. The bottoms of the columns rest on and are attached to the footings through fabricated metal column bases and bolts or lag screws. Column caps may be all wood, all metal, or a combination of wood and metal with bolts and timber connectors to provide for the transfer of loads and stresses from upper columns, girders, floor beams, or trusses. For multistoried buildings it is common practice to provide a joint for each column at each floor level. Figure 11-5 shows typical details for column bases. Figure 11-6 shows typical detail for floor-beam and column connections. Figure 11-7 shows typical details for roof-beam and column connections.

CARPENTRY

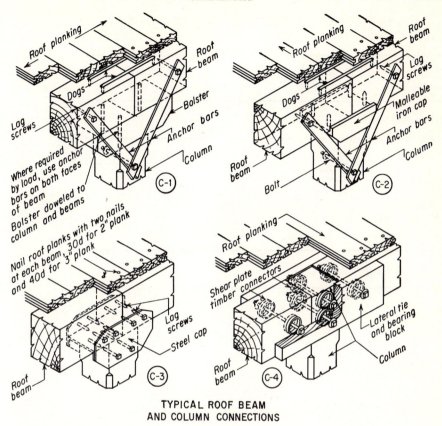

TYPICAL ROOF BEAM
AND COLUMN CONNECTIONS

Fig. 11-7 Typical details for roof-beam and column connections. (*National Lumber Manufacturing Co.*)

Beams and girders. If the floor loads are not extremely heavy and the spacings of the columns are not too great, it will be safe and economical to use beams only to support the floors and the roof. The beams are connected to and supported by the columns. If the floor loads are heavy and the column spacings are large, it will be necessary to use a beam-and-girder type of construction to support the floors and roof.

Floor. The floors may consist of tongue-and-groove planks, laid flat, or S4S planks laid on edge, designated as laminated floors. The tongue-and-groove planks are securely nailed to the floor beams. The laminated floor is constructed by laying the planks on edge, with the sides of adjacent planks brought in close contact by nails, driven horizontally, with spacings of 12 to 24 in. along each plank laid. The laminated floor may or may not be fastened

TABLE 11-26   Quantities of Lumber Required for 1 Sq Ft of
Wood Flooring

Nominal size plank, in.	Dressed size plank, in.	Percent side waste	FBM lumber per sq ft of floor*
Laminated flooring			
2 × 4	1½ × 3½	33.3	5.33
2 × 6	1½ × 5½	33.3	8.00
2 × 8	1½ × 7¼	33.3	10.67
3 × 6	2½ × 5½	20.0	7.20
3 × 8	2½ × 7¼	20.0	9.60
4 × 6	3½ × 5½	14.3	6.85
4 × 8	3½ × 7¼	14.3	9.15
Tongue-and-groove flooring			
2 × 4	1½ × 3	33.3	2.67
2 × 6	1½ × 5	20.0	2.40
2 × 8	1½ × 6¾	18.5	2.37
2 × 10	1½ × 8¾	14.3	2.29
2 × 12	1½ × 10¾	11.6	2.23
3 × 4	2½ × 3	33.3	4.00
3 × 6	2½ × 5	20.0	3.60
3 × 8	2½ × 6¾	18.5	3.55
3 × 10	2½ × 8¾	14.3	3.43
3 × 12	2½ × 10¾	11.6	3.25

* These quantities do not include any waste for end sawing lumber.

to the floor beam. End joints between the planks may be staggered over the floor beams, or they may be made between the floor beams, depending on the specifications for the structure.

Table 11-26 gives the quantities of lumber required for wood flooring, using various sizes of S4S planks and D and M flooring.

Table 11-27 gives the weight of common nails required for 100 sq ft of laminated-wood flooring or dressed and matched flooring.

Labor fabricating and erecting columns.   The labor operations required to fabricate and erect wood columns include sawing the two ends square, chamfering the four corners, drilling boltholes, and attaching bases and caps. The time will vary with the size and length of the columns and the kind of bases and caps used. For an 8-in. by 8-in. by 12-ft 0-in.-long column, using

TABLE 11-27 Quantities of Common Nails in Pounds Required for 100 Sq Ft of Wood Flooring

Laminated flooring

Size nails	Spacing of nails, in.		
	12	18	24

For 2-in.-nominal-thickness plank

16d	15	10	8
20d	24	16	12
30d	31	21	16
40d	41	28	21

For 3-in.-nominal-thickness plank

20d	15	10	8
30d	19	13	10
40d	26	18	13
60d	42	28	21

For 4-in.-nominal-thickness plank

40d	19	13	10
50d	22	15	11
60d	30	20	15

D and *M* Flooring

Nominal size flooring	Size nails	Nails per support	Spacing of supports, ft				
			4	6	8	10	12
2 × 4	20d	2	13.0	8.5	6.3	5.0	4.2
2 × 6	20d	2	7.8	5.0	3.8	3.0	2.5
2 × 8	20d	2	5.7	3.7	2.8	2.2	1.9
2 × 10	20d	3	6.7	4.4	3.3	2.6	2.2
2 × 12	20d	3	5.4	3.5	2.7	2.1	1.8
3 × 4	40d	2	22.4	14.7	10.8	8.6	7.3
3 × 6	40d	2	13.4	8.6	6.6	5.2	4.3
3 × 8	40d	2	9.8	6.4	4.9	3.8	3.3
3 × 10	40d	3	11.6	7.6	5.7	4.5	3.8
3 × 12	40d	3	9.3	6.1	4.7	3.6	3.2

power equipment for fabricating, the time should be about as follows:

Operation	Time, min
End sawing, 2 × 2 =	4
Chamfering, 40 ft × 1.2 =	48
Drilling holes, 6 × 1.7 =	10
Attaching base =	9
Attaching cap =	9
Erecting =	40
Total time =	120

It will be necessary to provide approximately one laborer for two carpenters in fabricating and erecting columns, the actual ratio depending on

TABLE 11-28   Approximate Labor-hours Required to Fabricate and Erect Timbers, Flooring, and Roofing for a Mill-type Building

Member	Size	Labor-hours per M fbm lumber	
		Carpenter	Helper
Columns	8 in. × 8 in. × 12 ft 0 in.	32	16
	10 in. × 10 in. × 12 ft 0 in.	21	10
	12 in. × 12 in. × 12 ft 0 in.	16	8
	12 in. × 12 in. × 16 ft 0 in.	13	7
Girders	8 in. × 12 in. × 12 ft 0 in.	8	8
	10 in. × 12 in. × 12 ft 0 in.	8	8
	12 in. × 16 in. × 16 ft 0 in.	6	6
	12 in. × 24 in. × 20 ft 0 in.	4	4
Floor beams and purlins	4 in. × 8 in. × 12 ft 0 in.	7	7
	4 in. × 12 in. × 18 ft 0 in.	6	6
	6 in. × 12 in. × 16 ft 0 in.	5	5
Laminated flooring	2 in. × 4 in.	8	5
	2 in. × 6 in.	7	5
	2 in. × 8 in.	6	5
	3 in. × 8 in.	5	4
	4 in. × 8 in.	4	4
	4 in. × 10 in.	4	4
D and M flooring and roofing	2 in. × 6 in.	8	4
	2 in. × 8 in.	7	4
	3 in. × 8 in.	5	3

the extent to which hand hoisting above the first floor is necessary. Thus for the column just mentioned, the total labor required for fabricating and erecting at ground level would be:

	Time, hr
Carpenter	2.1
Helper	1.0

Table 11–28 gives the approximate time required for fabricating and erecting wood columns.

Labor fabricating and erecting girders and floor beams. The labor operations required to fabricate and erect wood girders and floor beams include sawing the two ends, drilling boltholes and notching, if necessary, hoisting into position, and attaching the connectors. If an accurate method of locating the boltholes is used, it will be possible to drill all holes with power equipment prior to erecting. Otherwise some drilling will be necessary after the members are hoisted into position. Such drilling is usually done with a portable electric drill. The time will vary with the size and length of the member, the kind of connections used, and the extent of hoisting necessary. For a 10-in. by 14-in. by 12-ft 0-in.-long beam, using power equipment for fabricating, the time should be about as follows:

Operation	Time, min
End sawing, 2 × 9	18
Drilling holes, 4 × 2	8
Hoisting and connecting	40
Total time	66

It will be necessary to provide approximately one laborer for each carpenter in fabricating and erecting beams and girders above the ground floor.

Table 11-28 gives the approximate time required to fabricate and erect timbers, flooring, and roofing for a mill-type building.

Example Estimate the direct cost only for furnishing and erecting the structural members, flooring, and roofing in place for the mill building illustrated in Fig. 11-8. The structure will be erected on a concrete floor with the anchor bolts already in place. The column bases will be type A-2 of Fig. 11-5, and the column caps will be type B-2 of Fig. 11-6. All columns will be type B-2 of Fig. 11-6. All columns will be chamfered on four corners. The columns will be 10 by 10 in. The floor beams will be 10 by 14 in. The roof beams will be 10 by 10 in. The flooring will be 2 by 4 in. laminated. The roofing will be 3- by 8-in. D and M lumber, whose net width will be 6-¾ in. All lumber will be structural grade southern yellow pine.

The lumber will be fabricated with power equipment. The lumber will be hoisted into position by hand.

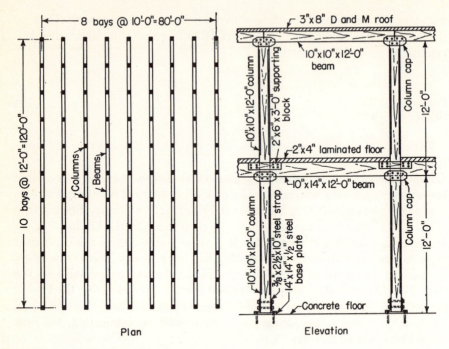

Fig. 11-8 Details for the framing for a mill building.

The cost will be

Materials f.o.b. the job:

Columns, 198, 10 in. × 10 in. × 12 ft 0 in. = 19,800 fbm @ $0.26	=	$ 5,148.00
Base plates, steel, 99, 14 × 14 × ½ in. = 2,583 lb @ $0.24	=	619.92
Steel straps, 198, ⅜ × 2½ × 10 in. = 526 lb @ $0.26	=	136.76
Shear plates. 2⅝ in., 396 @ $0.34	=	134.64
Bolts, ½ × 13 in., 198 @ $0.48	=	95.04
Steel caps, 198 @ $4.55	=	900.90
Lag screws, ½ × 4 in., 1,560 @ $0.16	=	249.60
Supporting blocks, 792 pc, 2 in. × 6 in. × 3 ft 0 in. = 2,376 fbm @ $0.20	=	475.20
Floor beams, 90, 10 in. × 14 in. × 12 ft 0 in. = 12,600 fbm @ $0.26	=	3,276.00
Roof beams, 90, 10 in. × 10 in. × 12 ft 0 in. = 9,000 fbm @ $0.26	=	2,340.00
Flooring, 9,600 sq ft × 5.33 fbm per sq ft = 51,168 fbm @ $0.20	=	10,233.60
Nails, 30d × 18-in. spacing, 9,600 sq ft × 21 lb per 100 sq ft = 2,016 lb @ $0.24	=	483.84
Roofing, 9,600 sq ft × 3.55 fbm per sq ft = 34,080 fbm @ $0.26	=	8,860.00
Nails, 9,600 sq ft × 3.8 lb per 100 sq ft 365 lb @ $0.24	=	87.60
Total cost of materials	=	$33,041.10

Labor fabricating and erecting:

Columns

Carpenter, 19,800 fbm × 21 hr per M fbm = 418 hr @ $7.67	=	$ 3,206.06
Helper, 19,800 fbm × 10 hr per M fbm = 198 hr @ $5.79	=	1,146.42

Floor beams
    Carpenter, 12,600 fbm $\times$ 8 hr per M fbm = 102 hr @ \$7.67   =   782.34
    Helpers, 12,600 fbm $\times$ 8 hr per M fbm = 102 hr @ \$5.79   =   590.58
Roof beams
    Carpenters, 9,000 fbm $\times$ 8 hr per M fbm = 72 hr @ \$7.67   =   552.24
    Helpers, 9,000 fbm $\times$ 8 hr per M fbm = 72 hr @ \$5.79   =   416.88
Flooring
    Carpenters, 51, 168 fbm $\times$ 8 hr per M fbm = 410 hr @ \$7.67   =   3,144.70
    Helpers, 51,168 fbm $\times$ 5 hr per M fbm = 256 hr @ \$5.79   =   1,482.24
Roofing
    Carpenters, 34,080 fbm $\times$ 5 hr per M fbm = 174 hr @ \$7.67   =   1,306.97
    Helpers, 34,080 fbm $\times$ 3 hr per M fbm = 102 hr @ \$5.79   =   590.58
Foreman, based on 6 carpenters, = 1,172.4 hr $\div$ 6 = 195.3 hr @
\$8.50   =   1,660.05

    Total cost of labor   = \$14,879.06
Equipment, 195 hr @ \$3.25   = \$ 632.75
Sundry tools and supplies   =   80.00

    Total cost of equipment   = \$ 712.75
Summary of costs:
    Materials   = \$33,041.10
    Labor   = 14,879.06
    Equipment   =   712.75

    Total cost   = \$48,632.91

ROOF TRUSSES

General information. Wood roof trusses are used in buildings which require a clear floor area, without columns or other interior supporting structures. Trusses have been used for spans varying from 25 ft 0 in. to 200 ft 0 in. Spacing of the trusses may vary from 12 ft 0 in. to 20 ft 0 in., with a 16-ft 0-in. spacing commonly used.

The introduction of metal timber connectors in the construction of trusses has increased the use of wood trusses.

Types of roof trusses Several types of trusses are used: bowstring, Fink, Belgian, Pratt, and Warren among the open types; and the glued laminated arches among the solid-member types. The selection of the particular type of truss for a building depends on the span, the need to economize, and the desired appearance within the building.

Trusses are supported at the two ends, which may bear upon steel columns, wood columns, or pilasters. Pratt trusses are frequently supported by end columns which are made an integral part of the trusses. The roof is supported by joists or purlins. For roofing, 1-in.-thick sheathing is generally used with joists and 2-in.-thick sheathing with purlins.

Estimating the cost of wood roof trusses The builder may completely fabricate all but the glued laminated trusses at the job; or he may purchase

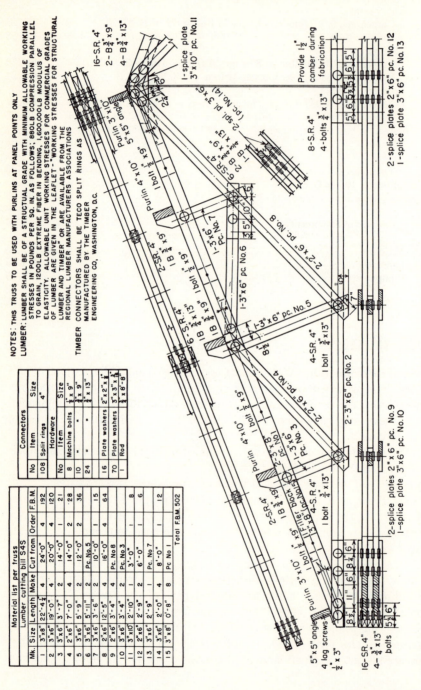

Fig. 11-9 Details of a 40-ft 0-in. Fink roof truss. (*Timber Engineering Company.*)

them completely fabricated and partly assembled; or in some instances he may purchase them completely fabricated and assembled, depending on the size of the trusses, the length of haul to the job, and transportation clearances along the haul road. Trusses are generally assembled into complete units prior to erection, regardless of where they are fabricated.

If the trusses are to be prefabricated and assembled at a shop for shipment to the job, the estimator should obtain the cost from the fabricator. The cost of transportation to the job should be added to obtain the cost at the job. To these costs should be added the cost of erecting. If the trusses are to be fabricated at the job, the estimator must determine the costs of all materials required, of fabricating and erecting equipment, and of labor fabricating, assembling, and erecting. Small trusses may be erected by hand using gin poles, but the larger trusses should be erected with power cranes.

Labor fabricating and assembling wood roof trusses   Table 11-29 gives the approximate labor-hours required to perform various operations in fabricating members for a truss. In assembling the members into a truss, considerable time will be saved if a jig or pattern is set up in a large working area, to assist in bringing the separate members into the correct position for the assembly. If all the members are prefabricated, the assembling operation will consist in placing the members in the correct positions, installing the metal timber connectors, if used, installing bolts and washers, and tightening the nuts. If the trusses are assembled at the job, this should be done as near the points of erection as possible, preferably on the floor of the building.

The labor-hours required to assemble a truss after the members are fabricated can be determined by estimating the time required to assemble each member of the truss. The sum of the separate times will give the time for the truss. Experienced estimators may determine the time on the basis of the quantity of lumber and an estimated rate of assembling, such as hours per M fbm of lumber.

Using the truss illustrated in Fig. 11-9, the approximate time required to fabricate and assemble a truss under favorable conditions should be as given in Table 11-29. The rates are based on using two carpenters and one helper with power equipment for sawing, drilling holes, and grooving for the connectors.

Table 11-30 gives the approximate labor-hours required to fabricate and assemble wood trusses. The rates given are based on using power equipment for sawing and boring holes and on performing the operations under favorable conditions.

Hoisting wood roof trusses   Wood roof trusses are usually hoisted into position with a gin pole, hand-operated derrick, or power crane, the equipment used being based on the weight, length, and number of trusses and the location of the job. For trusses whose weight does not exceed approximately 1,000 lb, a gin pole or a hand derrick may be used to lift the entire truss in a single operation. For trusses weighing approximately 1,000 to 2,500 lb, a gin pole or

TABLE 11-29   Approximate Time Required to Fabricate and Assemble Wood Members for a 40-ft Fink Truss

Mark	Size, in.	Pieces	Sawing			Drilling			Total time, hr	Total time for carpenter, hr
			No. cuts	Time per cut, min	Total time, hr	No. grooves	Time per groove, min	Total time, hr		
1	3 × 8	4	8	0.9	0.12	58	1.0	0.97	1.09
2	3 × 6	4	8	0.8	0.11	40	1.0	0.67	0.78
3	3 × 6	2	4	0.8	0.06	4	1.0	0.07	0.13
4	2 × 6	4	8	0.7	0.10	8	0.8	0.11	0.21
5	3 × 6	2	4	0.8	0.06	8	1.0	0.14	0.20
6	3 × 6	2	4	0.8	0.06	8	1.0	0.14	0.20
7	3 × 6	2	4	0.8	0.06	8	1.0	0.14	0.20
8	2 × 6	4	8	0.7	0.10	16	0.8	0.22	0.32
9	2 × 6	4	8	0.7	0.10	16	0.8	0.22	0.32
10	3 × 6	2	4	0.8	0.06	16	1.0	0.27	0.33
11	3 × 10	1	4	1.0	0.07	8	1.0	0.14	0.21
12	2 × 6	2	4	0.7	0.05	4	0.8	0.06	0.11
13	3 × 6	1	2	0.8	0.03	4	1.0	0.07	0.10
14	3 × 6	4	8	0.8	0.11	12	1.0	0.20	0.31
15	3 × 8	8	16	0.9	0.24	8	1.0	0.14	0.38
Total time to fabricate, hr								· · · ·	4.89

Mark	Pieces	Time to assemble, hr	
		Per piece	Total time
1	4	0.10	0.40
2	4	0.10	0.40
3	2	0.04	0.08
4	4	0.04	0.16
5	2	0.04	0.08
6	2	0.06	0.12
7	2	0.04	0.08
8	4	0.06	0.24
9	4	0.04	0.16
10	2	0.04	0.08
11	1	0.06	0.06
12	2	0.04	0.08
13	1	0.04	0.04
14	4	0.04	0.16
15	8	0.04	0.32

Total time to assemble, hr = 2.46
Number of carpenter-hr to fabricate and assemble = 7.35
Quantity of lumber per truss = 502 fbm

Carpenter-hr per M fbm, $\dfrac{7.35 \times 1,000}{502} = 14.7$

Helper-hr per M fbm = 7.3

TABLE 11-30 Approximate Labor-hours Required per M FBM of Lumber to Fabricate and Assemble Wood Roof Trusses

Span, ft	FBM per truss	Labor-hour	
		Carpenter	Helper
Pitched truss			
20	120	17.8	8.9
30	220	16.8	8.4
40	500	14.7	7.3
50	800	13.9	7.0
60	1,150	12.7	6.3
80	2,060	11.1	5.5
Bowstring truss with glued laminated top chord*			
50	750	12.4	6.2
80	1,450	9.7	4.9
100	2,200	9.0	4.5
120	3,130	8.5	4.3
140	4,120	8.3	4.2
160	5,380	7.8	3.9

* For nailed top chord, reduce the labor approximately 10 percent.

a hand derrick may be used to lift one end at a time. Trusses weighing more than 2,000 lb should be hoisted with a power crane, if one is available.

About 1½ hr will be required to hoist and erect a truss, by hand, if the entire truss is hoisted in one operation. If it is necessary to hoist each end separately, about 2 hr will be required to hoist and erect a truss. The gang should include a foreman, four or five laborers, and two carpenters.

In hoisting trusses with power equipment about 1 to 2 hr will be required to hoist and erect a truss, depending on the equipment, size of the trusses, and available working area.

Table 11-31 gives the approximate labor-hours required to hoist and erect wood trusses of various types and sizes. It is assumed that the trusses are prefabricated and assembled near the points where they will be hoisted. The trusses will be lifted 14 to 18 ft above the floor.

Example

Pitched Truss Estimate the total direct cost of furnishing all materials, equipment, and labor required to fabricate, assemble, and erect the wood roof truss illustrated in Fig. 11-9.

TABLE 11-31   Approximate Labor-hours Required per M FBM of
Lumber to Hoist and Erect Wood Trusses

Span	FBM per truss	Hoisting* time per truss	Labor-hr per M fbm of lumber		
			Foreman	Carpenter	Helper
Pitched truss—hand hoisting					
20	120	1.0	8.3	16.6	16.6
30	220	1.5	5.0	10.0	15.0
40	500	2.0	2.4	4.5	9.6
Pitched truss—power hoisting					
50	800	1.0	1.25	2.50	2.50
60	1,150	1.1	1.00	2.00	2.00
Bowstring truss					
80	1,450	1.3	0.90	1.80	1.80
100	2,200	1.2	0.55	1.10	1.10
120	3,130	1.4	0.45	0.90	0.90
140	4,120	1.7	0.41	0.82	0.82
160	5,380	2.0	0.37	0.74	0.74

* One power crane is used for the 50-, 60-, and 80-ft trusses and two cranes for
the 100-, 120-, 140-, and 160-ft trusses.

The project requires six trusses 16 ft 0 in. apart. Lumber will be S4S structural grade
southern yellow pine. The lumber will be fabricated at the job, using power equipment.
The trusses will be hoisted with a power crane.
    The cost per truss will be

Materials f.o.b. the job:		
Lumber, 502 fbm @ $0.26	=	$130.52
Split rings, 4 in., 108 @ $0.43	=	46.44
Machine bolts, ½ × 9 in., 8 @ $0.18	=	1.44
Machine bolts, ¾ × 9 in., 10 @ $0.38	=	3.80
Machine bolts, ¾ × 13 in., 24 @ $0.50	=	12.00
Steel washers, 2 × 2 × ⅛ in., 16 @ $0.11	=	1.76
Steel rod, ¾ in. × 8 ft 8 in., 1 @ $4.20	=	4.20
Nails, 6 lb @ $0.24	=	1.44
Total cost of materials	=	$201.60
Equipment:		
Power crane, with operators, rented, ¾ hr @ $22.00 per hr	=	$ 16.50
Power saws, 2 hr @ $0.75	=	1.50

Power drill, 2 hr @ $0.65	=	1.30
Sundry equipment and tools	=	1.80
Total cost of equipment	= $ 21.10	

Labor:
  Fabricating and assembling:

Carpenters, 502 fbm × 14.7 hr per M fbm = 7.5 hr @ 7.67	=	$ 57.52
Helpers, 502 fbm × 7.3 hr per M fbm = 3.7 hr @ $5.79	=	21.43
Foreman, based on 4 carpenters, 1.9 hr @ $8.50	=	16.15

  Erecting:

Carpenters, 4 × ¾ hr per truss = 3 hr @ $7.67	=	23.01
Helpers, 2 × ¾ hr per truss = 1.5 hr @ 5.79	=	8.69
Foreman, based on 4 carpenters, 0.75 hr @ $8.50	=	6.37
Total cost of labor	= $133.17	

Summary of costs:

Materials	=	$201.60
Equipment	=	21.10
Labor	=	$133.17
Total cost per truss	= $355.87	

Example

Bowstring Truss Estimate the total direct cost of furnishing all materials, equipment, and labor required to fabricate, assemble, and erect the bowstring truss illustrated in Fig. 11-10. The project requires seven trusses spaced 15 ft 6 in. apart. The lumber will be *S4S* structural grade southern yellow pine. The trusses will be fabricated and assembled at a shop and then transported to the job and hoisted into position using a truck-mounted power crane.

The cost per truss will be

Materials f.o.b. the job:

Lumber, 748 fbm @ $0.26	=	$194.48
Split rings, 2½ in., 112 @ $0.27	=	30.24
Shear plates, 4 in., 16 @ $0.72	=	11.52
Bolts, ¾ × 13 in., 8 @ $0.50	=	4.00
Bolts, ½ × 17 in., 16 @ $0.30	=	4.80
Bolts, ½ × 14 in., 4 @ $0.26	=	1.04
Bolts, ½ × 13 in., 26 @ $0.24	=	6.24
Bolts, ½ × 11 in., 4 @ $0.22	=	0.88
Bolts, ½ × 9 in., 18 @ $0.19	=	3.24
Washers, 2½ in., 136 @ $0.035	=	4.76
Steel straps, 2 @ $22.70	=	45.40
Glue, 90 lb liquid mix per 1,000 sq ft of glued area = 13 lb @ 0.65	=	8.45
Total cost of materials	= $315.05	

Equipment:

Power crane, with operators, rented, 1.25 hr per truss @ $26.00	=	$ 32.50
Power saws, 5 hr @ $0.70	=	3.50
Power planer, 3 hr @ $0.85	=	2.55
Power drill, 4 hr @ $0.70	=	2.80

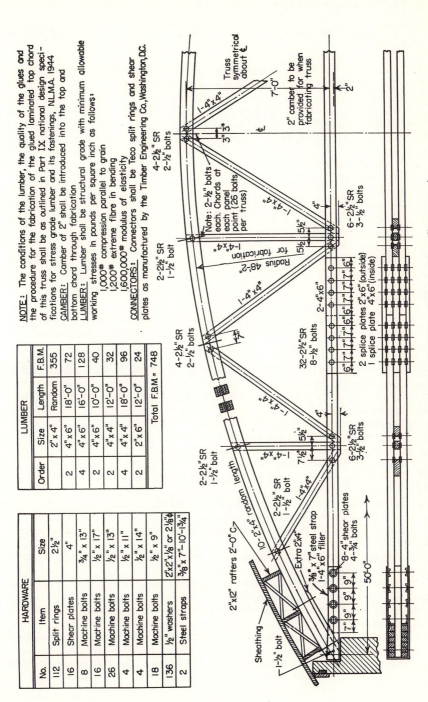

NOTE: The conditions of the lumber, the quality of the glues and the procedure for the fabrication of the glued laminated top chord of this truss shall be as outlined in Part IX national design specifications for stress grade lumber and its fastenings, NLMA. 1944

CAMBER: Camber of 2" shall be introduced into the top and bottom chord through fabrication

LUMBER: Lumber shall be structural grade with minimum allowable working stresses in pounds per square inch as follows:[1]

    1,000# compression parallel to grain
    1,200# extreme fibre in bending
    1,600,000# modulus of elasticity

CONNECTORS: Connectors shall be Teco split rings and shear plates as manufactured by the Timber Engineering Co., Washington, D.C.

HARDWARE

No.	Item	Size
112	Split rings	2½"
16	Shear plates	4"
8	Machine bolts	¾" x 13"
16	Machine bolts	½" x 17"
26	Machine bolts	½" x 13"
4	Machine bolts	½" x 11"
4	Machine bolts	½" x 14"
18	Machine bolts	½" x 9"
136	½" washers	2½"x²⅛" or 2⅛"⌀
2	Steel straps	⅜" x 7"—10'—1¾"

LUMBER

Order	Size	Length	F.B.M.
	2" x 4"	Random	355
2	4" x 6"	18'-0"	72
4	4" x 6"	16'-0"	128
2	4" x 6"	10'-0"	40
2	4" x 4"	12'-0"	32
4	4" x 4"	18'-0"	96
2	2" x 6"	12'-0"	24

Total F.B.M. = 748

Fig. 11-10 Details of a 50-ft 0-in. bowstring roof truss. (*Timber Engineering Company.*)

Jigs and clamps		=	4.50
Sundry equipment and tools		=	1.80
Total cost of equipment		=	$ 47.65

Labor:

Fabricating and assembling:

Carpenters, 748 fbm × 12.4 hr per M fbm = 9.3 hr @ $7.67		=	$ 71.33
Helpers, 748 fbm × 6.2 hr per M fbm = 4.7 hr @ $5.79		=	27.21

Erecting:

Carpenters, 2 men × 1.25 hr per truss = 2.5 hr @ $7.67		=	19.18
Helpers, 2 men × 1.25 hr per truss = 2.5 hr @ $5.79		=	14.48
Foreman, fabricating and erecting, 2.30 + 1.25 = 3.55 hr @ $8.50		=	30.18
Total cost of labor		=	$161.98

Summary of costs:

Materials	=	$315.05
Equipment	=	47.65
Labor	=	161.98
Transporting to project	=	18.50
Total cost per truss	=	$543.18

PROBLEMS

For all of these problems use the national average wage rates for labor, as given in the text. Assume that *S4S* lumber will cost $200.00 per M fbm at the job and that common nails will cost $0.25 per lb, including an allowance for waste. Omit the cost of a foreman.

11-1 Estimate the total direct cost and the cost per M fbm for furnishing and installing 96 floor joists size 2 in. by 8 in. by 17 ft 4 in., with the joists to rest on wood sills at each end of the joist and at the midpoint of the joist. Each joist will be fastened to the sills with twelve 16d common nails.

Both ends of each joist must be sawed square.

11-2 Estimate the total direct cost and the cost per M fbm for furnishing and installing diagonal subflooring for a floor area of 2,240 sq ft. Use 1 by 6-in. *D* and *M* lumber.

The joists will be spaced 16 in. apart.

11-3 Estimate the cost of Prob. 11-2 if the subflooring is installed perpendicular to the joist.

11-4 The total length of the exterior and interior walls of a residence is 464 lin ft. Studs, size 2 in. by 4 in. by 8 ft 0 in. long, will be spaced not over 16 in. on centers for all walls and partitions. There will be eight outside corners and 12 inside corners, each requiring three studs.

The studs will rest on a 2- by 4-in. plate and will be capped at the tops with two 2- by 4-in. plates for the full length of the walls.

Neglecting the door and window openings, prepare a bill of materials for the studs and the bottom and top plates, showing the number of pieces of each length required.

Estimate the total direct cost and the cost per M fbm for furnishing and installing the studs and plates.

11-5 Estimate the total direct cost per M fbm of lumber for furnishing and installing 2- by 8-in. rafters, spaced 24 in. on centers for a rectangular building whose outside dimensions are 28 ft 6 in. wide and 74 ft 8 in. long. The roof will be a gable type, with no dormers and with ¼ pitch. The rafters are to extend 18 in., measured horizontally, outside the outer surfaces of the walls.

The rafters will be sawed as illustrated in Fig. 11-3.

11-6 Estimate the total direct cost per M fbm of lumber for furnishing and installing diagonal sheathing on the outside walls of a building. The walls to be covered will be 8 ft 6 in. high and 324 ft 8 in. long. Use 1- by 8-in. *S4S* sheathing.

The studs will be spaced 16 in. on centers.

11-7 Estimate the total direct cost of furnishing and installing 2,480 sq ft, net area, 1- by 6-in. *D* and *M* siding, using first-grade workmanship. The siding will be installed horizontally on studs spaced 16 in. on centers.

The cost of materials will be

Lumber, $380.00 per M fbm
Nails, $0.30 per lb

11-8 Estimate the total direct cost and the cost per net 100 sq ft of floor area for furnishing and laying 1- by 3-in. *D* and *M* hardwood flooring on a wood subfloor.

Use first-grade workmanship.

The cost of materials will be

Flooring, $380.00 per M fbm
Nails, $0.32 per lb

11-9 Estimate the total direct cost and the cost per square for furnishing and laying wood shingles for a roof whose area will be 3,620 sq ft, consisting of irregular hips and gables. Use 18-in.-long shingles, in random widths, with 5½ in. exposed to the weather.

The roof, which is for a one-story home, will have a ¼ pitch.

The cost of the materials will be

Shingles, $39.00 per square
Nails, $0.30 per lb

12
Interior Finish, Millwork, and Wallboard

Interior finish and millwork include such items as base, molding, chair rails, cornices, panel strips, window and door frames, windows and doors, casings, trim, wood cabinets, etc. The cost of these items will vary considerably with the kind and grade of materials and the grade of workmanship required.

In the tables giving the labor-hours required to install finish and millwork, the lower values are for ordinary workmanship and the higher values are for first-grade workmanship. Because of the more rigid tolerances when first-grade workmanship is specified, the time required may be 25 to 50 percent greater than for ordinary workmanship.

Cost of interior finish and millwork The most dependable method of estimating the cost of interior finish and millwork is to list the quantity and cost of each item separately, together with the quantity and cost of the labor required to install each item. Finish material, such as molding, base, chair rail, and strip paneling, may be priced by the 100 lin ft, while window and door frames, sash, doors, cabinets, etc., usually are priced by the unit. The time required to install these items should be estimated by the 100 lin ft or by the unit, to conform with the method of pricing the materials.

Labor required to install interior finish When molding, base, chair rails, etc., are installed in rooms with wood walls, they are nailed directly to the walls. However, if they are installed in rooms with plastered walls, they should be nailed to grounds set in and flush with the surface of the plaster.

TABLE 12-1 Labor-hours Required to Install 100 Lin Ft of Interior Wood Finish

Item	Labor-hours	
	Carpenter	Helper
Picture molding	3.0–4.5	0.5–0.8
Chair rail	3.0–4.5	0.5–0.8
Base, 1 member	3.0–4.5	0.5–0.8
Base, 2 member	5.5–7.5	0.8–1.0
Base, 3 member	7.0–9.0	0.9–1.1
Panel strips	3.5–4.5	0.5–0.8

Table 12-1 gives the labor-hours required to install various types of interior finish. Use the lower rates for ordinary workmanship and the higher rates for first-grade workmanship.

Wood window frames Prefabricated window frames may be shipped from the mill unassembled, or they may be completely assembled with the exception of the interior casing. For the former condition, it is necessary to assemble the frames at the job prior to setting them in the openings.

Setting window frames involves carrying them to the openings, installing, plumbing, and bracing them in position. The time required will vary with the height to which they must be hoisted and the sizes of the frames, such as for single, double, or triple windows.

The interior trim for window frames usually is installed after the plaster walls are completed.

TABLE 12-2 Labor-hours Required to Install a Wood Window Frame

Operation	Labor-hours per unit	
	Carpenter	Helper
Assembling frame, single opening	1.0–1.5	0.2–0.3
Assembling frame, double opening	1.5–2.0	0.2–0.3
Assembling frame, triple opening	2.0–2.5	0.3–0.4
Setting frame, single opening	0.8–1.0	0.2–0.3
Setting frame, double opening	0.9–1.1	0.3–0.4
Setting frame, triple opening	1.0–1.2	0.4–0.5
Installing interior trim	1.0–2.0	0.2–0.4

TABLE 12-3 Labor-hours Required to Fit and Hang a Pair of Wood Sash Windows

Operation	Labor-hours	
	Carpenter	Helper
Fitting double-hung sash	0.5–0.8	0.1–0.2
Hanging double-hung sash	0.6–0.8	0.1–0.2
Fitting casement sash	0.6–1.0	0.1–0.2
Hanging casement sash	0.8–1.2	0.2–0.3

Table 12-2 gives the labor-hours required to perform various operations in installing wood window frames and the trim, for ordinary and first-grade workmanship.

Fitting and hanging wood sash Window sash most commonly used are of two types, double hung, which are raised and lowered vertically, and casement, which open inward or outward. Double-hung sash are fitted with sash cord, or chain and weights, or counterbalances. Some fitting and finishing is required before a sash can be installed.

Table 12-3 gives the labor-hours required to perform the necessary operations in fitting and hanging wood sash windows. Use the lower rates for ordinary workmanship and the higher rates for first-grade workmanship.

Wood door frames and jambs The time required to set door frames and jambs will vary with the size of the opening, the type of frame or jamb set,

TABLE 12-4 Labor-hours Required to Set a Wood Door Frame Jambs, and Trim

Operation	Maximum size of opening, ft	Labor-hours	
		Carpenter	Helper
Set frame for exterior door	3 × 7	1.0–1.5	0.2–0.3
	6 × 7	1.5–2.0	0.3–0.4
	8 × 8	1.5–2.0	0.3–0.4
Set frame for interior door	3 × 7	0.5–1.0	0.1–0.2
	6 × 7	1.5–2.0	0.3–0.4
Install interior trim, two sides:			
Single casing	3 × 7	1.0–2.0	0.2–0.4
	6 × 7	1.5–2.5	0.3–0.5
Two-member trim	3 × 7	1.5–2.5	0.3–0.5
	6 × 7	2.0–3.0	0.4–0.6

TABLE 12-5   Labor-hours Required to Set the Frame or Jambs, Install Stops and Hardware, and Fit and Hang a Door

Type of door and trim	Maximum size of opening, ft	Labor-hours	
		Carpenter	Helper
Exterior, single casing	3 × 7	3.0–6.0	0.6–1.0
	6 × 7	5.0–10.0	1.0–2.0
Exterior, two-member trim	3 × 7	3.5–6.5	0.7–1.3
	6 × 7	6.0–10.5	1.2–2.1
Interior, single casing	3 × 7	2.5–5.0	0.5–1.0
	6 × 7	4.5–9.0	0.9–1.8
Interior, two-member trim	3 × 7	3.0–6.0	0.6–1.2
	6 × 7	5.5–10.0	1.1–2.0
Sliding doors for closets, install track, two doors	5 × 8	6.0–8.0	1.2–1.6
Screen doors	3 × 7	1.0–1.5	0.2–0.3
	6 × 7	1.5–2.5	0.3–0.5

and the grade of workmanship. The units, which are furnished by a mill, may require some fitting before they are set, plumbed, and fastened in position. The trim is installed after the walls are finished.

Table 12-4 gives the labor-hours required to set door frames, jambs, and trim. Use the lower rates for ordinary workmanship and the higher rates for first-grade workmanship.

Fitting and hanging wood doors   Exterior doors are usually thicker and heavier than interior doors and require more time to fit and hang than interior doors. The operations include fitting the door to the opening in the frame, routing the jamb and door for butts, mortising the door for the lock, installing the butts and lock, and hanging the door. Fitting, routing, and mortising may be done with hand tools or with power tools. Production rates are considerably higher when power tools are used.

Table 12-5 gives the labor-hours required to set the frame or jambs, install the stops, casing, butts, and lock, and fit and hang one door. The rates are based on using hand tools. If power tools are used to fit, route, and mortise, reduce the rates by about 10 percent. Use the lower rates for ordinary workmanship and the higher rates for first-grade workmanship.

WALLBOARDS

Wallboards include such items as gypsum boards, fiberboards, pressed wood, plywood, asbestos, and cement boards, etc., which are sold under a variety of trade names. They are used for walls, ceilings, and wainscotings in many buildings, especially houses. They may be installed on studs, joists, furring strips, plaster, or masonry, using nails or cement to secure them in place.

The cost of wallboards in place will include the cost of materials and labor. The cost of materials can be obtained readily from the manufacturer or jobber. The cost of labor installing wallboards will vary considerably with the size, shape, complexity of the area to be covered. Where areas are large and plain, the waste of materials will be low, and labor can install them rapidly, but where areas are small or irregular, with many openings, which require considerable cutting and fitting, the waste may be large and the cost of labor will be higher. The area to be covered should be carefully examined prior to preparing an estimate.

Gypsum wallboards Gypsum wallboards, which are made by molding gypsum between two sheets of paper, are used on walls and ceilings in many houses and other buildings. They are manufactured in sheets having the dimensions given in the accompanying table.

Thickness, in.	Width, in.	Length, ft	Approx weight per 100 sq ft, lb
¼	48	6–12	270
⅜	48	6–12	400
½	48	6–12	530
⅝	48	6–12	660

Installing gypsum wallboards Where gypsum wallboards are installed on wood studs and ceiling joists, the studs and joists should be spaced 12, 16, or 24 in. on centers as these dimensions are divisible into the 48-in. width. For walls, the boards are usually installed in vertical positions, with the length equal to the height of the wall, up to 12 ft. For ceilings, they may be installed

TABLE 12-6 Quantities of Nails Required per 1,000 Sq Ft of Gypsum Wallboard

Thickness, in.	Size nail	Quantity of nails, lb		
		Spacing of studs or joists		
		12 in.	16 in.	24 in.
¼	4d	5	4	3
⅜	4d	5	4	3
½	5d	8	6	4
⅝	6d	9	7	5

TABLE 12-7 Labor-hours Required to Install Gypsum Wallboard and Perforated Tape

Operation	Unit	Labor-hours		
		Spacing of studs or joists		
		12 in.	16 in.	24 in.
Install wallboard:				
Large areas	100 sq ft	1.7	1.5	1.2
Medium-size rooms	100 sq ft	2.8	2.5	2.0
Small rooms	100 sq ft	4.0	3.5	3.0
Install perforated tape and sand the surface	100 lin ft	4.0	4.0	4.0

perpendicular to or parallel with the joists. Some specifications require the use of nailing strips, such as 1- by 4-in. planks, installed perpendicular to the ceiling joists on 12-in. centers, to which the boards are fastened with nails.

The boards are fastened with plasterboard nails, 4d, 5d, or 6d, with $\frac{5}{16}$- or $\frac{3}{8}$-in. heads, depending on the thickness of the board. Recommended nail spacings are 8 in. for wall and 7 in. for ceiling installation. Table 12-6 gives recommended sizes of nails and the approximate quantities required per 1,000 sq ft of wallboard.

The joints between adjacent boards are filled with a special cement, which is spread with a putty knife to cover a strip about $1\frac{1}{2}$ in. wide on each side of the joint. After the cement is applied, the joint is covered with a perforated fiber tape about 2 in. wide, and another layer of cement is applied and spread over the tape. After the cement dries, it is sanded to a uniformly smooth surface, which should eliminate all evidence of a joint, especially after the surface is painted. The nailheads are likewise covered with the cement and sanded flush with the surface of the wallboard.

Labor installing gypsum wallboard The labor required to install gypsum wallboard will vary considerably with the size and complexity of the area to be covered. For large wall or ceiling areas, which require little or no cutting and fitting of the boards, two carpenters should install a board 48 in. wide and 8 to 9 ft long in about 10 to 15 min. This is equivalent to 1 to $1\frac{1}{2}$ man-hours per 100 sq ft. However, where an area contains numerous openings, it is necessary to mark the boards and cut them to fit the openings. For such areas the labor required may be as high as 3 to 4 man-hours per 100 sq ft. Cutting is done by scoring one side of a board with a curved knife, which is drawn along a straightedge. A slight bending force will break the board along the scored line.

Table 12-7 gives representative labor-hours required to install gypsum wallboard and perforated-tape joint system.

PROBLEMS

For these problems use the national average wage rates for labor, as given in this book. Assume that the cost of nails will be $0.30 per lb.

12-1 Estimate the total direct cost and the cost per unit for furnishing, assemblying, and setting 24 double-opening wood window frames in wood walls, using first-grade workmanship.

The sizes of the frames will be 2 ft 8 in. wide and 5 ft 2 in. high per opening.

The cost of the frames per double opening will be $25.60.

12-2 Estimate the total direct cost per pair of sashes for furnishing, fitting, and hanging wood sash for the window frames of Prob. 12-1, using first-grade workmanship.

The cost of the glazed sash will be $11.20 per opening.

12-3 Estimate the direct cost per opening for furnishing and setting an interior wood door frame and installing, single-casing trim on two sides for doors whose dimensions are 2 ft 8 in. wide by 6 ft 8 in. high. The frames will be furnished to the job with stops installed.

The cost of materials per unit, including the casing, will be $14.40.

12-4 Estimate the direct cost per opening for furnishing and setting a wood frame, installing stops, and fitting and hanging an inside door, with two-member trim, using first-grade workmanship. The door size will be 2 ft 8 in. wide by 6 ft. 8 in. high.

The cost of materials will be as follows:

Door frame, stop, and casing, $15.60

Door, good-grade birch, $18.75

Hardware, butts, and lock, $12.65

13
Lathing and Plastering

This chapter discusses methods of estimating the cost of materials and labor for placing lath and plaster, which are frequently used for finished surfaces for walls, ceilings, etc., in buildings.

Several types of materials are used for lath, as discussed under lathing.

Various materials are used for plaster, as discussed under plastering.

Estimating the cost of lathing and plastering The cost of lathing and plastering may be estimated by the square yard, square foot, or for some classes of work by the linear foot.

Within the lathing and plastering trades, the policy used to determine the area covered varies considerably. Some estimators do not deduct the areas of openings such as doors, windows, etc., when determining the area for cost purposes. Some estimators deduct one-half the areas of all openings larger than a specified dimension, such as 22 sq ft, but make no deductions for smaller areas. Still other estimators deduct the areas of all openings, regardless of size, to give the net area of the surface covered. When specifying a cost per unit area, the method of arriving at the total area used for payment purposes should be clearly stated.

When payment is made by the linear foot, it is also necessary to state clearly the basis on which the number of units will be determined.

Grades of workmanship Two grades of workmanship are commonly used in specifying the quality of work required, namely, ordinary and first grade.

The cost of the work will be influenced considerably by the grade of workmanship specified.

Ordinary workmanship permits surfaces, grounds, corners, etc., to vary from true vertical or plane positions by as much as $\frac{1}{8}$ to $\frac{3}{16}$ in.

First-grade workmanship requires that grounds, corners, etc., be in true alignment, both horizontally and vertically, and that areas of the finished surfaces shall not vary from a true plane by more than $\frac{1}{32}$ to $\frac{1}{16}$ in.

LATHING

Types of lath Plaster may be applied to masonry surfaces or to several types of lath, classified as follows:

1 Masonry
   a Brick, stone, concrete, tile, and concrete and gypsum blocks
2 Wood lath
3 Metal lath
   a Flat, self-furring, or ribbed lath
   b Exposed or sheet metal
   c Flat perforated sheet lath
   d Wire lath, plain or stiffened
4 Gypsum and fiber lath and board

Wood lath Dimensions of wood lath commonly used are $\frac{1}{4}$ or $\frac{3}{8}$ in. thick, 1 or $1\frac{1}{2}$ in. wide, and 32 or 48 in. long, the 48-in. length being more popular. They are nailed to wood studs, joists, or furring strips, usually spaced 12 or 16 in. on centers, with $\frac{1}{4}$- or $\frac{3}{8}$-in.-wide spaces between them to permit the forming of plaster keys for anchorage.

Wood lath should be thoroughly saturated with water prior to applying the plaster in order to prevent subsequent swelling, which will crack the plaster.

Table 13-1 gives quantities of materials required for installing wood lath. If the net surface area is used, with the area of all openings deducted, 5 to 10 percent should be added to the quantity of lath for waste. The lath, which are packed in bundles of 100, are sold by the 1,000 pieces.

Labor applying wood lath Wood lath may be applied by experienced lathers or by carpenters. If carpenters are used, the time required will be about 25 percent more than when lathers are used.

Table 13-2 gives representative rates for labor applying wood lath.

Metal lath Metal lath has substantially replaced wood lath as a base for plaster. Table 13-3 gives the properties of the more popular types of metal lath.

Flat expanded and rib metal lath are available in sheets 24 and 27 in. wide and 96 in. long, whose areas are 1.78 and 2.0 sq yd, respectively. The 24-in.-wide sheets are packed in bundles of nine sheets, containing 16 sq yd,

TABLE 13-1   Quantities of Materials Required per 100 Sq Yd of Net Area for Wood Lath

Size of lath, in.	Nail spacings, in.	Number of lath required	Nails required, lb	
			Per 1,000 lath	Per 100 sq yd
1 × 32	16	2,600	7	182
1 × 48	12	1,950	13	253
1 × 48	16	1,950	10	195
1½ × 48	12	1,450	13	189
1½ × 48	16	1,450	10	145

TABLE 13-2   Labor-hours Required by Experienced Lathers to Apply Wood Lath

Size of lath, in.	Nails spacings, in.	Labor-hours required	
		Per 1,000 lath	Per 100 sq yd
1 × 32	16	4.0–5.0	10.5–13.0
1 × 48	12	5.0–6.0	10.0–12.0
1 × 48	16	4.5–5.5	9.0–11.0
1½ × 48	12	5.0–6.0	7.3– 8.7
1½ × 48	16	4.5–5.5	6.5– 8.0

TABLE 13-3   Properties of Metal Lath

Type of lath	Weight, lb per sq yd	Maximum spacing of supports, in.	
		Vertical	Horizontal
Flat expanded metal lath	2.5	16	0
	3.4	16	16
⅛-in.-rib metal lath	2.75	16	16
	3.4	24	19
⅜-in.-rib metal lath	3.4	24	24
	4.0	24	24
¾-in.-rib metal lath	5.4	30	30
Sheet-metal lath	4.5	24	24
Stiffened wire lath	3.3	24	19

and the 27-in.-wide sheets are packed in bundles of ten sheets, containing 20 sq yd.

Applying metal lath on wood studs, joists, and furring strips Metal lath is fastened to wood supports with staples, large-head roofing nails, or common nails which are driven part way into the wood, then bent over the lath, spaced not more than 8 to 9 in. apart. The sheets should be applied with the lengths perpendicular to the supports.

Side laps should be 1 in., with two strands of No. 18 gauge soft wire used to tie them together at spacings of about 6 in. End laps should be at least $1\frac{1}{2}$ in., with wire used to lace the adjacent sheets together at the lap. About 6 to 10 percent should be added for laps and waste on flat areas.

Applying metal lath on steel studs for partition walls Where plastered partition walls, with total thicknesses of about $2\frac{1}{2}$ in., are to be constructed, metal lath is applied on both sides of lightweight $\frac{3}{4}$-in.-steel channels, spaced 12 to 16 in. on centers. These channels, which serve as studs, are anchored to the floor and ceiling, using floor and ceiling grounds for anchorage.

Metal lath is applied on each side of the channels, using wire at about 6-in. spacings to fasten the lath to the channels. The sides and ends of the lath are lapped 1 and $1\frac{1}{2}$ in., respectively, and fastened with wire ties spaced about 6 in. apart.

If the channels are cut a full 9 ft 0 in. long and spaced 16 in. on centers for a wall 9 ft 0 in. high and 100 ft 0 in. long, whose total area is 100 sq yd,

$$\frac{100 \times 12}{16} = 75$$ channels will be required. The total length of the channels will

be $75 \times 9 = 675$ lin ft. Table 13-4 gives the quantities of materials required for 100 sq yd of wall area, considering one side of the wall only. If the opposite

TABLE 13-4 Quantities of Materials Required for 100 Sq Yd of Metal-lath Partition, Considering One Side of Wall Only

Item	Channel spacing, in.	Quantity
Channels	12	900 lin ft*
	16	675 lin ft*
Metal lath	..	106 sq yd
Tie wire	..	10 lb

* These quantities include no allowance for waste.

TABLE 13-5   Sizes and Maximum Spacings of Runners and Furring for Suspended Ceilings

Size and kind of runners	Max spacing of hanger wires, in.
1½-in. channel	48
2-in. channel	60

| Runners, max spacing, in. | Furring | |
	Size and kind	Max spacing, in.
24	¾-in. channel	24*
30	¾-in. channel	24*
36	¾-in. channel	24
42	¾-in. channel	19
48	¾-in. channel	16
54	1-in. channel	19
60	1-in. channel	12

* These spacings may be increased if a metal lath is used that will permit greater spacing.

side of the wall is plastered, only metal lath and wire need be added, since the single row of channels will serve as studs for both sides.

Applying metal lath for suspended ceilings   Where metal lath is used in constructing suspended ceilings, it is necessary to support the lath below the ceiling. A commonly used method of supporting the lath is to install hanger

TABLE 13-6   Quantities of Materials Required for 100 Sq Yd of Suspended Metal Lath Ceiling (Includes Waste)

Item	Quantity
Metal lath, ⅛-in. rib, 2.75 lb	108 sq yd
Furring, ¾-in. channels	710 lin ft
Runners, 1½-in. channels	240 lin ft
Hangers, 8 gauge wires	60 each
Tie wire, 16 gauge	12 lb

wires in the concrete ceiling at spacings of 3 to 5 ft each way. These wires support steel runners, such as 1½- or 2-in. channels, spaced 3 to 5 ft apart. Cross furring, consisting of ¾- to 1-in.-steel channels, spaced 12 to 16 in. on centers, are placed under and wired to the main runners to support lath.

Table 13-6 gives the quantities of materials required for 100 sq yd of suspended ceiling, using ¾-in.-thick plaster, whose weight will be about 75 lb

---

TABLE 13-7   Labor-hours Required to Apply Metal Lath

| | | Labor-hours per unit | |
| | | Spacing of supports | |
Class of work	Unit	12 in.	16 in.
On wood furring			
	Sq yd		
Walls and partitions	100	9–12	7–10
Flat ceilings	100	9–13	7–11
Beams and girders	100	22–32	18–25
Columns	100	22–30	20–28
Simple coves and cornices	100	20–25	15–20
Arch and groined ceilings	100	15–20	12–17
On metal furring			
	Sq yd		
Partitions, including furring, lath one side only	100	25–30	20–25
Flat suspended ceilings, including furring and runners	100	30–35	25–30
Beams and girders	100	25–32	20–25
Columns	100	25–32	20–25
Simple coves and cornices	100	25–30	20–25
Arch and groined ceilings	100	16–24	12–18
Accessories			
	Lin ft		
Corner bead, molding, base	100	3–4	
Corner reinforcing	100	2–3	

per sq yd. Hanger wires will be spaced 48 in. each way. Use 1½-in. channels for main runners and ¾-in. channels, spaced 16 in. apart, for furring.

The weight on one hanger wire will be

    Plaster, 1.78 sq yd × 75 lb   = 134 lb
    Lath, 1.78 sq yd × 2.75 lb   =   5 lb
    Furring, 12 lin ft × 0.3 lb   =   4 lb
    Runners, 4 lin ft × 1.32 lb   =   5 lb
                                    ─────
    Total weight   = 148 lb

Use 8 gauge wire for hangers.

Labor applying metal lath   The labor required to apply metal lath will vary considerably with the size and complexity of the areas covered, the type of lath used, the type of base to support the lath, and the method of fastening. When an area contains openings, it usually is necessary to cut the lath to fit around the openings, which requires additional time.

Sufficient labor should be included to cover the installation of corner beads, moldings, bases, reinforcing for inside corners, etc., if such items are required.

Wood grounds are usually installed by carpenters and need not be considered here.

Table 13-7 gives the labor-hours required to apply metal lath under different conditions.

Example   Estimate the cost of applying metal lath, furring, and runners for 100 sq yd of suspended ceiling in rooms whose average dimensions are 16 ft 0 in. by 18 ft 0 in. The hanger wires are already in place.

The following materials will be used:

Runners, 1½-in. channels @ 48 in.

Furring, ¾-in. channels @ 16 in.

Metal lath, ⅛-in.-rib, 2.75 lb per sq yd

Tie wire, 16 gauge

Use the information in Table 13-6 for quantities of materials required. The cost will be

    Runners, 240 lin ft @ $0.11   = $ 26.40
    Furring, 710 lin ft @ $0.085   =   60.35
    Metal lath, 108 sq yd @ $0.88   =   95.04
    Tie wire, 12 lb @ $0.46   =   5.52
    Lather, 28 hr @ $7.25   =   203.00
    Scaffolds and tools   =   3.00
       Total cost   = $393.31

PLASTERING

Plaster is applied to masonry, wood lath, metal lath, and plasterboard inside a building to provide walls, ceilings, ornamental surfaces, etc. When it is applied on the outside of a building it is called stucco. Several materials are used to make plaster and stucco, as more fully described hereafter.

The cost of plaster is usually estimated by the square yard, except for certain trim and novelty uses, whose cost may be estimated by the linear foot or square foot. There is no uniform method of determining the area of a surface to be plastered. Some contractors and estimators make no deductions for the areas of openings, while others deduct areas larger than specified sizes, and still others deduct the areas of all openings. An estimate should clearly state what method is used; otherwise there may be a misunderstanding when computing the quantity for payment purposes. Net areas are used in this book.

The labor required to apply plaster will vary considerably with the class of work specified. Two classes are usually recognized, ordinary workmanship and first-grade workmanship. Ordinary workmanship permits surface variations from a true plane by as much as $\frac{1}{8}$ to $\frac{3}{16}$ in., while first-grade workmanship permits variations not greater than $\frac{1}{32}$ to $\frac{1}{16}$ in. The labor required for first-grade workmanship may be 30 to 50 percent more than for ordinary workmanship.

Number of coats applied Plaster may be applied in one, two, or three coats, the number depending on the base against which it is applied, the desired thickness, and the requirements of the specifications.

Scratch coat When plaster is applied to wood and metal lath, it is common practice to apply three coats. The first coat is called the scratch coat. It should be thick enough to flow through all joints and openings in the lath and form keys which will securely bond it to the lath. A thin covering of the lath is necessary in order that the surface may be scratched prior to final hardening to provide a good bond with the second coat. The total thickness should be $\frac{1}{8}$ to $\frac{1}{4}$ in., with about $\frac{1}{8}$ in. coverage over the lath.

Brown coat The second coat, called the brown coat, is applied to the scratch coat after the latter has dried. The brown coat contributes the greatest thickness to a plastered surface, usually being $\frac{3}{8}$ to $\frac{1}{2}$ in. thick. Since this coat will determine the quality of workmanship for the job, it is necessary to finish it as carefully as the specifications require. Screeds, which span across the grounds, may be used to enable a plasterer to bring the surface to the desired smoothness. The surface is wood floated to provide a good bond with the finish coat.

Finish coat This coat, which is also called the white coat, is the final coat. It is usually $\frac{1}{16}$ to $\frac{1}{8}$ in. thick. After it is applied, it may be finished with a steel trowel to produce a hard smooth surface, or it may be finished with a sanded surface.

Total thickness of plaster The total thickness of plaster may vary from ⅜ to 1 in., depending on the base against which it is applied.

TABLE 13-8 Quantities of Materials Required for 1 Cu Yd of Plaster (No Waste Included)

Kind of plaster	Proportions by volume	Quantity Cementing material, sacks*	Sand, cu yd
Scratch or brown coat			
Hydrated lime†	1:2	13.5	1
	1:3	9.0	1
Gypsum	1:2	13.5	1
	1:3	9.0	1
Finish coat			
Gypsum	1:0	27.0	0
Exterior stucco			
Hydrated lime	1:2	13.5	1
	1:3	9.0	1
Portland cement‡	1:2	13.5	1
	1:3	9.0	1

* Sacks of hydrated lime and gypsum should contain 1 cu ft each, weighing 50 and 100 lb, respectively.

† If quicklime putty is used instead of hydrated lime, the following weight ratios should be used, compared to hydrated lime:

For lump quicklime   = 0.65
For pulverized quicklime = 0.57

‡ For portland-cement stucco, add 8 lb or ⅙ cu ft of hydrated lime for each sack of cement.

Where three coats are applied to metal lath, with the scratch coat $\frac{1}{4}$ in. thick, the brown coat $\frac{3}{8}$ in. thick, and the finish coat $\frac{1}{8}$ in. thick, the total thickness will be $\frac{3}{4}$ in.

> Where plaster is applied to brick, tile, concrete blocks, or plasterboard, it is common practice to omit the scratch coat, which will give a total thickness of $\frac{1}{2}$ to $\frac{5}{8}$ in.
>
> Grounds having the required thickness should be used.

Materials used for plaster The materials used for plaster include lime, gypsum plaster, cement, sand, hair, or fiber, and sometimes special cements or plasters to produce desired effects.

Plaster may be designated as lime, gypsum, or cement, depending on the chief cementing agent used.

Hair, asbestos, or manila fibers may be added to the scratch and brown coats to give them extra strength. Hair is usually packed in bags containing 1 bushel and weighing 7 to 8 lb.

Lime is usually sold by the ton and may be obtained in bulk or in sacks. Hydrated lime is available in 50-lb sacks, which contain 1 cu ft. Gypsum plaster is available in 100-lb sacks, which contain 1 cu ft. Sand may be assumed to weigh 100 lb per cu ft.

Table 13-8 gives the quantities of materials required for 1 cu yd of plaster. The mix proportions are by volume. The first number designates the quantity of cementing agent and the second number the quantity of sand.

Covering capacity of plaster Neglecting waste, the quantity of plaster required to cover 100 sq yd of area with a $\frac{1}{8}$-in.-thick coat will be $(100 \times 9)/(8 \times 12) = 9.4$ cu ft, or 0.35 cu yd. The quantities for other thicknesses will be in the ratio of the thickness divided by $\frac{1}{8}$. Table 13-9 gives the quantities of plaster required to cover 100 sq yd of area for various thicknesses.

TABLE 13-9 Quantities of Plaster Required to Cover 100 Sq Yd of Area (No Waste Included)

	Thickness of plaster, in.								
	$\frac{1}{16}$	$\frac{1}{8}$	$\frac{1}{4}$	$\frac{3}{8}$	$\frac{1}{2}$	$\frac{5}{8}$	$\frac{3}{4}$	$\frac{7}{8}$	1
Quantity, cu ft	4.7	9.4	18.8	28.2	37.6	47.0	56.4	65.8	75.2
Quantity, cu yd	0.18	0.35	0.70	1.05	1.40	1.74	2.10	2.44	2.80

Labor applying plaster The labor required to apply plaster will vary considerably with the class of workmanship specified, as previously stated. Large flat surfaces will require less labor than small irregular surfaces. The number of helpers needed will vary with the speed at which the plaster can be applied and the height to which the plaster must be hoisted. If high scaffolds are required, additional labor will be needed.

Table 13-10 gives the labor-hours required to apply 100 sq yd of plaster, based on net areas. Use the lower rates for ordinary workmanship and the higher rates for first-grade workmanship.

Example Estimate the cost of applying a ¾-in.-thick three-coat gypsum plaster on 100 sq yd of plain wall and ceiling, using ordinary workmanship. The use of fiber is required in the scratch coat. The thicknesses of the coats will be as follows: scratch coat, ⅛ in.; brown coat, ⅜ in.; finish coat, ⅛ in. The plaster will be applied on metal lath. The scratch and brown coats will consist of gypsum and sand mixed in the ratio 1:3, respectively.

Assume that 10 percent of the sand delivered to the job for this use will be wasted.
The cost will be

Scratch coat:
Gypsum, 0.7 × 9 = 6.3 sk @ $1.95	= $ 12.29
Sand, 0.7 × 1.1 = 0.8 cu yd @ $4.20	= 3.36
Fiber, 0.7 × 2 = 1.4 lb @ $1.20	= 1.68
Plasterer, 5 hr @ $7.69	= 38.45
Helper, 5 hr @ $5.79	= 28.95

TABLE 13-10 Labor-hours Required to Apply 100 Sq Yd of Plaster

	Labor-hours	
Class of work	Plasterer	Helper
Scratch coat, 3-coat work:		
Plain surface	5–6	5–6
Irregular surface	8–12	5–6
Brown coat, 3-coat work:		
Plain surface	8–12	5–6
Irregular surface	10–15	5–6
Brown coat, 2-coat work:		
Plain surface	8–12	5–6
Irregular surface	10–15	5–6
Finish coat	8–12	4–5
Sand finish coat	9–13	4–5

Brown coat:
Gypsum, 1.05 × 9 = 9.5 sk @ $1.95	= 18.53
Sand, 1.05 × 1.1 = 1.2 cu yd @ $4.20	= 5.04
Plasterer, 8 hr @ $7.69	= 61.52
Helper, 5 hr @ $5.79	= 28.95

Finish coat:
Gypsum, 0.35 × 27 = 9.5 sk @ $1.95	= 18.53
Plasterer, 8 hr @ $7.69	= 61.52
Helper, 4 hr @ $5.79	= 23.16
Scaffolds and sundry equipment	= 3.00

Total cost	= $304.98
Cost per sq yd, $304.98 ÷ 100	= 3.05

PROBLEMS

In solving the following problems the wage rates should be

Lathers, per hr	=	$7.69
Plasterers, per hr	=	7.69
Helpers, per hr	=	5.79

13-1   Estimate the direct cost of furnishing and installing 100 sq yd of flat metal lath weighing 3.4 lb per sq yd on wood studs spaced 16 in. on centers, using bent nails to secure the lath to the studs.

The cost of materials will be

Metal lath, per sq yd	=	$0.98
Tie wire, per lb	=	0.32
Nails, per lb	=	0.25

13-2   Estimate the direct cost of furnishing and installing 100 sq yd of ⅜-in.-rib metal lath weighing 3.4 lb per sq yd to two sides of lightweight ¾-in.-steel channels, spaced 12 in. on centers.

The cost of materials will be

Metal lath, per sq yd	=	$0.96
Tie wire, per lb	=	0.32

13-3   Estimate the total cost and the cost per sq yd for furnishing and applying a ¾-in.-thick three-coat gypsum plaster on 3,688 sq yd of irregular surface wall and ceiling, using first-grade workmanship. The plaster will be applied on metal lath.

The thicknesses of the coats will be: scratch coat, ¼ in.; brown coat, ⅜ in.; and finish coat, ⅛ in. For the scratch and brown coats, mix the gypsum and sand in the ratio 1:3, respectively, with no fiber required. The finish coat will be gypsum only.

The cost of materials will be

Gypsum, per sk	=	$ 1.95
Sand, per cu yd	=	4.20
Scaffolds and sundry equipment	=	10.00

14
Painting

General information Painting is the covering of surfaces of wood, plaster, masonry, metal, and other materials with a compound for protection or for the improvement of the appearance of the surface painted. Many kinds of paint are used, some of which will be described briefly.

As various practices are used in determining the area of the surfaces to be painted, it is desirable to state what methods are used. Some estimators make no deductions for the areas of openings unless they are quite large, whereas other estimators deduct all areas that are not painted. The latter method seems to be more accurate. Practices vary when estimating the areas for trim, windows, doors, metal specialties, masonry, etc.

Most paints are liquids, which contain pigments such as titanium dioxide, aluminum, etc., suspended in linseed oil, tung oil, or an acrylic latex and alkyd resin emulsion. The vehicle carries the pigment during painting; and when it solidifies, it holds the pigment on the surface to provide the desired protection or appearance.

Paints may be applied in one, two, three, or more coats, with sufficient time allowed between successive coats to permit the prior coat to dry thoroughly. Paint may be applied with a brush, roller, or spray gun. The first coat, which is usually called the prime coat, should fill the pores of the surface, if such exist, and bond securely to the surface to serve as a base for the other coats.

The cost of painting includes materials, labor, and sometimes equipment, especially when the paint is applied with a spray gun. Because the cost of the paint is relatively small when compared with the other costs, it is not good economy to purchase cheap paint.

Materials   It is beyond the scope of this book to discuss all the materials used for paints, but the ones that are most commonly used will be discussed briefly.

Ready-mixed paints   These paints, which are mixed by the manufacturers, may be purchased with all of the ingredients combined into a finished product.

Colored pigments   Colored pigments are added to paints by the manufacturer or at the job to produce the desired color.

Linseed oil   Linseed oil is frequently used as a vehicle for paints. The oil is processed from flax seed. When the oil dries, it forms a hard film on the surface to which it is applied.

Boiling the oil before using it will speed its rate of drying and hardening. Raw oil may be used for inside painting, but a commercial drier should be added to increase the rate of drying.

Turpentine   Turpentine, which is obtained by distilling the gum from pine trees, may be used to thin oil paints, especially the priming coat in order to obtain better penetration into the wood.

Varnish   Varnish, which is a solution of gums or resins in linseed oil, turpentine, alcohol, or other vehicles, is applied to produce hard transparent surfaces. Spar is a special varnish which is used on surfaces that may be exposed to water for long periods of time.

Shellac   Shellac is a liquid consisting of a resinous secretion from several trees dissolved in alcohol. It may be applied to knots and other resinous areas of lumber prior to painting to prevent bleeding of the resinous substance through the paint.

Stains   Stains, which are liquids of different tints, are applied to the surfaces of wood to produce desired color and texture effects. The vehicle may be oil or water.

Putty   Putty, which is a mixture of powdered chalk or commercial whiting and raw linseed oil, is used to fill cracks and joints and to cover the heads of countersunk nails. It should be applied following the application of the priming coat of paint. If it is applied directly on wood, the wood will absorb most of the oil before the putty hardens, and the putty will not adhere to the wood.

Covering capacity of paint   The covering capacity of paint, which is generally expressed as the number of square feet of area covered per gallon for one coat, will vary with several factors, including the following:

1   The kind of surface painted, wood, masonry, metal, etc.
2   The porosity of the surface.

3 The extent to which paint is spread as it is applied.

4 The extent to which a thinner is added to the paint.

5 The temperature of the air. Thinner coats are possible during warm weather, resulting in greater coverage.

Table 14-1 gives representative values for the covering capacities of various paints, varnishes, and stains when applied to different surfaces.

Preparing a surface for painting The operations required to apply a complete paint coverage will vary with the kind of surface to be painted, the number of coats to be applied, and the kind of paint used.

When painting new wood surfaces, it may be necessary to cover all resinous areas with shellac before applying the priming coat of paint. After

TABLE 14-1 Number of Square Feet of Surface Area Covered by 1 Gal of Paint, Varnish, Stain, Etc.

Material	Surface area covered per gal, sq ft
Applied on wood	
Prime coat paint	450–550
Oil base paint	450–550
Latex paint	450–550
Varnish, flat	500–600
Varnish, glossy	400–450
Shellac	600–700
Enamel paint	500–600
Aluminum paint	550–650
Oil stain on wood	500–700
Oil paint on floors	400–500
Stain on shingle roofs	125–225
Applied on plaster and stucco	
Size or sealer	600–700
Flat-finish paint	500–550
Calcimine on smooth plaster	225–250
Caesin paint on smooth plaster	400–450
Caesin paint on sanded plaster	400–450
Oil paint on stucco, first coat	150–160
Oil paint on stucco, second coat	350–375

this coat is applied, all joints, cracks, and nail holes may be filled with putty, which should be sanded smooth before applying the second coat of paint.

It may be necessary to apply a filler to the surfaces of certain woods, such as oak, before they are painted. The surfaces may be sanded before and after applying the filler and sometimes following the application of each coat of paint or varnish, except the last coat.

Before applying paint to plaster surfaces, it may be necessary to apply a sealer to close the pores and neutralize the alkali in the plaster.

The surfaces of new metal may be covered with a thin film of oil which must be removed with warm water and soap prior to applying the priming coat of paint.

Sometimes it is necessary to place masking tape, such as a strip of kraft paper with glue on one side, over areas adjacent to surfaces to be painted. This is especially true when paint is applied with a spray gun.

Labor applying paint   The labor required to apply paint may be expressed in hours per 100 sq ft of surface area, per 100 lin ft for trim, cornices, posts, rails, etc., or sometimes per opening for windows and doors. If the surfaces to be painted are properly prepared, and the paint is applied with first-grade workmanship, more time will be required than when an inferior job is permitted. An estimator must consider the kinds of surfaces to be painted and the requirements of the specifications before he can prepare an accurate estimate for a job.

The following factors should be considered when preparing an estimate for painting:

1   Treatment of the surface prior to painting, removing old paint, sanding, filling, etc.

2   Kind of surface, wood, plaster, masonry, etc.

3   Size of area, large flat, small, irregular.

4   Height of area above the floor or ground.

5   Kind of paint. Some flows more easily than others.

6   Temperature of air. Warm air thins paint and permits it to flow more easily.

7   Method of applying paint, with brushes or spray guns. Spraying may be five times as fast as brushing.

Table 14-2 gives representative labor-hours required to apply paint. The lower values should be used when the paint is applied under favorable conditions and the higher values when it is applied under unfavorable conditions or when first-grade workmanship is required.

Since the heavy pigment will settle to the bottom of a paint container, it may be necessary for a painter to spend 5 to 15 min per gal mixing paint before it can be used.

TABLE 14-2   Labor-hours Required to Apply a Coat of Paint and Perform Other Operations

Operation	Unit	Hours per unit
Exterior work		
Oil paint on wall siding and trim	100 sq ft	0.60–1.00
Oil paint on wood-shingle siding	100 sq ft	0.65–0.90
Stain on wood-shingle roofs	100 sq ft	0.60–0.75
Oil paint on wood floors	100 sq ft	0.30–0.40
Painting doors, windows, and blinds	Each	0.50–0.75
Oil paint on stucco	100 sq ft	0.60–1.00
Oil paint on steel roofing and siding	100 sq ft	0.50–0.60
Oil paint on brick walls	100 sq ft	0.60–0.90
Interior work		
Size or seal plaster and wallboard	100 sq ft	0.40–0.50
Calcimine on plaster and wallboard	100 sq ft	0.30–0.40
Casein paint on plaster and wallboard	100 sq ft	0.20–0.30
Trim, mold, base, chair rail:		
Sanding	100 lin ft	0.80–1.00
Varnishing	100 lin ft	0.50–0.75
Enameling	100 lin ft	0.50–0.75
Flat painting	100 lin ft	0.50–0.75
Staining	100 lin ft	0.40–0.60
Painting doors and windows	Each	0.50–0.75
Floors:		
Sanding with power sander	100 sq ft	0.75–1.00
Filling and wiping	100 sq ft	0.50–0.60
Shellacking	100 sq ft	0.25–0.30
Varnishing	100 sq ft	0.40–0.50
Waxing	100 sq ft	0.40–0.50
Polishing with power polisher	100 sq ft	0.30–0.40
Structural steel		
Brush painting:		
Beams and girders, 150–250 sq ft per ton	Ton	0.80–1.00
Columns, 200–250 sq ft per ton	Ton	0.90–1.00
Roof trusses, 275–350 sq ft per ton	Ton	1.25–1.75
Bridge trusses, 200–250 sq ft per ton	Ton	1.00–1.25
Spray painting:		
Beams and girders	Ton	0.20–0.25
Columns	Ton	0.20–0.25
Roof trusses	Ton	0.25–0.35
Bridge trusses	Ton	0.20–0.30

Equipment required for painting The equipment required for painting with brushes will include brushes, ladders, sawhorses, scaffolds, and foot boards. Inside latex paints may be applied with rollers. Where painting is done with spray guns, it will be necessary to provide one or more small air compressors, paint tanks, hose, and spray guns, in addition to the equipment listed for brush painting.

The cost of painting This example illustrates a method of determining the cost of furnishing materials and labor in painting a surface.

Example Estimate the cost of furnishing and applying paint for a three-coat application on 100 sq ft of exterior new wood wall for a residence, using ordinary workmanship. The top of the wall will be 12 ft above the ground. The temperature of the air will be about 70 deg. Ready-mixed oil paint will be used.

The cost of the three coats will be

Priming coat:
Paint, 100 sq ft ÷ 500 sq ft per gal = 0.2 gal @ $8.00 = $ 1.60
Painter, 100 sq ft × 0.9 = 0.9 hr @ $7.24 = 6.52
Second coat:
Paint, 100 sq ft ÷ 500 sq ft per gal = 0.2 gal @ $8.75 = 1.75
Painter, 100 sq ft × 0.8 = 0.8 hr @ $7.24 = 5.79
Third coat:
Paint, 100 sq ft ÷ 550 sq ft per gal = 0.18 gal @ $8.75 = 1.58
Painter, 100 sq ft × 0.8 = 0.8 hr @ $7.24 = 5.79
Painter mixing and stirring paint, 0.2 hr @ $7.24 = 1.45
Ladders, brushes, etc., 2.5 hr @ $0.30 = 0.75

Total cost = $25.23

PROBLEMS

For these problems use the national average wage rates, as given in this book.

14-1 Estimate the total cost of furnishing and applying a prime coat and two finish coats of oil base paint using brushes, on 3,380 sq ft of exterior wood surface, for a one-story house.

The cost of paint will be

Prime coat, $8.25 per gal
Finish paint, $9.25 per gal

14-2 Estimate the total direct cost of furnishing and applying one sealing coat and two coats of flat wall paint on 3,680 sq ft of smooth plaster walls and ceilings.

The cost of materials will be

Sealer, $6.80 per gal
Paint, $8.25 per gal

14-3 Estimate the total direct cost of sanding, filling, wiping, applying four coats of varnish, waxing, and polishing 3,640 sq ft of oak flooring, using first-grade workmanship,

The cost of equipment and materials will be

Power sander, $0.75 per hr
Polisher, $0.20 per hr
Filler, $8.25 per gal
Varnish, $10.25 per gal
Wax, $7.75 per gal

15
Glass and Glazing

It is common practice to furnish wood sash with the glass already installed by the manufacturer. Steel and aluminum sash may be furnished with the glass installed for use in residences, but when such sash are furnished for industrial and commercial buildings, it is common practice to install the glass after the sash are set in the walls, usually just before completing a building in order to reduce the danger of breakage.

The installation of glass is called glazing, and the individuals who install it are glaziers.

An estimate covering the cost of furnishing and installing glass should include a detailed list of the quantity for each size, kind, and grade of glass required and for each the cost of the glass, putty, and glaziers.

Glass   There are many kinds and grades of glass, not all of which will be covered in this book. Glass used for glazing purposes is called window glass, and may be divided into sheet glass and plate glass.

Sheet glass   Sheet glass is manufactured in several grades, such as AA, A, and B, with AA the highest quality. The thickness of ordinary window glass is single strength and double strength.

Plate glass   Plate glass used for glazing purposes is available in two grades, second silvering and glazing quality.

TABLE 15-1 Thicknesses and Weights of Clear Sheet Glass

Designation	Thickness, in.		Average weight per sq ft	
	Min.	Max	Oz	Lb
Single strength	0.085	0.100	18.5	1.16
Double strength	0.115	0.133	24.5	1.53
3/16 in.	0.182	0.205	39.0	2.44
7/32 in.	0.205	0.230	45.5	2.85
1/4 in.	0.240	0.255	52.0	3.25
3/8 in.	0.312	0.437	78.0	4.87
1/2 in.	0.438	0.556	104.0	6.50

Figured sheet glass Many kinds of figured sheet glass are available, with various patterns or figures rolled or otherwise produced on them.

The thicknesses and weights of clear sheet glass are given in Table 15-1.

Window glass is designated by size and grade, such as 20- by 36-in. B double strength. The first number represents the width and the second number the height of the sheet.

Glass is sold by the square foot in sheets whose dimensions vary in steps of 2 in. Fractional dimensions are figured for the next standard dimensions. thus a sheet whose actual size is 17⅝ by 14⅝ in. is classified for cost purposes as 18 by 16 in.

Costs and certain operations related to glass are frequently based on the united inches of a sheet of glass, which is the sum of the width and height of a sheet.

Glass for steel and aluminum sash Steel and aluminum sash are manufactured in standard units which require glass having designated sizes. Each pane of glass in a sash is called a light. Thus a sash that requires three lights for width and four lights for height is called a 12-light sash.

Table 15-2 lists representative net prices for various sizes and grades of flat glass used for steel and aluminum sashes effective in October, 1973. All dimensions are given according to standard pricing procedures, even though the actual sizes of the lights may be smaller than the dimensions listed. For example, a sash may require lights whose actual dimensions are 12½ by 20-¾ in. Table 15-2 lists this glass as size 14 by 22 in., because this is the smallest standard size from which it can be cut.

The net cost of glass is obtained by applying applicable discounts to a schedule of semipermanent list prices. The list price may remain constant for several years, but the discounts may change frequently and may vary with

TABLE 15-2   Representative Costs of Flat Window Glass per Box of 50 Sq Ft

Glass size, in.	Lights per box	Single strength Grade		Double strength Grade	
		A	B	A	B
8 × 12	75	$15.60	$14.15	$22.50	$19.80
8 × 16	56	15.60	14.15	22.50	19.80
9 × 12	67	15.60	14.15	22.50	19.80
9 × 14	57	15.60	14.15	22.50	19.80
9 × 16	50	15.60	14.15	22.50	19.80
10 × 22	33	15.60	14.15	22.50	19.80
10 × 24	33	15.60	14.15	22.50	19.80
12 × 16	38	15.60	14.15	22.50	19.80
12 × 18	33	15.60	14.15	22.50	19.80
12 × 24	25	15.60	14.15	22.50	19.80
14 × 18	29	15.60	14.15	22.50	19.80
14 × 24	21	15.60	14.15	22.50	19.80
16 × 18	25	15.60	14.15	22.50	19.80
16 × 24	19	15.60	14.15	22.50	19.80
16 × 32	14	16.20	14.30	24.50	21.30

the quality and quantity of glass purchased and the location of the purchase. *Putty required to glaze steel and aluminum sashes*   Putty and glazing compound are used to glaze glass in steel and aluminum sashes. The compound must be used with aluminum sashes and is frequently used with steel sashes. Putty will cost about $0.60 per lb, whereas the glazing compound will cost about $.75 per lb.

TABLE 15-3   Representative Costs for Cutting a Box of Glass of 50 Sq Ft to Odd Sizes or Irregular Shapes

Operation	Cost per box
Cut to 1 odd size	$2.50
Cut to 2 odd sizes	4.50
Cut circles or ovals	7.50

TABLE 15-4   Labor-hours Required to Glaze 100 Window Lights

Size of glass, in.	Putty, lb	Labor-hours	Size of glass, in.	Putty, lb	Labor-hours
Set window or plate glass using wood stops					
12 × 14	. . .	18	30 × 36	. . .	46
14 × 20	. . .	24	36 × 42	. . .	54
20 × 28	. . .	36	40 × 48	. . .	62
Set glass in wood sash using putty					
12 × 14	54	12	30 × 36	138	25
14 × 20	72	15	36 × 42	162	32
20 × 28	100	20	40 × 48	184	40
Set glass in steel or aluminum sash					
24 × 10	114	18	18 × 14	107	17
30 × 10	134	24	24 × 14	127	20
38 × 10	160	29	30 × 14	147	24
			38 × 14	173	30
9 × 12	70	16			
16 × 12	94	18	18 × 16	114	18
18 × 12	100	19	20 × 16	120	19
24 × 12	120	20	22 × 16	127	20
30 × 12	140	23	32 × 16	160	29
38 × 12	167	28	38 × 16	180	32
42 × 12	180	32	46 × 16	204	42

Before installing a glass into a metal sash, a quantity of putty or glazing compound should be spred uniformly against the metal of the rabbet to hold the glass away from the metal. After the glass is installed, additional putty is applied to complete the glazing. A pound of putty or compound will glaze about 5 lin ft of rabbet for steel and aluminum sash and about 8 lin ft for wood sash.

Labor required to install glass   Table 15-4 gives the labor-hours required to glaze windows during warm weather when the glazing is done from inside the building.

If glazing is done during cold weather, the putty or glazing compound will be more difficult to work, which will require about 25 percent more labor time than when glazing is done during warm weather.

If glazing is done from the outside of a building, it will require about 20 percent more time than when it is done from inside a building.

Example Estimate the direct cost of glass, glazing compound, and labor required to glaze 240 lights in aluminum sashes. The actual size of the lights will be 15½ by 31½ in. This will require lights whose standard sizes are 16 by 32 in., which will require two cuttings per light.

The cost will be

No. boxes of glass, 240 ÷ 14 = 17.2
Add for possible breakage   = 0.8
                                 _____
    Total no. boxes   = 18.0
Use double strength grade A glass
Glass, 18 boxes @ $24.50   = $ 441.00
Cutting to size, 18 boxes @ $4.50   =   81.00
Glazing compound, 240 × 160/100 = 384 lb @ $0.75   =   288.00
Glazier, 240 × 29/100 = 69.6 hr @ $6.19   =   430.82
                                                               _____
    Total cost   = $1,240.82

PROBLEMS

For all these problems use the national average wage rates as given in this book.

15-1 Estimate the total direct cost for furnishing and installing grade B double-strength glass in 120 aluminum sashes. Each sash will require two sheets of glass size 15¼ by 31-¾ in. Use glazing compound to set the glass. The installation will be done from outside the building during the summer.

Use the costs of materials given in this book.

15-2 Estimate the cost of furnishing and installing the glass of Prob. 15-1 when the glazing is done from the inside of the building during cold weather.

15-3 Estimate the cost of furnishing and installing 62 plate-glass lights size 36 by 42 in. in wood frames using wood stops. The glass, which is ¼ in. thick, will cost $2.15 per sq ft.

15-4 Estimate the total direct cost of furnishing and installing 280 size 11½ by 23¼ in. single-strength grade A lights in aluminum sashes, using glazing compound. The glazing will be done from inside the building during the summer.

Use the material prices given in this book.

16
Roofing and Flashing

Roofing materials Roofing refers to the furnishing of materials and labor to install coverings for roofs of buildings. Several kinds of materials are used, including but not limited to the following:

1 Shingles
   a Wood
   b Asphalt
   c Asbestos
2 Slate
3 Built-up
4 Clay tile
5 Metal
   a Copper
   b Aluminum
   c Galvanized steel
   d Tin
6 Corrugated asbestos

Area of a roof Several methods are used when measuring the area of a roof for estimating purposes. Some estimators make no deduction for the areas of openings containing less than 100 sq ft but deduct one-half the areas of openings containing 100 to 500 sq ft. A method which may be used for measuring the area of slate roofs is to make no deductions for skylights, chimneys, etc., whose dimensions are less than 4 ft square, deduct one-half the areas whose dimensions are between 4 and 8 ft square, and deduct all the areas whose dimensions are larger than 8 ft square. An estimator should determine what method will be used before preparing an estimate.

When determining the area of a pitched roof, measure along the full length of the roof; and for the width, measure along the slope of the roof. Be sure that every dimension includes the maximum length or width to be covered, including eaves, overhang, etc.

The unit of area most commonly used for roofing is the square, which is 100 sq ft.

Materials used to cover ridges and valleys should be measured by the linear foot, with the width specified.

Roof pitch   The pitch of a roof indicates its steepness. The terms frequently used and their values are as follows:

Pitch	Rise, in. per ft
$\frac{1}{8}$	3
$\frac{1}{4}$	6
$\frac{1}{3}$	8
$\frac{1}{2}$	12
$\frac{2}{3}$	16

Roofing felt   When certain materials such as asphalt and asbestos-cement shingles, slate, and tile are used for roofing, the specifications may require the application of roofing felt to the entire roof prior to applying the covering material. This material is sold in rolls containing 108, 216, and 432 sq ft, which will cover 100, 200, and 400 sq ft, respectively, with a 2-in. lap. Weights

TABLE 16-1   Quantities of Asphalt Shingles and Roofing Nails per Square

Style shingle	Size, in.	No. per square	Length exposed, in.	No. nails per shingle	Nails per square, lb
Individual	12 × 16	226	5	2	2.0
Strip, 3 tab	12 × 36	80	5	6	1.5
Hexagon strip	11$\frac{1}{3}$ × 36	86	4$\frac{2}{3}$	6	1.5
Double coverage	18 × 20	120	12	4	2.0

TABLE 16-2   Representative Prices per Square for Asphalt Shingles

Style shingle	Size, in.	Weight per square, lb	Cost per square
Individual	12 × 16	215	$ 9.45
Strip, 3 tab	12 × 36	235	12.45
Hexagon strip	12 × 36	170	7.85
Random shakes	12 × 36	345	19.25
Hip and ridge, 50 lin ft	9 × 12		5.60

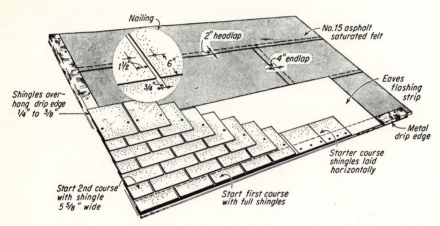

Fig. 16-1 Individual asphalt shingles applied by the American method. (*The Ruberoid Company.*)

TABLE 16-3   Labor-hours Required to Lay a Square of Asphalt Shingles, Using Carpenters

Style shingle	Labor-hours*	
	Carpenter	Helper
Individual	3.5–5.5	1.0–1.2
Strip, 3 tab	2.0–2.6	1.0–1.2
Hexagonal strip	2.0–2.6	1.0–1.2
Double coverage	2.5–3.5	1.0–1.2

* If experienced shinglers are used, reduce the hours by 15 to 20 percent.

of felt used are 15 and 30 lb per square. Galvanized nails about $\frac{3}{4}$ in. long, with $\frac{7}{16}$-in. heads, spaced about 6 in. apart along the edges, are used to hold the felt in place. About $\frac{3}{4}$ lb of $\frac{3}{4}$-in. nails will be required per square. Asphalt felt, weighing 15 lb per square, will cost about $0.65 per square and 30-lb felt about $1.30 per square.

The labor required to lay felt should be about 0.5 hr per square.

WOOD SHINGLES

Wood shingles are discussed in Chap. 11 under Carpentry.

ASPHALT SHINGLES

Asphalt shingles are manufactured in several styles, colors, and sizes, including individual shingles, strips containing three shingles, hexagons, double coverage, etc.

The shingles are fastened to the wood decking with galvanized roofing nails, 1 to $1\frac{1}{2}$ in. long, starting at the lower edge of the roof. Asphalt starting strips should be laid along the eaves of a roof prior to laying the first row of shingles, or the first row may be doubled.

The net area of a roof should be increased about 10 percent for gable roofs, 15 percent for hip roofs, and 20 percent for roofs with valleys and dormers to provide for waste.

Table 16-1 gives the net quantities of asphalt shingles of various styles required to cover one square.

Cost of asphalt shingles The cost of asphalt shingles will vary with the style, materials, and weight per square for the shingles selected. The quoted

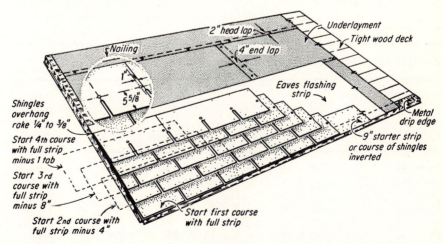

Fig. 16-2 Three-tab strip asphalt shingles. (*The Ruberoid Company.*)

price per square is based on furnishing enough shingles to cover 100 sq ft of roof in the manner indicated by the manufacturer, with no allowance for waste included. Table 16-2 gives representative prices for asphalt shingles.

Labor required to lay asphalt shingles Asphalt shingles may be laid by carpenters or experienced roofers. The latter should lay them more rapidly. The rate of laying shingles will be lower for simple areas than for irregular areas. Hips, valleys, gables, dormers, etc., require the cutting of shingles for correct fit, which will reduce the production rates. Table 16-3 gives the labor-hours required to lay a square of asphalt shingles. Use the lower rates for gable roofs and the higher rates for roofs with valleys, hips, dormers, etc.

ASBESTOS SHINGLES

Asbestos shingles are rigid shingles made of asbestos and portland cement in various styles, sizes, colors, and textures. The roof decking should be covered with 15- or 30-lb asphalt roofing felt prior to laying the shingles. The shingles are factory punched to permit fastening with roofing nails and storm anchors.

Table 16-4 gives properties and coverage of several popular styles of asbestos shingles when laid in different patterns.

Cost of asbestos shingles The cost of asbestos shingles will vary with the style, texture, color, and weight required per square. Shingles are sold by the square for a specified size and exposure. It will be necessary to allow about 10 percent for gable roofs, 15 percent for hip roofs, and 20 percent for roofs with valleys, hips, and dormers for waste. Table 16-5 gives representative prices for asbestos shingles.

Laying asbestos shingles After the roofing felt is laid and prior to laying the first course of shingles, a course of asbestos starter strips, 4 to 9 in. wide, should be laid along the eaves, with the edge extending ⅜ to ½ in. beyond the roof decking to form a drip. The shingles are then laid in whatever pattern is

---

TABLE 16-4   Properties and Coverage of Asbestos Shingles

Pattern	Size, in.	Exposure, in.	No. per square	Weight per square, lb	Nails per square, lb
American	8 × 16	7 × 8	260	570	3.0
American	12 × 24	7 × 24	86	275	1.0
American colonial	14 × 30	6 × 30	80	265	2.0
Dutch lap	12 × 24	9 × 20	80	260	1.0
Dutch lap	16 × 16	12 × 13	92	260	0.8
Dutch lap	16 × 16	10⅔ × 13	104	295	1.0
Hexagonal	16 × 16	13 × 13	86	265	1.0

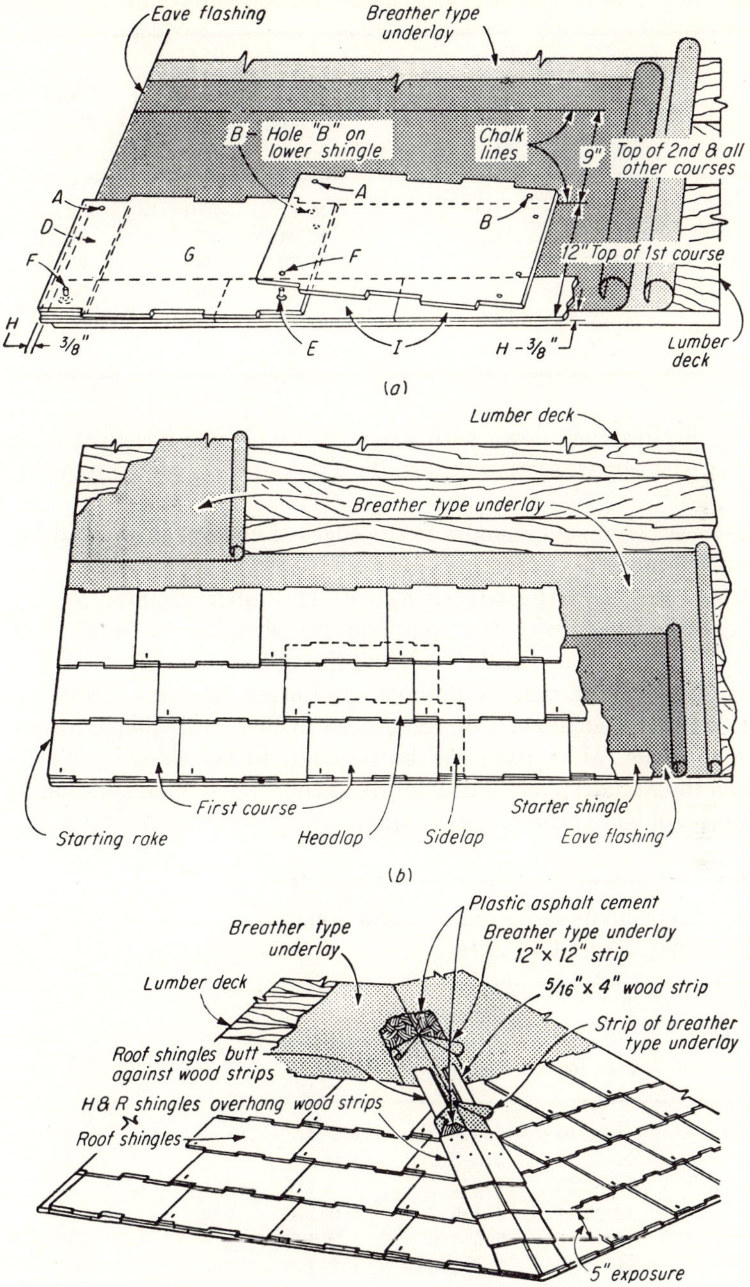

ESTIMATING CONSTRUCTION COSTS

(a)

(b)

Fig. 16-3 Methods of applying American thatch asbestos roof shingles. (*The Ruberoid Company.*)

TABLE 16-5 Representative Prices per Square for Asbestos Shingles

Style shingle	Size, in.	Thickness, in.	Exposure, in.	Weight per square, lb	Cost per square
American	8 × 16	¼	7 × 8	570	$66.75
American	12 × 24	5⁄32	7 × 24	275	32.80
American colonial	14 × 30	5⁄32	6 × 30	265	31.50
Dutch lap	16 × 16	5⁄32	12 × 13	260	30.60
Dutch lap	12 × 24	5⁄32	9 × 20	260	30.60
Hexagonal	16 × 16	5⁄32	13 × 13	265	31.00

specified. They are fastened down with galvanized nails and in addition, in some instances, with copper storm anchors. Two to four nails will be required for each shingle.

It will be necessary to cut shingles at the edges of valleys, at hips, ridges, and at walls or dormers.

Labor required to lay asbestos shingles The labor required to lay asbestos shingles will vary with the style and size shingles, the method of fastening them down, the slope of the roof, and the extent to which the roof is simple or cut up with hips, valleys, dormers, and other openings. Less time will be required to lay a square of strip shingles 24 to 30 in. wide than a square of shingles 8 in. wide. More time will be required to lay shingles by the hexagonal than the American method. If storm anchors are used in addition to nails, more time will be required to lay a square. More time is required to lay

TABLE 16-6 Labor-hours Required to Lay a Square
of Asbestos Shingles, Using Carpenters

Style shingle	Size, in.	Labor-hours*	
		Carpenter	Helper
American	8 × 16	4.5–6.0	2.2–3.0
American	12 × 24	2.0–3.0	1.0–1.5
American colonial	14 × 30	1.8–2.7	1.0–1.5
Dutch lap	12 × 24	2.1–3.2	1.0–1.5
Dutch lap	16 × 16	2.5–4.0	1.2–1.8
Hexagonal	16 × 16	2.5–4.0	1.2–1.8

* If experienced shinglers are used, reduce the hours by 25 to 30 percent.

shingles on a steep than on a flat roof. When it is necessary to cut shingles to fit valleys, hips, ridges, and dormers, the time required to lay a square may be increased considerably.

Table 16-6 gives representative labor-hours required to lay one square of asbestos shingles using carpenters. Use the lower values for plain roofs, such as gable, and the higher values for steep roofs with hips, valleys, and dormers.

SLATE ROOFING

Slate for roofing is made by splitting slate blocks into pieces having the desired thickness, length, and width. Most slates are produced in Maine, Maryland, Pennsylvania, and Vermont. Although slate is available in various thicknesses up to about 2 in., the most common thickness used is $\frac{3}{16}$ in. Sizes vary from 10 by 6 in. to 24 by 16 in., with the lengths varying in steps of 2 in. and the widths varying in steps of 1 in.

Colors and grades of slate The trade names of roofing slate, based on color, are as follows.

Black	Mottled purple and green
Blue black	Green
Gray	Purple variegated
Blue gray	Red
Purple	

The trade grades of slate are as follows:

No. 1 clear	No. 1 ribbon
Medium clear	No. 2 ribbon

Quantities of slate Slate is priced by the square, with sufficient pieces furnished to cover a square with a 3-in. head lap over the second course under the given course. Thus for slates 16 in. long, the length exposed to weather will be $(16 - 3)/2 = 6.5$ in. This same calculation will apply to any length for which the head lap is 3 in. If the slates are 8 in. wide, the area covered will be $8 \times 6.5 = 52$ sq in. The number of slates required to cover a square will be $(100 \times 144)/52 = 277$. For other sizes the number required to cover a square can be determined in the same manner. An area should be increased 10 to 25 percent to allow for waste.

Weight of slate roofing Slate weighs 170 to 180 lb per cu ft. The weights per square on a roof with a standard 3-in. head lap are given in Table 16-7.

Nails for slate roofing The weight of nails required for slate roofing will depend on the number, length, gauge, and type. The number required will be twice the number of slates laid. Table 16-8 gives the number of copper nails per pound.

TABLE 16-7 Weights of
Slate Roofing per Square
for a 3-in. Lap

Thickness, in.	Weight per square, lb
3⁄16	700
1⁄4	900
3⁄8	1,400
1⁄2	1,800
3⁄4	2,700
1	4,000

Laying slate   Roofing slate should be laid on asphalt felt, weighing 30 lb per square. The first course of slate should be doubled, with the lower end laid on a wood strip to give the slate the proper cant for the succeeding courses. Joints should be staggered. Each slate should be fastened with two nails driven through prepunched holes.

Hips and valleys will require edge mitering of the adjacent slates. The lengths of slates laid in courses along ridges must be reduced.

Labor required to apply slate roofing   The labor operations required for slate roofing will include laying the felt, applying slates, and installing snow

TABLE 16-8   Number of Copper Nails per Pound

Length, in.	Copper-wire slating nails		Copper-wire common nails	
	Gauge	No. per lb	Gauge	No. per lb
1	12	310	15	575
1¼	10	144	14	400
	11	196		
	12	216		
1½	10	124	12	260
	11	160		
	12	192		
1¾	10	112	11	140
			12	156
2	10	100	10	114
			11	128
2½	...	...	10	84
3	...	...	9	57

TABLE 16-9 Labor-hours Required to Apply a Square of Felt and Roofing Slate

Size slate,* in.	Labor-hours	
	Slater	Helper
12 × 8	5.0–7.0	2.5–3.5
16 × 8	4.0–6.0	2.0–3.0
16 × 12	3.5–5.2	1.8–2.7
18 × 12	3.2–5.0	1.6–2.5
20 × 10	2.9–4.5	1.4–2.2
20 × 12	2.8–4.3	1.4–2.1
22 × 12	2.5–3.9	1.2–2.0
24 × 12	2.3–3.6	1.2–1.8
24 × 16	2.0–3.2	1.0–1.6

* The first number is the length and the second number is the width of a slate.

guards, if they are required. Production rates will vary considerably with the size and thickness of slates, slope of the roof, kind of pattern specified, and the complexity of the areas.

Table 16-9 gives labor-hours to apply a square of felt and slate under various conditions. Use the lower rates for plain roofs, such as gable roofs, and the higher rates for steep roofs with valleys, hips, and dormers.

BUILT-UP ROOFING

Built-up roofing consists in applying alternate layers of roofing felt and hot pitch or asphalt over the area to be covered, with gravel, crushed stone, or slag applied uniformly over the top layer of pitch or asphalt. This type of roofing may be applied to wood sheathing, concrete, poured gypsum, precast concrete tiles, precast gypsum blocks, book tile, and approved insulation.

The quality of built-up roofing is designated by specifying the weight and number of plies of felt, the weight and number of applications of pitch or asphalt, and the weight of gravel or slag used. The unit of area is a square.

Felt Pitch-impregnated felt should be used with pitch cement and asphalt-impregnated felt with asphalt. Felt weighing 15 or 30 lb per square may be used, the 15-lb felt being more commonly specified. Felt is available in rolls, 36 in. wide, with gross areas of 108, 216, and 432 sq ft.

Pitch and asphalt Pitch should be applied at a temperature not exceeding 400°F and asphalt at a temperature not exceeding 450°F. Applications are made with a mop to the specified thickness or weight. Pitch and asphalt will weigh about 10 lb per gal.

Gravel and slag   After the final layer of pitch or asphalt is applied, and while it is still hot, gravel or slag is spread uniformly over the area. The aggregate should be ¼ to ⅝ in. in size and thoroughly dry. Application rates are about 400 lb per square for gravel and 300 lb per square for slag.

Laying built-up roofing on wood decking   While specifications covering the roofing laid on wood decking will vary, the method described below is representative of common practice.

First   Over the entire surface lay two plies of 15-lb asphalt felt, lapping each sheet 19 in. over the preceding one and turning these felts up not less than 4 in. along all vertical surfaces. Nail as often as necessary to secure until the remaining felt is laid.

Second   Over the entire surface embed in asphalt two plies of asphalt felt, lapping each sheet 19 in. over the preceding one, rolling each sheet immediately behind the mop to ensure a uniform coating of hot asphalt so that in no place shall felt touch felt. Each sheet shall be nailed 6 in. from the back edge at intervals of 24 in. These felts shall be cut off at the angles of the roof deck and all walls or vertical surfaces.

Third   Over the entire surface spread a uniform coating of asphalt into which, while hot, embed not less than 400 lb of gravel or 300 lb of slag per 100 sq ft of area. Gravel or slag must be approximately ¼ to ⅝ in. in size, dry and free from dirt. If the roofing is applied during cold weather or the gravel or slag is damp, it shall be heated and dried immediately before application.

Not less than the following quantities of materials shall be used for each 100 sq ft of roof area:

Material	Weight, lb	
	Gravel	Slag
4 plies of 15-lb asphalt felt	60	60
Roofing asphalt	100	100
Gravel or slag	400	300
Total weight	560	460

The application previously specified is designated as four-ply roofing, with two plies dry and two plies mopped.

Sometimes specifications require that the decking first be covered with a single thickness of sheathing paper, weighing not less than 5 or 6 lb per square, with the edges of the sheets lapped at least 1 in.

TABLE 16-10 Representative Costs
of Materials Required for Built-up
Roofing

Material	Unit	Cost per unit
Sheathing paper	Square	$ 1.00
Asphalt felt, 15 lb	Square	1.10
Asphalt felt, 30 lb	Square	2.20
Concrete primer	Gallon	1.05
Roofing asphalt	55 gal	45.00

Laying built-up roofing on concrete Before applying built-up roofing on a concrete deck, the concrete should be cleaned and dried, after which a coat of concrete primer should be applied cold at a rate of about 1 gal per square. After the primer has dried, embed in hot asphalt two, three, four, or five plies of asphalt felt, lapping each sheet enough to give the required number of plies. Roll each sheet immediately behind the mop to ensure a uniform coating of hot asphalt, so that in no place shall felt touch felt. Over the entire area spread a uniform coating of hot asphalt into which, while hot, embed not less than 400 lb of dry gravel or 300 lb of dry slag per square.

TABLE 16-11 Labor-hours Required to Perform Operations and Lay a Square of Built-up
Roofing

Operation	Labor-hours*
Apply sheathing paper on wood deck	0.10–0.15
Apply primer on concrete	0.10–0.15
Lay 1 ply of roofing felt	0.10–0.15
Apply asphalt with mop	0.10–0.15
Apply asphalt and gravel	0.50–0.75
Apply 2-ply roofing on wood deck	1.3–2.0
Apply 3-ply roofing on wood deck	1.5–2.2
Apply 4-ply roofing on wood deck	1.7–2.5
Apply 5-ply roofing on wood deck	1.8–2.7
Apply 3-ply roofing on concrete	1.6–2.3
Apply 4-ply roofing on concrete	1.8–2.6
Apply 5-ply roofing on concrete	1.9–2.8

* The labor-hours for the last eight applications include three men on the roof and two men heating and supplying hot asphalt.

For a three-ply roofing the quantities of materials given in the following table might be used per square.

Material	Weight, lb	
	Gravel	Slag
Concrete primer	10	10
3 plies of 15-lb asphalt felt	45	45
Roofing asphalt	125	125
Gravel or slag	400	300
Total weight	580	480

Cost of materials   Table 16-10 gives representative costs of materials required for built-up roofing.

Labor laying built-up roofing   The operations required to lay built-up roofing will vary with the type of roofing specified. If the time required to perform each operation is estimated, the sum of these times will give the total time for a roof or for completing a square. If a building is more than two or three stories high, additional time should be allowed for hoisting materials.

A typical crew for laying roofing on buildings up to three stories high would include one man each tending the kettle, handling hot asphalt or pitch, laying felt, rolling felt, and mopping, with a foreman supervising all operations.

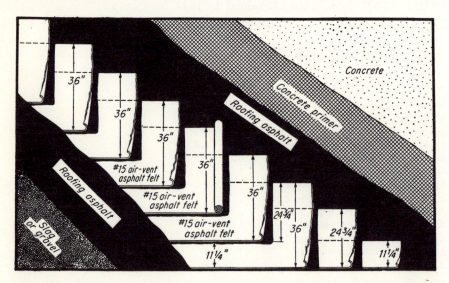

Fig. 16-4   Method of laying built-up roofing on concrete. (*The Ruberoid Company.*)

On some jobs one man may be able to tend the kettle and handle the asphalt or pitch.

Table 16-11 gives the labor-hours per square required to perform each operation and to complete all operations for various types of built-up roofing. Use the lower values for large plain roofs and the higher values for roofs with irregular areas.

FLASHING

Flashing is installed to prevent water from passing into or through areas such as valleys and hips on roofs, where roofs meet walls, or where openings are cut through roofs. Materials used for flashing include sheets of copper, tin, galvanized steel, aluminum, lead, and sometimes mopped layers of roofing felt.

Flashing is usually measured by the linear foot for widths up to 12 in. and by the square foot for widths greater than 12 in.

Metal flashing   When metal flashing requires nails or other metal devices to hold it in place, it is essential that the fastener and the flashing be of the same metal; otherwise galvanic action will soon destroy one of the metals.

Flashing roofs at walls   Where a roof and a parapet wall join, it is common

ESTIMATING CONSTRUCTION COSTS

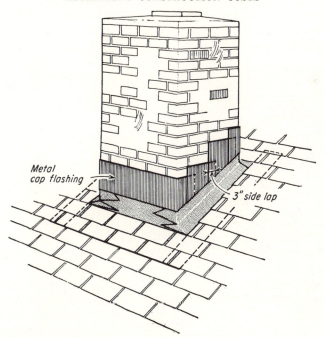

Fig. 16-5 Metal cap flashing applied over base flashing. (*The Ruberoid Company.*)

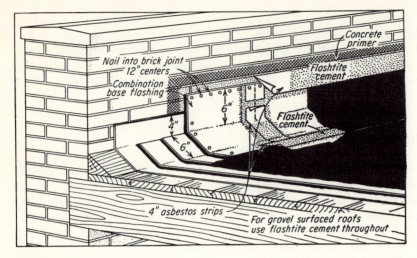

Fig. 16-6 Method of flashing roof at parapet wall. (*The Ruberoid Company.*)

practice to extend the layers of built-up roofing 4 to 8 in. up the wall, with mopping applied to the wall and the felt, with no metal flashing underneath the felt. A metal counter flashing, whose upper edge is bent and inserted in a raggle or slot in the mortar joint between bricks about 12 in. above the roof, is installed to cover the portion of the felt flashing attached to the wall. If

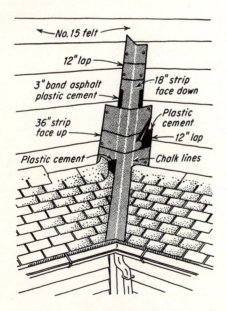

Fig. 16-7 Use of roll roofing for valley flashing. (*The Ruberoid Company.*)

TABLE 16-12 Costs of Materials Used for Flashing per 100 Lin Ft

Galvanized steel:	
Plain ridge roll, 8 in. wide	$ 17.36
Plain ridge roll, 10 in. wide	18.50
Plain ridge roll, 12 in. wide	20.65
Roll flashing, weight per 100 lin ft:	
8 in. wide, 45 lb	13.50
10 in. wide, 56 lb	16.56
12 in. wide, 67 lb	19.80
14 in. wide, 78 lb	21.26
16 in. wide, 89 lb	24.25
24 in. wide, 134 lb	36.25
Flashing shingles, 5 × 7 in., per 1,000	50.30
Aluminum flashing:	
4 in. wide	9.60
5 in. wide	12.60
6 in. wide	15.10
20 in. wide	42.00
Copper flashing, valley, etc., 16 oz	
12 in. wide	152.00
16 in. wide	192.00
20 in. wide	240.00

metal is installed under the felt and extended up the wall, it is called base flashing. Base and counter flashing should be soldered at all end joints.

Flashing valleys and hips When roofs are covered with shingles, it is necessary to flash the valleys and hips, usually with metal flashing 10 to 30 in. wide. The flashing may be soldered at end joints, or it may be lapped enough to eliminate danger of leakage. The quantity is measured by the linear foot.

Cost of flashing Table 16-12 lists prices for certain types of metal flashing materials.

TABLE 16-13 Labor-hours Required to Install Flashing

Class of work	Unit	Labor-hours
Metal base around parapet wall	100 lin ft	5.0– 6.0
Metal counter flashing around walls	100 lin ft	5.0– 6.0
Metal along roof and wood walls	100 lin ft	4.0– 5.0
Metal valleys and hips	100 lin ft	5.0– 6.0
Metal shingles along chimneys	100 each	8.0–10.0

Labor required to install flashing Flashing is installed by tinners. The labor required will vary with the material used, the type of flashing required, and the specifications covering the installation. Table 16-13 gives the labor-hours required to install flashing.

PROBLEMS

For these problems use the national average wage rates and the costs of materials given in this book.

16-1 Estimate the total cost and the cost per square for furnishing and laying 15-lb asphalt felt and three-tab strip asphalt shingles, size 12 by 36 in., on a wood roof decking whose area is 3,436 sq ft., for a building one-story high. The roof will have numerous valleys, hips, and dormers. Use carpenters to lay the roofing.

The nails will cost $0.35 per lb.

16-2 Estimate the total cost and the cost per square for furnishing and laying 12- by 24-in. American-style asbestos shingles weighing 275 lb per square. Lay one ply of 15-lb roofing felt over the wood decking prior to applying the shingles. Use carpenters to apply the felt and shingles.

The roof, whose area is 2,640 sq ft, is a simple gable type with a ¼ pitch.

The nails will cost $0.35 per lb.

17
Plumbing

Although general contractors usually subcontract the furnishing of materials and the installation of plumbing in a building, they should have a reasonably good knowledge of the costs of plumbing. The fact that general contractors do not prepare detailed estimates for plumbing does not eliminate the need for estimating, it simply transfers the preparation to another party.

Plumbing involves the furnishing of materials, equipment, and labor to bring gas and water to a building, the furnishing and installation of fixtures, and the removal of the water and waste from the building. Water usually is obtained from a water main, and the waste is discharged into a sanitary sewer line. Materials include various types of pipe, fittings, valves, and fixtures, which will be more fully described hereafter. Consumable supplies such as lead, oakum, gasoline, etc., must be included in an estimate.

Plumbing requirements   Some plans and specifications clearly define the types, grades, sizes, and quantities and furnish other information required for a complete plumbing installation, while other plans and specifications furnish limited information and place on the plumbing contractor the responsibility for determining what is needed to satisfy the owner and the local plumbing ordinance.

All cities have ordinances which require that plumbing installations must conform with the plumbing code for the city in which the project will be constructed. A contractor must obtain a plumbing permit prior to starting an

TABLE 17-1 Minimum Requirement for Plumbing-fixture Facilities (One Fixture for Each Designated Group)

Type of building	Water closet	Urinal	Lavatory	Drinking fountain	Shower	Bathtub	Kitchen sink
Dwellings and apartment houses	Each family	....	Each family	....		Choice of 1 per family	Each family
Places of employment, such as mercantile and office buildings, work-shops, and factories where 5 or more persons work	25 males 20 females	25 males	15 persons	75 persons			
Foundries, mines, and places where exposed to dirty or skin irritating materials where 5 or more persons work	25 males 20 females	25 males	5 persons	75 persons	15 males 15 females		
Schools	20 males 15 females	25 males	20 persons	75 persons			
Dormitories	10 males 8 females	25 males	6 persons	50 persons	8 males 10 females	40* males 35* females	

* Half may be additional showers.

TABLE 17-2   Fixture-unit Values and Minimum-size Traps and Outlets Required

Fixture	Fixture unit value as load factor	Minimum size trap and outlet connection, in.
Bathroom group consisting of water closet, lavatory and bathtub or shower stall and		
Tank with water closet	6	3
Flush-valve water closet	8	3
Bathtub with or without overhead shower	2	1½
Bidet	3	1½
Combination sink and tray	3	1½
Combination sink and tray with food disposal unit	4	1½
Dental unit or cuspidor	½	1¼
Dental lavatory	1	1¼
Drinking fountain	½	1
Dishwasher, domestic	2	1½
Floor drains	1	2
Kitchen sink, domestic	2	1½
Kitchen sink, domestic, with food-disposal unit	3	1½
Lavatory	2	1½
Lavatory, barber, beauty parlor	2	1½
Lavatory, surgeon's	2	1½
Laundry tray, 1 or 2 compartment	2	1½
Shower stall, domestic	2	2
Showers, group, per head	3	Varies
Sinks:		
Surgeon's	3	1½
Flushing rim, with valve	8	3
Service, standard trap	3	3
Service, with P trap	2	2
Pot, scullery, etc.	4	1½
Urinal, pedestal, syphon jet blow out	8	3
Urinal, wall lip	4	1½
Urinal, stall, washout	4	2
Urinal trough, each 2-ft section	2	1½
Wash sink, circular or multiple, each set of faucets	2	1½
Water closet:		
Tank operated	4	3
Valve operated	8	3

installation, and the work is checked by a plumbing inspector for compliance with the code. Although plumbing codes in different localities are similar, there will be variations which make it necessary for a plumbing contractor to be fully cognizant of the requirements of the code which will apply to a given project.

Plumbing code Some but not all of the requirements of the plumbing code for a major city are given here. The information is intended to serve as a guide in demonstrating the requirements of city codes for plumbing. When preparing an estimate of the cost of furnishing and installing items for plumbing services, the estimator should use the appropriate code for the area in which the project will be constructed.

Permit fee A fee, which is generally paid by the plumbing contractors and included in their estimate, is charged by a city for issuing a plumbing permit. This fee may vary with the type and size of the project and with the location.

Plumbing-fixture facilities Table 17-1 lists the minimum requirements for plumbing-fixture facilities for a representative city. The requirements may differ in other cities.

Minimum-size trap and outlet Table 17-2 gives the minimum-size trap and outlet permitted for the indicated fixture.

Fixture-unit values Table 17-2 gives the value of each fixture unit when determining the relative load factors of different kinds of plumbing fixtures and in estimating the total load carried by soil and waste pipe.

TABLE 17-3   Maximum Number of Fixture Units Permitted for Soil and Waste Pipe

| Pipe size, in. | Fall or slope, in. per ft | | | |
	$\frac{1}{16}$	$\frac{1}{8}$	$\frac{1}{4}$	$\frac{1}{2}$
2½	. . . . .	2	24	31
3	. . . . .	20*	27+	36+
4	. . . . .	180	216	250
5	. . . . .	390	480	575
6	. . . . .	700	840	1,000
8	1,400	1,600	1,920	2,300
10	2,500	2,900	3,500	4,200
12	3,900	4,800	5,600	6,700

* Water closet not permitted.
+ Not over two water closets permitted.

TABLE 17-4   Dimensions and Weights of Black and Galvanized Standard Steel Pipe

Size, in.	Diameter, in.		Thickness, in.	Internal area, sq in.	Weight per lin ft, lb	Threads per in.
	External	Internal				
⅛	0.405	0.269	0.068	0.057	0.24	27
¼	0.540	0.364	0.088	0.104	0.42	18
⅜	0.675	0.493	0.091	0.191	0.56	18
½	0.840	0.622	0.109	0.304	0.84	14
¾	1.050	0.824	0.113	0.533	1.12	14
1	1.315	1.049	0.133	0.861	1.67	11½
1¼	1.660	1.380	0.140	1.496	2.25	11½
1½	1.900	1.610	0.145	2.036	2.68	11½
2	2.375	2.067	0.154	3.356	3.61	11½
2½	2.875	2.467	0.203	4.780	5.74	8
3	3.500	3.066	0.217	7.383	7.54	8
3½	4.000	3.548	0.226	9.886	9.00	8
4	4.500	4.026	0.237	12.730	10.67	8
5	5.563	5.045	0.259	19.985	14.50	8
6	6.625	6.065	0.280	28.886	18.76	8
7	7.625	7.023	0.301	38.734	23.27	8
8	8.625	7.981	0.322	50.021	28.18	8
9	9.625	8.937	0.344	62.72	33.70	8
10	10.750	10.018	0.336	78.82	40.07	8
12	12.750	12.000	0.375	113.09	48.99	8

Sizes of soil and waste pipe   Table 17-3 gives the maximum number of fixture units which may be connected to any given size of drain, soil, or waste pipe serving a building.

Steel pipe   Black and galvanized steel pipes are available in standard, extra-strong, and double-extra-strong weights, with standard weights most commonly used for plumbing purposes. Table 17-4 gives the dimensions and weights of standard steel pipe.

Brass and copper pipe   Brass and copper pipe are frequently used instead of steel pipe for water. They are available in standard and extra-strong grades. The external diameters and number of threads per inch are the same as for steel pipe. Table 17-5 gives the dimensions of brass and copper pipe.

Copper tubing   Copper tubing is frequently used instead of steel pipes for water. It is furnished in both hard and soft tempers. Both tempers are available in straight 20-ft lengths, and the soft temper is furnished in 30-, 45-, and 60-ft coils in sizes to 1¼ in. diameter. The tubes are joined by sweating the ends into special fittings, using a solder or brazing alloy, which is heated with a blowtorch. Because of the flexibility, the tubing can be bent easily to change directions, thus reducing the number of fittings and the amount of labor required. This tubing is frequently used instead of lead pipe for connections from water mains into buildings.

Table 17-6 gives the dimensions and weights of soft- and hard-temper copper tubing.

PVC water-supply pipe   This pipe, made of polyvinyl chloride, is a rigid plastic pipe capable of withstanding internal water pressures up to 160 or 200 lb per sq in. It is light in weight and highly resistant to corrosion and has low resistance to the flow of water and can be installed rapidly and economically.

Joints between the pipe and fittings are made by applying a coating of solvent cement around the end of the pipe and then inserting it into the fitting.

TABLE 17-5   Dimensions and Weights of Standard Brass and Copper Pipe

Size, in.	Diameter, in.		Thickness, in.	Weight, lb per lin ft	
	External	Internal		Brass	Copper
⅛	0.405	0.281	0.0620	0.246	0.259
¼	0.540	0.375	0.0825	0.437	0.460
⅜	0.675	0.494	0.0905	0.612	0.644
½	0.840	0.625	0.1075	0.911	0.959
¾	1.050	0.822	0.1140	1.235	1.299
1	1.315	1.062	0.1265	1.740	1.831
1¼	1.660	1.368	0.1460	2.558	2.692
1½	1.900	1.600	0.1500	3.038	3.196
2	2.375	2.062	0.1565	4.018	4.228
2½	2.875	2.500	0.1875	5.832	6.136
3	3.500	3.062	0.2190	8.316	8.75
3½	4.000	3.500	0.2500	10.85	11.42
4	4.500	4.000	0.2500	12.30	12.94
4½	5.000	4.500	0.2500	13.74	14.46
5	5.563	5.062	0.2500	15.40	16.20
6	6.625	6.125	0.2500	18.45	19.41
7	7.625	7.062	0.2815	23.92	25.17
8	8.625	8.000	0.3125	30.06	31.63
9	9.625	8.937	0.3440	36.95	38.88
10	10.750	10.019	0.3655	43.93	46.22
12	12.750	12.000	0.3750	53.71	56.51

TABLE 17-6   Dimensions and Weights of Soft- and Hard-temper Copper
Tubing

Nonimal size, in.	Outside diameter, in.	Type K soft		Type M hard	
		Wall thickness, in.	Weight, lb per lin ft	Wall thickness, in.	Weight, lb per lin ft
¼	⅜	0.032	0.133	0.025	0.106
⅜	½	0.049	0.269	0.025	0.144
½	⅝	0.049	0.344	0.028	0.203
⅝	¾	0.049	0.418		
¾	⅞	0.065	0.641	0.032	0.328
1	1⅛	0.065	0.839	0.035	0.464
1¼	1⅜	0.065	1.04	0.042	0.681
1½	1⅝	0.072	1.36	0.049	0.940
2	2⅛	0.083	2.06	0.058	1.46
2½	2⅝	0.095	2.92	0.065	2.03
3	3⅛	0.109	4.00	0.072	2.68
3½	3⅝	0.120	5.12	0.083	3.58
4	4⅛	0.134	6.51	0.095	4.66
5	5⅛	0.160	9.67	0.109	6.66
6	6⅛	0.192	13.87	0.122	8.91
8	8⅛	0.271	25.90	0.170	16.46
10	10⅛	0.388	40.26	0.212	25.57
12	12⅛	0.405	57.76	0.254	36.69

This pipe is not suitable for use with hot water.

Table 17-7 lists representative sizes of PVC pipe and costs.

Table 17-8 lists representative PVC fittings for use with the pipe, with current costs.

Indoor CPVC plastic water pipe This pipe, made of chlorinated polyvinyl chloride, may be used inside a building for hot or cold water, if local building codes permit its use. It will withstand internal water pressures up to 100 lb per sq in., or more, at temperatures up to 180°F. Joints between pipes and fittings are made by applying coatings of solvent cements to the ends of the pipes and then inserting them into the fittings.

Labor installing plastic water pipe Installing plastic water pipe involves cutting standard lengths of pipe to the desired lengths, if necessary using a hack saw, applying a solvent cement to the ends, and then inserting the ends into plastic fittings. This forms a solid joint, which should be free of any leakage.

TABLE 17-7 Representative Costs of PVC Plastic Water Pipe Class 160 Lb per Sq In.

Pipe size, in.	Cost per lin ft
½	$0.08
¾	0.09
1	0.11
1¼	0.17
1½	0.20
2	0.28

TABLE 17-8 Representative Costs of Fittings for PVC Plastic Water Pipe

Fitting	Price each by size, in.				
	½	¾	1	1-¼	1-½
90° elbow	$0.29	$0.39	$0.69	$0.80	$1.00
45° elbow	0.55	0.59	0.65	0.95	1.30
Tee	0.30	0.50	0.89	0.99	1.20
Coupling	0.10	0.11	0.29	0.40	0.55
Male adapter	0.15	0.25	0.40	0.55	0.65
Female adapter	0.17	0.27	0.35	0.50	0.60

TABLE 17-9 Representative Time Required to Install Plastic Water Pipe and Fittings

Operation	Time, hr, by size of pipe, in.				
	½	¾	1	1¼	1½
Cut pipe	0.08	0.10	0.11	0.12	0.13
Apply cement and join pipe and fittings*	0.05	0.05	0.06	0.07	0.08
Join plastic pipe to steel pipe with adapter	0.20	0.20	0.25	0.25	0.30

* This is the time required for making one joint. A coupling and an elbow require two joints and a tee requires three joints.

Connections between plastic pipes and standard steel pipes can be made using plastic male or female adapters, which are threaded at one end to match the threads of the steel pipe.

Table 17-9 gives representative times required to perform various operations in installing plastic pipe.

Soil, waste, and vent pipes The pipes which convey the discharge liquids from plumbing fixtures to the house drain are called soil and waste pipes. Pipes which receive the discharge from water closets are called soil pipes, while the pipes that receive the discharge from other fixtures are called waste pipes. Pipes which provide ventilation for a house drainage system and prevent siphonage of water from traps are called vent pipes.

Cast-iron soil pipe is furnished in 5-ft lengths, with single or double hubs, in standard and extra-heavy weights. The weights are given in Table 17-10.

The joints for cast-iron soil pipe are usually made with lead and oakum. The oakum, which is wrapped around the spigot end of the pipe, should be well calked into the hub of the joint to center the spigot and to prevent molten lead from flowing into the pipe. An asbestos runner is wrapped around the pipe adjacent to the hub, with an opening at the top, into which molten lead is poured to fill the joint in one operation. After the lead solidifies, the runner is removed, and the lead is heavily calked.

Table 17-11 gives the approximate quantity of lead and oakum required for a joint in cast-iron soil pipe.

House drain pipe The cast-iron soil pipe must extend a specified distance outside a building, depending on the plumbing code. The balance of the drain

TABLE 17-10 Approximate Weights of Soil Pipe, in Pounds per Linear Foot

Size, in.	Standard, single hub	Extra heavy	
		Single hub	Double hub
2	3.6	5.0	5.2
3	5.2	9.0	9.4
4	7.0	12.0	12.6
5	9.0	15.0	15.6
6	11.0	19.0	20.0
8	17.0	30.0	31.4
10	23.0	43.0	45.0
12	33.0	54.0	57.0

TABLE 17-11 Approximate Quantities of Lead and Oakum Required for Cast-iron Soil Pipe

Size pipe, in.	Weight per joint, lb	
	Lead	Oakum
2	1.5	0.13
3	2.5	0.16
4	3.5	0.19
5	4.3	0.22
6	5.0	0.25
8	7.0	0.38
10	9.0	0.50
12	11.0	0.70

TABLE 17-12 Check List for Estimating the Cost of Roughing-in Plumbing

Description	Quantity	Unit cost	Total cost
Permits			
Tapping fees			
Removing and replacing pavement			
Water meter and box, if required			
Steel pipe			
Copper pipe and tubing			
Pipe fittings			
Valves			
Soil pipe			
Water pipe			
Vent pipe and stacks			
Traps, cleanouts			
Cast-iron fittings			
Floor drains			
Roof flashings			
Vitrified-clay sewer pipe and fittings			
Concrete sewer pipe and fittings			
Catch basins and covers			
Manholes and covers			
Lead, solder, oakum, gasoline, etc.			
Cement and sand			
Brackets, hangers, supports, etc.			
Other items			

pipe extending to the sanitary sewer main may be vitrified clay, concrete sewer pipe, or sometimes a composition pipe of fiber and bitumen. Joints for clay and concrete pipes may be made with cement mortar or an asphaltic jointing compound.

Fittings Fittings for black and galvanized-steel pipe should be malleable iron.

Fittings for copper and brass pipe and tubing should be copper and brass, respectively, threaded or sweat type.

Fittings for plastic pipe should be of the same material as the pipe. If different types of plastic are used, the solvent cement will not be effective on both types of materials.

Fittings for cast-iron pipe should be cast iron.

Adapters are available which may be used in joining one type of pipe or fitting to another type.

Valves Valves are installed to shut off the flow of water. Several types are used including globe, gate, check, sill cocks, drain, etc. They are usually made from bronze and brass.

Traps Traps are installed below the outlets from fixtures to retain water as a seal to prevent sewer gases from entering a building. Vent pipes installed into the drain pipes below the traps prevent the water in the traps from being siphoned out.

Roughing-in plumbing Roughing-in includes the installation of all water pipes from the meter into and through a building, soil pipe, waste pipe, drains, vents, traps, plugs, cleanouts, etc., but does not include the installation of plumbing fixtures, which is called finish plumbing.

Estimating the cost of roughing-in plumbing When preparing a detailed estimate covering the cost of furnishing materials, equipment, and labor for roughing-in plumbing, a comprehensive list of items required should be used as a check and for establishing all costs. Each item should be listed separately by description, grade, size, quantity, and cost, both unit and total. If there are other costs such as tapping fees, permits, removing and replacing pavement, they must be included in an estimate. Table 17-12 may be used as a guide in preparing a list for estimating purposes.

Cost of materials for rough plumbing While the costs of materials for rough plumbing will vary with grades, quantities, locations, and time, the costs given in the tables in this book may be used as a guide in preparing an approximate estimate. The actual costs that will apply should be determined and used when preparing an estimate for bid purposes

Cost of lead, oakum, and solder Ingot or pig lead used for joints with cast-iron pipe will cost about $0.35 per lb.

Oakum will cost about $0.55 per lb.

TABLE 17-13   Representative Costs of ABS Plastic Drainage and Waste Pipe and Fittings (See Fig. 17-1)

Number	Item	Size in.	Unit	Unit cost
1	ABS pipe	1½	Lin ft	$0.19
	ABS pipe	2	Lin ft	0.26
2	ABS pipe	3	Lin ft	0.58
3	Roof flashing, Neoprene	1½	Each	3.10
		2	Each	3.40
		3	Each	3.90
4	Coupling	1½	Each	0.17
		2	Each	0.25
		3	Each	0.54
5	Sanitary tee	1½	Each	0.38
		2	Each	0.70
		3	Each	1.85
	Sanitary tee reducer	3 × 1½	Each	1.75
6	90° elbow	1½	Each	0.38
		2	Each	0.55
7	90° elbow	3	Each	1.50
8	Sanitary tee with two 1½ in. side outlets	3	Each	6.50
9	Slip plug	1½	Each	0.35
10	Reducing closet flange	4 × 3	Each	2.30
11	45° elbow	1½	Each	0.32
		2	Each	0.45
		3	Each	1.20
12	P trap with union	1½	Each	1.25
		2	Each	2.30
13	Trap adapter, 1½-in. pipe to 1¼-in. waste	1¼	Each	0.95
14	Trap adapter, 1½-in. pipe to 1½-in. waste	1½	Each	1.05
15	Wye branch	1½	Each	0.55
		2	Each	1.00
		3	Each	2.40

TABLE 17-13—(Continued)

Number	Item	Size in.	Unit	Unit cost
16	Cleanout adapter	1½	Each	0.35
		2	Each	0.45
		3	Each	1.00
17	Threaded cleanout plug	1½	Each	0.18
		2	Each	0.20
		3	Each	0.36
18	Combination Y and ⅛ bend	3	Each	3.25
19	Male iron pipe adapter	1½	Each	0.27
		2	Each	0.50
20	Adapter, plastic to iron hub	1½ × 2	Each	1.15
		2 × 2	Each	1.05
	Adapter, plastic to plastic	3 × 4	Each	1.65

Wire solder used to sweat joints for copper pipe and tubing will cost about $1.70 per lb.

Plastic drainage pipe and fittings Many building codes permit the use of plastic drainage pipe and fittings, such as ABS-DWV ASTM-2661-68, or later, for use in draining liquids from buildings. These pipes are generally available in sizes 1¼ in. in diameter and larger, and in lengths of 10 or 20 ft, or more. Joints are made by applying a solvent cement to the spigot end of a pipe before inserting it into a fitting. Adapters are available to permit this pipe to be connected into cast-iron pipe.

Figure 17-1 illustrates an assembly of pipes and fittings representing a typical drainage system for a home or other building. Table 17-13 lists the items illustrated in Fig. 17-1, together with representative costs of the items.

Labor required to rough-in plumbing Plumbers frequently work in teams of two men, a plumber and a helper. However, since this arrangement is not always followed, the labor time given in the tables is expressed in labor-hours, which includes the combined time for a plumber and a helper.

The operations required to install water pipe will include cutting and threading the pipe and screwing it together with the appropriate fittings. Cutting and threading may be done with hand or power tools. The time required will vary with the size of the pipe, tools used, and working conditions at the job. If power tools can be moved along with the work, a minimum

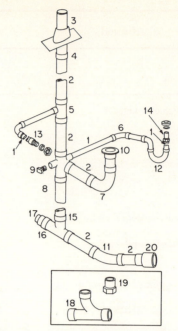

Fig. 17-1 Representative piping and fittings for a drain system.

amount of time will be required, but if the pipe must be carried some distance to the tools, a great deal of time will be consumed in walking. If the pipe is suspended from hangers, additional time will be required to install the hangers. Some specifications require that all pipe cut on the job shall be reamed to remove burrs. This will consume additional time.

TABLE 17-14 Representative Cost of Standard-weight Steel Pipe

Pipe size, in.	Cost per lin ft	
	Black	Galvanized
½	$0.24	$0.29
¾	0.31	0.37
1	0.43	0.52
1¼	0.54	0.66
1½	0.64	0.78
2	0.82	0.99
2½	1.31	1.58
3	1.65	2.05
4	2.62	3.17

TABLE 17-15   Cost of Malleable Fittings for Steel Pipe
per Fitting

Fitting	Size, in					
	½	¾	1	1¼	1½	2
Black						
Couplings	$0.19	$0.26	$0.35	$0.45	$0.57	$0.82
Elbows	0.19	0.26	0.58	0.87	1.08	1.57
Reducers	0.27	0.31	0.44	0.61	0.72	1.04
Tees	0.19	0.26	0.68	1.08	1.30	1.89
Unions	0.61	0.65	0.99	1.43	1.71	2.48
Galvanized						
Couplings	0.21	0.29	0.39	0.50	0.64	0.92
Elbows	0.20	0.29	0.65	0.98	1.20	1.73
Reducers	0.29	0.35	0.49	0.69	0.80	1.15
Tees	0.21	0.29	0.75	1.20	1.45	2.08
Unions	0.68	0.72	1.10	1.59	1.90	2.73

Copper tubing is furnished in coils up to 60 ft long in all sizes to 1¼ in. Since the tubing is longer than steel pipe and changes in direction are obtained by bending the tubing, fewer fittings are required and the labor time will be less than for steel pipe. Joints usually are made by sweating the tubing into fittings, using solder and a blowtorch. Some cutting of the tubing with a hack saw will be necessary.

TABLE 17-16   Cost of Copper
Pipe and Tubing per Lin Ft

Pipe size, in.	Type L hard or soft	Type K hard or soft
⅜	$0.38	$0.50
½	0.52	0.62
¾	0.78	1.04
1	1.07	1.29
1¼	1.43	1.63
1½	1.80	2.03
2	2.80	3.10

TABLE 17-17   Cost of Copper Solder Fittings

Fitting	Size, in.					
	½	¾	1	1¼	1½	2
Couplings	$0.15	$0.24	$0.62	$0.92	$1.20	$ 2.50
Elbows	0.15	0.30	0.92	1.15	1.55	3.00
Tees	0.20	0.54	1.82	2.50	3.75	8.00
Unions	0.75	1.15	2.40	3.15	4.00	10.00
Gate valves	2.95	3.40	4.65			

TABLE 17-18   Cost of 125-lb Brass Screwed Valves per Valve

Type valve	Size in.					
	½	¾	1	1¼	1½	2
Globe and angle	$3.10	$3.75	$5.60	$8.05	$11.80	$12.90
Gate and check	3.10	3.65	5.40	8.10	11.90	14.30
Stop cock	3.00	4.15	4.85	5.60	6.25	
Hose bib	1.69	1.98				

TABLE 17-19   Cost of Cast-iron Soil Pipe per Lin Ft

Pipe size, in.	Single hub		Double hub	
	Standard	Extra heavy	Standard	Extra heavy
2	$0.65	$0.85	$0.98	$ 1.15
3	0.75	1.05	1.05	1.40
4	0.90	1.20	1.20	1.70
5	1.35	1.50	1.45	1.95
6	1.80	2.25	2.20	2.65
8	2.95	4.15	4.75	5.75
10	5.00	6.90	6.90	8.95
12	7.00	9.50	9.50	11.75

TABLE 17-20 Cost of Single-hub Fittings for Cast-iron Pipe, per Fitting

Fitting	Size, in.				
	2	3	4	6	8
Quarter bend, standard	$0.86	$1.40	$1.90	$ 3.50	$ 9.75
Extra heavy	0.96	1.70	2.50	4.10	10.90
Eighth bend, standard	0.68	1.15	1.70	2.90	8.75
Extra heavy	0.88	1.40	1.95	3.40	9.75
T and Y branch, standard	1.30	2.25	3.40	5.90	19.50
Extra heavy	1.45	2.75	3.90	7.25	22.50
Cross, standard	1.90	3.15	4.45	12.50	24.50
Extra heavy	2.15	3.60	5.25	13.40	26.50
Running trap, with hub vent	2.25	3.90	4.95		
Extra heavy	2.45	4.40	5.75		
Plain trap, S and P, standard	1.30	2.50	3.95	14.50	
Extra heavy	1.60	3.25	4.75	16.75	

Cast-iron soil, waste, and vent pipes are joined with molten lead and oakum joints. Many of these joints can be made with the pipe or fittings standing in an upright position, which will reduce the time required. Line joints are made by pouring the lead with the pipe in place, using an asbestos runner to hold the lead in the joint. After the lead has solidified, it must be calked. Time should be estimated by the number of joints.

TABLE 17-21 Cost of Vitrified-clay Sewer Pipe, per Lin Ft

Pipe size, in.	ASTM C-700-71	ASTM C-425-71 Rubber seal
4	$ 0.53	$ 0.53
6	0.80	0.76
8	1.25	1.15
10	1.75	1.65
12	2.25	2.22
15	4.28	4.07
18	5.95	5.87
24	12.50	12.30

TABLE 17-22   Cost of Vitrified-clay Sewer Pipe Fittings
with Rubber Seal

| Pipe size, in. | Cost each | | | |
| | Fitting | | | |
	Wyes & tees	⅛, ¼, ½ bends	Cleanout wyes	Stoppers
4	$ 2.52	$ 2.52	. . . .	$0.51
6	4.39	3.68	$5.00	0.71
8	6.67	5.65	7.69	1.04
10	10.60	9.87	. . . .	1.56
12	14.34	13.33	. . . .	2.08

Vitrified-clay sewer pipe is manufactured in lengths varying from $2\frac{1}{2}$ to 4 ft. Joints are made with oakum and cement mortar or a heated asphaltic joint compound.

Many plumbing contractors add a percentage of the cost of materials to cover the cost of labor required to rough-in plumbing. The rates used vary from 40 to 80 percent of the cost of materials. While this is a simple operation, it does not necessarily produce an accurate estimate.

Table 17-23 gives representative time in labor-hours required to rough-in plumbing. The values are based on an analysis of the rates reported by a substantial number of plumbing contractors whose individual rates varied considerably. The rates given in the table include the combined time for plumbers and helpers. Thus, if a plumber and a helper work together, one-half of the labor-hours should be assigned to each man.

FINISH PLUMBING

Finish plumbing includes fixtures such as lavatories, bathtubs, shower stalls, water closets, sinks, urinals, water heaters, etc., and the accessories which usually are supplied with the fixtures. The prices of finish plumbing vary a great deal with the type, size, and quality specified. No book of this kind can give a complete listing of fixtures and prices. An estimator should list each item required for a building and then refer to a current catalog for prices.

Table 17-24 gives representative prices for plumbing fixtures in the medium-grade range, which may be used as a guide only.

Labor required to install fixtures   The labor required to install plumbing fixtures will vary with the kind and quality of fixture, the kind of accessories used, the type of building, and to some extent the building code. The simplest

TABLE 17-23 Labor-hours Required for Roughing-in Plumbing

Class of work	Labor-hours
Hand cut, thread, and install steel pipe, per joint:	
½- and ¾-in. pipe	0.6–0.7
1- and 1¼-in. pipe	0.8–0.9
1½- and 2-in. pipe	1.3–1.5
2½- and 3-in. pipe	1.8–2.4
4-in. pipe	2.2–2.5
Machine cut, thread, and install steel pipe, per joint:	
½- and ¾-in. pipe	0.4–0.7
1- and 1¼-in. pipe	0.5–0.6
1½- and 2-in. pipe	0.7–0.8
2½- and 3-in. pipe	0.9–1.0
4-in. pipe	1.2–1.4
Install copper tubing, per joint:	
½- and ¾-in. tubing	0.3–0.4
1- and 1¼-in. tubing	0.4–0.5
1½- and 2-in. tubing	0.5–0.6
2½- and 3-in. tubing	0.6–0.7
4-in. tubing	0.7–0.8
Install cast-iron soil pipe and fittings, per joint:	
2-in. diameter	0.3–0.4
3-in. diameter	0.4–0.5
4-in. diameter	0.5–0.6
6-in. diameter	0.7–0.9
8-in. diameter	1.0–1.3
Install vitrified-clay or concrete pipe, per joint:	
4-in. diameter	0.2–0.3
6-in. diameter	0.2–0.3
8-in. diameter	0.3–0.4
10-in. diameter	0.5–0.6
12-in. diameter	0.7–0.9
Rough-in for fixtures:	
Bathtub	10–18
Bathtub with shower	16–24
Floor drain	4–6
Grease trap	5–10
Kitchen sink, single	8–14
Kitchen sink, double	10–16
Laundry tub, 2-compartment	8–12
Lavatory	10–15
Shower with stall	12–18
Slop sink	8–12
Urinal, pedestal type	8–12
Urinal with stall	10–14
Urinal, wall type	7–10
Water closet	9–18
Water heater, 30–50 gal automatic	10–12
Water heater, 50–100 gal automatic	12–15

TABLE 17-24 Representative Costs of Plumbing Fixtures and Accessories

Fixture	Cost per unit
Bathtub, porcelain on cast iron, flat bottom	
5 ft 0 in.	$175.00
5 ft 6 in.	185.00
Kitchen sink, enamel on cast iron, single, 24 × 21 × 8 in.	80.00
Kitchen sink, enamel on cast iron, double, 32 × 21 × 8 in.	110.00
Lavatory, porcelain on cast iron, 18 × 24 in.	95.00
Urinal stall, vitreous china	210.00
Water closet, vitreous china, close coupled	75.00
Garbage disposal unit, $\frac{1}{2}$ hp	125.00

fixtures will require the least time, and the more elaborate fixtures will generally require the most time.

A lavatory to be attached to the wall will be delivered to a job packed in a crate, complete with wall brackets, faucets, pop-up stopper, and other accessories, unassembled. It is necessary to remove it from the crate, attach

TABLE 17-25 Labor-hours Required to Install Plumbing Fixtures and Accessories

Fixtures	Labor-hours
Bathtub, leg type	10–16
Bathtub, flat bottom, no shower	14–20
Bathtub, flat bottom, with shower	18–26
Kitchen sink, enamel single	6–10
Kitchen sink, enamel, double	8–12
Laundry tub, double	10–14
Lavatory, wall type	5–8
Lavatory, pedestal type	6–9
Urinal, wall type	8–12
Urinal, pedestal type	9–13
Urinal with stall	16–20
Water closet	6–9
Garbage-disposal unit	6–8
Drinking fountain	5–8

the brackets, then attach and level the lavatory, connect the hot- and cold-water lines, and connect the trap and drain into the waste pipe. Additional time will be required to install legs and towel bars.

More time is required to install a combination bathtub with a shower than to install a bathtub only. Also, considerably more time is required to install a built-in bathtub, especially in a bathroom with tile floors and walls, than to install a tub supported on legs. All factors that affect the rates of installation must be considered when preparing an estimate.

Table 17-25 gives representative labor-hours required to install plumbing fixtures, based on reports furnished by a substantial number of plumbing contractors. Use the lower rates for simple fixtures installed under favorable conditions and the higher rates for more complicated fixtures installed under more difficult conditions.

PROBLEMS

For these problems use the national average wage rates, as given in this book. Use a plumber and a helper as a team in performing the work. Use the costs of materials given in this book.

17-1 Estimate the direct cost of furnishing and installing 340 lin ft of 1-in. galvanized-steel pipe, whose average length per joint prior to cutting will be 21 ft 0 in. It will be necessary to cut and thread the pipe to install ten galvanized tees, six galvanized elbows, four galvanized unions, and three brass gate valves. The pipe will be cut and threaded by hand.

17-2 Estimate the cost of furnishing and installing 420 lin ft of ¾-in. type K soft-copper tubing, including 12 couplings, 28 tees, 18 ells, and 4 unions, using sweated joints.

17-3 Estimate the cost of furnishing and installing 360 lin ft of 1½-in. type L copper pipe, including 14 couplings, 12 tees, 4 unions, and 6 elbows, using sweated joints.

17-4 Estimate the cost of furnishing and installing 286 lin ft of 4-in. single-hub, standard weight cast-iron soil pipe, including six wyes, six eighth bends, and three P traps, all standard fittings, under average conditions. All joints will be made with oakum and lead.

17-5 Estimate the cost of furnishing and installing 280 lin ft of 1½-in. diameter ABS drain pipe supplied in 20-ft lengths. The following fittings will be required, with each fitting requiring that the pipe be cut:

   6 sanitary tees
   5 90° elbows

The cost of a 1-pint can of solvent cement will be $1.60. This is enough cement for making 50 joints with 1½-in. pipe.

18
Electric Wiring

The installation of electric wiring is generally done by a subcontractor who specializes in this type of work, with all materials and labor furnished under the contract. This procedure does not eliminate the need for estimating a job; it simply transfers the operation from the general to the electrical contractor.

Approximate estimates, and sometimes estimates for bid purposes, are prepared by counting the number of outlets required for a job and then applying a unit price to each outlet for the total cost. While this method will permit estimates to be prepared quickly, it will not always give an accurate cost for a job, as conditions may vary a great deal between two jobs. When this method is used, each switch, plug, and fixture served will count as an outlet.

Factors which affect the cost of wiring The cost of wiring a building will be affected by the type, size, and arrangement of the building and the materials used in constructing the building. Wiring for a frame building may be attached to the framing members or run through holes drilled in the members. Wiring for a building constructed of masonry may be placed in conduit attached to or concealed within the walls. Wiring for concrete structures is placed in conduit installed in the concrete, which will require the installation of the conduit prior to placing the concrete.

The cost per outlet will be affected by the lengths of runs between outlets, the kind of installation, and the sizes of wires required. Job specifications may require no conduit, rigid or flexible conduit, armored cable, or nonmetallic cable. The costs of these materials and the labor required to install them may vary considerably. Unless an estimator is able to apply a dependable cost factor, based on job conditions, to a particular job, the outlet method may result in a substantial variation from the true cost.

The best method of preparing a dependable estimate is to prepare an accurate material take-off which lists the items separately by type, size, quantity, quality, and cost. When working from plans, it is good practice to indicate on the plans, using a colored pencil, each item as it is transferred to the material list. This procedure will reduce the danger of omitting items or counting them more than once. The application of appropriate costs for materials and labor to these items will give a dependable estimate.

Items included in the cost of wiring   When preparing a detailed estimate covering the cost of wiring a project, it is good practice to start with the first item required to bring electric service to the project. This may be the service wires from the transmission lines to the building, unless they are furnished by the utility. The listing should proceed from the first item to and including the last ones, which may be the fixtures. The following list may be used as a guide:

1   Service wires, if required
2   Meter box
3   Entrance switch box and fuses
4   Wire circuits
   a   Conduit
   b   Wires
5   Junction boxes
6   Outlets for
   a   Switches, single pole, double pole, three way
   b   Plugs, wall and floor
   c   Fixtures
7   Bell systems, with transformers
8   Sundry supplies, such as solder, tape, gasoline, etc.

Types of wiring   The types of wiring used in buildings may be governed by the National Electrical Code, published by the National Board of Fire Underwriters, or in some locations by state or municipal codes, which specify minimum requirements for materials and workmanship. Codes require that electric wiring be protected against physical damage and possible short circuits by one or more methods such as installing them in conduit or by the use of wires that are furnished with factory-wrapper coverings.

Rigid conduit   Rigid conduit is made of light galvanized pipe. Heavy conduit, which may be used for outside or inside installations, is threaded at each end. As the walls of light conduit are too thin for threads, it is necessary

TABLE 18-1 Size Conduit Required for Rubber-covered Electric Wires

Size of wire, B & S gauge	Maximum number of wires in one conduit								
	1	2	3	4	5	6	7	8	9
	Size conduit required, in.								
14	½	½	½	½	¾	¾	¾	1	1
12	½	½	½	¾	¾	1	1	1	1¼
10	½	½	¾	¾	1	1	1	1¼	1¼
8	½	¾	1	1	1¼	1¼	1¼	1¼	1½
6	½	1	1¼	1¼	1½	1½	2	2	2
5	½	1¼	1¼	1¼	1½	2	2	2	2
4	½	1¼	1¼	1½	2	2	2	2	2½
3	¾	1¼	1¼	1½	2	2	2	2½	2½
2	¾	1¼	1¼	1½	2	2	2½	2½	2½
1	¾	1½	1½	2	2	2½	2½	3	3
0	1	1½	2	2	2½	2½	3	3	3
00	1	2	2	2½	2½	3	3	3	3½
000	1	2	2	2½	3	3	3	3½	3½
0000	1¼	2	2	2½	3	3	3½	3½	4

TABLE 18-2 Resistance and Current Capacity of Copper Wire

B & S gauge	Resistance, ohms per M ft	Capacity, amp	
		Rubber covered	Weather-proof
18	6.374	3	5
16	4.009	6	10
14	2.527	15	20
12	1.586	20	25
10	0.997	25	30
8	0.627	35	50
6	0.394	50	70
5	0.313	55	80
4	0.248	70	90
3	0.197	80	100
2	0.156	90	125
1	0.124	100	150
0	0.098	125	200
00	0.078	150	225
000	0.062	175	275
0000	0.049	225	325

to use special clamping fittings when making joints or entering boxes. The conduit extends into outlet or other boxes, where it is securely fastened with locknuts and bushings or with clamps. The conduit must be adequately supported and fastened in position along its run. The electric wires are pulled through the conduit at any desirable time after the conduit is installed.

Table 18-1 gives the sizes of conduit required for the indicated sizes and numbers of wires, as specified by the National Electrical Code.

Flexible metal conduit   This conduit consists of an interlocking spiral steel armor, which is constructed in such a manner that it permits considerable flexibility. It may be used where frequent changes in direction are necessary. End connections must be made with special fittings.

Armored cable   Armored cables or conductors are made with or without a lead sheath under the armor. Special fittings are required to connect this cable into boxes, outlets, etc.

Nonmetallic cable   The wires of this cable are covered with a flexible insulated braiding impregnated with a moisture-resisting compound. It is furnished in wire sizes 14 to 4 gauge. It is very popular for use in residences.

Electric wire   Electric wires are usually made of copper, either solid or standard. The wires may be rubber covered for use indoors or weatherproof for use outdoors. The size of a wire is designated by specifying the Brown and Sharpe, or B & S, gauge number, or for sizes larger than B & S gauge No. 0000, by the number of circular mils.

Table 18-2 gives the resistance in ohms per M ft, and the electric current capacity in amperes, as specified by the National Electrical Code, for copper wire.

TABLE 18-3 Representative Costs of Rigid Conduit

Size, in.	Cost per 100 ft	
	Heavy	Light
½	$ 29.70	$ 11.70
¾	38.80	15.50
1	55.70	22.30
1¼	69.50	27.83
1½	83.25	31.70
2	105.60	42.25
2½	164.27	66.70
3	224.80	91.35
3½	278.02	113.15
4	328.85	133.35

Accessories In addition to the conduit and wire, numerous accessories will be needed to install wiring. Among the items needed will be the following:

1 Meter box
2 Entrance cap and conduit
3 Entrance box with switch and fuses or circuit breaker
4 Couplings, elbows, locknuts, bushings, and straps for conduit
5 Special fittings and clamps for flexible conduit and cable
6 Outlet boxes
7 Junction boxes
8 Light sockets
9 Rosettes
10 Solder, tape, straps, etc.

Cost of materials The cost of materials for electric wiring will vary with the location, quality, quantities purchased, and time. Representative costs in effect in 1974 are given in Tables 18-3 to 18-8. The costs given in the tables should be used as a guide only. When estimating the cost of a project for contract purposes, the estimator should obtain current prices applicable to the particular project.

TABLE 18-4 Representative Cost per 100 Fittings for Rigid Conduit

| Fitting | Size, in. | | | | | |
	½	¾	1	1¼	1½	2
For heavy conduit						
Couplings	$ 12.32	$ 18.12	$ 26.55	$ 31.85	$ 43.40	$ 66.50
Elbows	48.80	63.70	98.10	138.50	185.70	285.25
Locknuts	2.46	3.12	5.08	6.60	8.20	14.40
Bushings	4.28	6.60	10.95	14.80	18.00	34.45
Straps, 2 hole	2.03	2.17	3.05	5.30	5.58	7.05
Condulets, LB, etc.	69.50	82.50	126.50	200.50	262.00	435.55
Condulets, T	87.20	104.60	157.00	235.50	313.50	487.50
Condulets, X	113.50	139.60	191.50	271.00	348.00	610.00
For thin-wall conduit						
Elbows	......	......	50.00	73.60	96.50	185.50
Connectors	18.90	26.15	40.70	74.10	109.00	155.50
Couplings	22.50	31.20	48.50	81.30	116.00	159.00

TABLE 18-5 Representative Cost of Flexible Metal Conduit

Size, in.	Cost per 100 lin ft
½	$14.34
¾	19.72
1	41.85
1¼	53.75
1½	66.90
2	86.70

Labor required to install electric wiring The labor required to install electric wiring is estimated by at least three methods: by assuming a certain cost per outlet, by assuming that labor will cost a certain percent of the cost of materials, and by assuming the time required to perform each operation of the work.

The first two methods may produce a sufficiently accurate estimate if dependable cost data are available from similar work previously done. If an estimate is prepared for a project that is unlike work previously done, it may be difficult to assume costs accurately.

The third method requires the preparation of a list of all materials needed, by quantity and quality. If the time required to install each item is known, the

TABLE 18-6 Representative Cost of Nonmetallic Cable

B & S gauge	Number of wires	Cost per 1,000 lin ft	
		Without ground wire	With ground wire
14	2	$ 70.50	$ 75.20
12	2	91.80	110.00
10	2	125.40	188.60
8	2	212.60	257.80
14	3	125.20	155.20
12	3	151.70	212.40
10	3	212.60	274.80
8	3	424.10	531.20
6	3	518.80	662.70

TABLE 18-7 Representative Cost of Single-strand Rubber-covered Copper Wire

B & S gauge	Cost per 1,000 lin ft
14	$ 41.25
12	56.80
10	82.60
8	161.65
6	222.90
4	316.60
2	441.00
1	624.30
0	740.80

total time for the job can be estimated. Electric wiring is usually installed by two men, an electrician and a helper, who work together as a team. The time per operation may be based on labor-hours or team-hours.

If the estimate is started with the service wires that enter a building, it should include any work required to bring the wires to the building. Frequently the utility company will bring the service wires to a building, and the contractor will install an entrance cap, a conduit, and wires to the meter box, then install conduit and wires from the meter to a service fuse panel, with the main switch located within the building. From this panel, as many separate circuits as are necessary are installed throughout the building. Frequently a single

TABLE 18-8 Representative Cost of Accessories

Item	Unit	Cost
Service fuse panel with switch:		
Four branches	Each	$ 11.00–15.00
Six branches	Each	15.00–22.00
Outlet boxes, 4 in.	100	35.00–47.00
Switch boxes	100	35.00–47.00
Wall plates	100	14.00–26.00
Switches:		
Toggle, single pole	100	35.00–55.00
Toggle, three way	100	50.00–65.00
Mercury, single pole	100	110.00–125.00
Receptacles	100	25.00–50.00

conduit may be used for several circuits for at least a portion of the building, with individual circuits coming out of junction boxes installed in the main conduit line.

Wires must be installed from a circuit to each outlet, switch, and plug.

Heavy rigid conduit is installed with threaded couplings, elbows, and locknuts and bushings where it enters boxes. It should be supported with pipe straps or in some other approved manner, depending on the type construction used for the building. This conduit is furnished in 10-ft lengths.

TABLE 18-9   Labor-hours Required to Install Electric Wiring

| Class of work | Unit | Labor-hours | |
		Electrician	Helper
Install service entrance cap and conduit	Each	0.5–1.0	0.5–1.0
Install conduit and fuse panel	Each	0.5–1.0	0.5–1.0
Install heavy rigid conduit with outlet boxes:			
½ and ¾ in.	100 lin ft	5.0–10.0	5.0–10.0
1 and 1¼ in.	100 lin ft	7.0–11.0	7.0–11.0
1½ in.	100 lin ft	9.0–13.0	9.0–13.0
2 in.	100 lin ft	12.0–17.0	12.0–17.0
2½ in.	100 lin ft	15.0–21.0	15.0–21.0
3 in.	100 lin ft	20.0–28.0	20.0–28.0
4 in.	100 lin ft	25.0–35.0	25.0–35.0
Install thin-wall conduit with outlet boxes:			
½ and ¾ in.	100 lin ft	4.0–6.0	4.0–6.0
1 in.	100 lin ft	4.3–7.0	4.3–7.0
1¼ in.	100 lin ft	4.5–7.5	4.5–7.5
1½ in.	100 lin ft	5.5–9.0	5.5–9.0
Install flexible conduit with outlet boxes:			
½ and ¾ in.	100 lin ft	3.0–5.0	3.0–5.0
1 and 1¼ in.	100 lin ft	4.0–6.0	4.0–6.0
Install nonmetallic cable with outlet boxes:			
14/2, 12/2, 10/2, and 8/2*	100 lin ft	3.0–6.0	3.0–6.0
14/3, 12/3, 10/3, and 8/3*	100 lin ft	3.5–6.5	3.5–6.5
6/3 and 4/3	100 lin ft	4.0–7.0	4.0–7.0
Pull wire through conduit and make end connections, per circuit:			
14, 12, and 10 gauge	100 lin ft	0.5–1.0	0.5–1.0
8 and 6 gauge	100 lin ft	1.0–1.5	1.0–1.5
4 gauge	100 lin ft	1.5–2.0	1.5–2.0
2 gauge	100 lin ft	2.2–3.2	2.2–3.2
1 gauge	100 lin ft	2.7–3.7	2.7–3.7

* The numbers 8/2 designate two wires, each No. 8 gauge in a cable.

Lightweight rigid conduit is installed with special couplings and connectors.

Flexible conduit, armored, and nonmetallic cable may be bent easily as it is installed. Usually only end connectors are required where it enters boxes. It can be installed more rapidly than rigid conduit.

Where conduit, boxes, and accessories are installed in a building under construction, especially a concrete building, it may be necessary to keep one or more electricians on the job most of the time, even though they are unable to work continuously. This possibility should be considered by an estimator.

When installing 100 ft of straight rigid conduit, the operations will include connecting 10 joints into couplings, plus the installation of any outlet boxes required. The installation of an outlet box requires the setting of a locknut and a bushing for each conduit connected to the box. More time will be required for large than for small conduit. Table 18-9 gives the labor-hours required for installing conduit.

The installation of a switch, fixture outlet, or plug requires a two-wire circuit from the main circuit to the outlet. The time required will vary with the length of the run and the type of building. For a frame building it may be necessary to drill one or more holes through lumber, whereas for a masonry building it may be necessary to install the conduit through cells in concrete blocks, then chip through the wall of a block to install the outlet box.

The wires usually are pulled through the conduit just before completing the building.

FINISH ELECTRICAL WORK

Under roughing-in electrical work the conduit, wires, and outlet boxes are installed. After the building is completed and the surfaces are painted, the fixtures are installed. The costs of fixtures vary so much that a listing of prices in this book would be of little value to an estimator. The costs should be obtained from a current catalog or a jobber.

TABLE 18-10 Labor-hours Required to Install Electrical Fixtures

Class of work	Labor-hours per fixture	
	Electrician	Helper
Install ceiling fixture	0.2–0.4	0.2–0.4
Install wall light	0.2–0.4	0.2–0.4
Install base or floor plug	0.1–0.2	0.1–0.2
Install wall switch	0.1–0.2	0.1–0.2
Install three-way switch	0.15–0.25	0.15–0.25

Labor required to install electrical fixtures Installing a fixture such as a switch or a plug involves connecting it to the wires which are already in the outlet box and attaching a cover plate. Installing a ceiling or wall fixture such as a light involves securing the fixture to a bracket and connecting it to the wires. Table 18-10 gives representative labor-hours required to install electrical fixtures.

PROBLEMS

For these problems use the national average wage rates and the costs of materials given in this book. Use an electrician and a helper to do the work.

18-1 Estimate the cost of furnishing and installing 420 lin ft of 1½-in. heavy rigid conduit, including 32 couplings, 10 elbows, and 15 T-type condulets under average conditions.

18-2 Estimate the cost of furnishing and installing 540 lin ft of 1-in. heavy rigid conduit, including 36 couplings, 18 elbows, and 12 T-type condulets when the installation is made in a concrete building under construction. Conditions will be such that the electrician and helper can work at only about 75 percent of normal efficiency.

18-3 Estimate the cost of furnishing and installing the following items in a frame residence:

   1 service fuse panel with switch for six branches

   160 lin ft of ¾-in. light conduit

   140 lin ft of ½-in. light conduit

   26 junction boxes, 4 in.

   28 switch boxes

   12 single-pole toggle switches

   16 flush receptacles, wall plugs

   8 ¾-in. elbows

   12 ¾-in. couplings

   24 ¾-in. connectors

   12 ½-in. elbows

   18 ½-in. couplings

   48 ½-in. connectors

   28 wall plates

   320 lin ft of No. 10 gauge rubber-covered copper wire, two wires per circuit

   280 lin ft of No. 12 gauge rubber-covered copper wire, two wires per circuit

18-4 Estimate the cost of furnishing and installing the wiring for a frame residence, including the following items:

   1 1¼-in. entrance cap, cost $2.20

   20 lin ft of 1¼-in. heavy rigid conduit

   4 1¼-in. elbows

   2 1¼-in. couplings

   4 1¼-in. locknuts and bushings

   1 service fuse panel with switch for six branches

980 lin ft of No. 12 gauge two-wire nonmetallic cable with ground wire
320 lin ft of No. 14 gauge two-wire nonmetallic cable with ground wire
26 junction boxes, 4 in.
36 switch boxes
36 wall plates
20 flush receptacles, wall plugs
16 single-pole mercury switches

19
Steel Structures

Types of steel structure Steel is used to erect such structures as multistory buildings, auditoriums, gymnasiums, theaters, churches, mill buildings, roof trusses, stadiums, bridges, wharfs, towers, etc. In addition to steel structures, steel members are frequently used for columns, beams, and lintels and for other purposes.

Materials used in steel structures In so far as it is possible, steel structures should be constructed with members which are fabricated from standard shapes, such as H columns, I beams, WF beams, channels, angles, and plates. Members made from standard rolled shapes are usually more economical than fabricated members. However, if standard shapes are not available in sufficient sizes to supply the required strength, it is necessary to fabricate the members from several parts such as standard shapes and plates or lattices.

Connections for structural steel In fabricating standard shapes to form the required members or in connecting the members into the structure, three types of connections are used: rivets, bolts, and welds. Each type of connection has its place in the field of structural-steel construction and will be discussed in greater detail in a later part of this chapter.

Estimating the costs of steel structures In estimating the cost of structural steel for a job, a contractor will submit a set of plans and specifications for the structure to a commercial steel fabricator for quotations. The fabricator will make a quantity take-off, including main members, details, and miscellaneous items, to which he will apply shop costs for fabricating, riveting, welding, painting, overhead, and profit as a basis for submitting a quotation to the general contractor. The cost of transporting the steel to the project must be added to the cost of the finished products at the shop. This procedure will establish the cost of the fabricated steel delivered to the job.

Most general contractors who erect buildings and similar structures subcontract the erection of the steel to contractors who specialize in this work. This practice is justified because the erection of steel is a highly specialized operation which should be performed by a contractor with suitable equipment and a well-trained erection crew. Because of these conditions, the general contractor can usually have the erection done more economically by a subcontractor than he can with his own equipment and employees. The charge for the erection is generally based on an agreed price per ton of steel in place, including riveting, bolting, or welding the connections.

When estimating the cost of structural steel in place, a building contractor will include in his estimate the cost of the steel delivered to the project, the cost of erection, and the cost of field painting as required. To these costs he will add his cost for job overhead, general overhead, and profit.

Items of cost in a structural-steel estimate The items of cost which should be considered in preparing a comprehensive detailed estimate for a steel structure include the following:

1 The cost of structural-steel shapes at the fabricating shop
2 The cost of preparing drawings for use by the shop in fabricating the steel
3 The cost of handling and fabricating the steel shapes into finished members
4 The cost of shop painting, if required
5 The cost of shop overhead, sales, and profit
6 The cost of transporting the steel to the job
7 The cost of erecting the steel, including equipment, labor, bolts, rivets, or welding
8 The cost of field painting the steel structure
9 The cost of job overhead, general overhead, insurance, taxes, and profit

The cost of any one or all of these items may vary considerably between two projects, and consequently the cost of each item must be estimated for a particular project. The costs of these items are discussed hereafter.

The cost of structural-steel shapes at the fabricating shop Structural-steel shapes are made in many sizes by rolling mills. These shapes are purchased by shops which specialize in fabricating steel members. The base price of the shapes at a fabricating shop varies with the price charged by the mills and the cost of transporting the steel to the shop.

If the steel members are furnished by a fabricating shop, the job estimator does not need to consider the mill price and the cost of transportation from the mill to the shop separately. Only the base price at the fabricating shop will concern him.

The actual price per pound of structural-steel shapes varies at a given mill or shop with the size and weight of the shape and the quantity of steel required.

Extra charges for size and section   As indicated in Table 19-1, the rolling mills which supply steel shapes to fabricating shops assess charges that are added to the base price of steel. Table 19-1 illustrates representative extra charges for only selected sizes and sections. A complete list of the extra charges may be obtained from mills which produce structural-steel shapes.

Extra charges for quantity   Quantity extra charges are determined by the total theoretical weight of the individual size, weight, gauge, or thickness of a structural section ordered of one grade or analysis on the same order, released and accepted for one mode of shipment to one destination at one time. Table 19-2 gives the extra charges for quantity effective as of January 1, 1973.

Example   Determine the cost per cwt for 3,160 lb of 6- by 6 by ½-in. structural-steel angles, including the base price, size extra, and quantity extra charges only.

$$
\begin{array}{llr}
\text{Base price at shop} & = & \$13.05 \text{ per cwt} \\
\text{Size extra charge} & = & 0.65 \text{ per cwt} \\
\text{Quantity extra charge} & = & 0.25 \text{ per cwt} \\
\hline
\text{Total cost} & = & \$13.95 \text{ per cwt}
\end{array}
$$

Estimating the weight of structural steel   In estimating the probable weight of structural steel for a job, the estimator should determine from the plans the total number of linear feet for each shape by size or weight. Structural-steel handbooks give the nominal weights of all sections. However, variations in weights, amounting to 2½ percent above or below the nominal weights, are permissible and may occur. The purchaser is charged for the actual weight furnished, provided that the weight does not fall outside the permissible variation.

The weight of the details for connections should be estimated and priced separately if a detailed estimate is desirable. In estimating the weight of a steel plate having an irregular shape, the weight of the rectangular plate from which the shape is cut should be used. Steel weighs 490 lb per cu ft.

The cost of preparing shop drawings   In preparing plans for a steel structure, the engineer or architect does not furnish drawings in sufficient detail to permit the shop to fabricate the members without additional

TABLE 19-1 Representative Size Extra Charges over the Base Prices for Structural-steel Shapes

Size, in.	Weight, lb per ft	Extra charge per cwt
Wide flange shapes		
W36	230–300	$0.55
W36	135–194	0.50
W30	172–210	0.50
W30	99–132	0.45
W24	68–160	0.45
W24	55–61	0.60
W21	55–142	0.45
W18	64–114	0.45
W18	35–40	0.65
W16	58–96	0.45
W16	36–50	0.55
W14	142–426	0.45
W14	61–136	0.45
W12	65–190	0.45
W12	40–58	0.50
W12	27–36	0.60
W10	49–112	0.50
W10	21–29	0.75
W10	15–19	1.10
W8	31–67	0.55
W8	17–20	0.90
W6	12–16	1.60
Standard beams		
S24	79.9–120	$0.80
S20	65.4–95	0.80
S18	....	0.80
S15	....	0.80
S12	31.8–35	0.75
S10	....	0.85
S8	....	1.00
S6	....	1.10
S4	....	1.30

TABLE   19-1   —(Continued)

Size, in.	Weight, lb per ft	Extra charge per cwt
Standard channels		
C15	All	0.70
C12	All	0.80
C10	All	0.90
C9	All	1.00
C8	All	1.10
C6	All	1.15
C4	All	1.30
Angle sections		
∟ 8 × 8	All	0.50
∟ 6 × 6	All	0.65
∟ 5 × 5	All	0.70
∟ 4 × 4	All	0.70
∟ 3 × 3	All	0.80
∟ 9 × 4	All	0.60
∟ 8 × 6	All	0.50
∟ 8 × 4	All	0.65
∟ 6 × 4	All	0.65
∟ 4 × 3*	. . .	0.80
∟ 3 × 2*	. . .	0.90

* Except for special thicknesses, for which other extra charges apply.

TABLE   19-2   Quantity Extra Charges for Structural Steel

Quantity	Extra charge per cwt
4,000 lb and over	None
2,000 lb to 3,999 lb	$0.25
1,000 lb to 1,999 lb	0.75
Under 1,000 lb	2.25

TABLE 19-3   Representative Costs of Shop Drawings for Structural Steel

| | Cost of shop drawings for various connections | | | |
| | Riveted and bolted | | Welded | |
	Per ton	Per cwt	Per ton	Per cwt
Steel skeleton, office buildings, etc.	$25.00–$45.00	$1.25–$2.25	$25.00–$45.00	$1.25–$2.25
Mill and factory buildings	40.00–75.00	2.00–3.75	40.00–85.00	2.00–4.25
Churches, museums, theaters, etc.	60.00–100.00	3.00–5.00	60.00–100.00	3.00–5.00
Roof trusses, per 24- × 36-in. sheet	125.00–175.00	. . . .	125.00–175.00	
Plate girders	18.00–35.00	0.90–1.75	20.00–45.00	1.00–2.25
Girders of rolled sections	20.00–40.00	1.00–2.00	22.50–45.00	1.13–2.25
Highway bridges, truss types, per lin ft	20.00–25.00	. . . .	20.00–25.00	

information. Fabricating shops maintain drafting departments, which prepare shop drawings in sufficient detail to permit the shops to fabricate the members. One job may require as few as one or two sheets, while another job may require more than 100 sheets. The cost of preparing the drawings is based on the complexity of the detailing and the number of sheets required. As the total cost of the drawings is charged to the steel supplied for a job, the cost per unit weight of steel will vary with the total cost of the drawings and the quantity of steel supplied. The cost per ton for lightweight roof trusses will be high, while the cost per ton for large beams fabricated from rolled sections will be considerably lower.

Shop drawings are usually prepared on 24- by 36-in. sheets. The cost of preparing a sheet may vary from $25.00 to $150.00, depending on the amount of detailing required. The cost of preparing a sheet must be charged to the finished steel members that are fabricated from the information given on the sheet. If only one member is fabricated, the cost per member is high; if a great many members are fabricated, the cost per member will be much lower.

Table 19-3 gives the approximate cost of shop drawings for various types of steel structures.

The cost of shop handling and fabricating structural steel   The cost of shop handling and fabricating structural steel will vary considerably with the operations performed, the sizes of the members, and the extent to which the operations are duplicated on similar members.

For riveted and bolted connections the fabricating operations include cutting, punching, milling, planing, and shop riveting.

For welded connections the fabricating operations include cutting, some punching for temporary bolt connections, milling, beveling, planing, and shop welding.

TABLE 19-4   Representative Costs of Fabricating Structural Steel Using Riveted and Bolted Connections

| | Cost of fabricating | |
Operation	Per ton	Per cwt
Punching beams and channels:		
6 to 10 in.	$28.00–$40.00	$1.40–$2.00
18 in. and larger	21.00–30.00	1.05–1.50
Punching and framing beams and channels with end connections:		
6 to 16 in.	52.00–72.00	2.60–3.60
18 in. and larger	35.00–52.00	1.75–2.60
Framing beams with plates and channels:		
6 to 16 in.	70.00–105.00	3.50–5.25
18 in. and larger	44.00–61.00	2.20–3.05
Fabricating plate and angle columns:		
Weight up to 40 lb per lin ft	105.00–140.00	5.25–7.00
Weight 40 to 80 lb per lin ft	70.00–95.00	3.50–4.75
Weight over 80 lb per lin ft	61.00–78.00	3.05–3.90
Fabricating columns from channels and lacings:		
Weight up to 40 lb per lin ft	150.00–170.00	7.50–8.75
Weight 40 to 80 lb per lin ft	105.00–140.00	5.25–7.00
Weight over 80 lb per lin ft	87.00–115.00	4.35–5.75
Fabricating built-up girders:		
Weight up to 200 lb per lin ft	115.00–130.00	5.75–6.50
Weight 200 to 300 lb per lin ft	95.00–120.00	4.75–6.00
Weight 300 to 400 lb per lin ft	87.00–115.00	4.35–5.75
Weight 400 to 600 lb per lin ft	79.00–96.00	3.95–4.80
Fabricating H columns, including bases, splices, and end milling:		
Up to 8 in.	80.00–115.00	4.00–5.75
Larger than 8 in.	62.00–87.00	3.10–4.35
Fabricating roof trusses*:		
Weight up to 1,200 lb	150.00–185.00	7.50–9.25
Weight 1,200 to 2,400 lb	140.00–168.00	7.00–8.40
Weight 2,400 to 3,600 lb	115.00–140.00	5.75–7.00
Weight over 3,600 lb	87.00–115.00	4.35–5.75

* These prices are for single trusses. If several identical trusses are fabricated, the prices will be less.

Tables 19-4 and 19-5 give ranges in the costs of various fabricating operations for structural steel. The cost per ton or per hundredweight is based on the weight of the finished members, including details, rivets, etc. The lower costs should be used when identical operations are performed on several members, while the higher costs should be used when the operations are performed on only a few members. This is necessary in order to provide for the time required to set up the fabricating equipment, which is fairly constant regardless of the number of operations performed.

The cost of applying a coat of shop paint to structural steel Specifications for structural steel frequently require the fabricator to apply a coat of paint after the fabricating is completed. A gallon of paint should cover about 400 sq ft of surface. Spray guns are generally used to apply the paint.

TABLE 19-5   Representative Costs for Fabricating Structural Steel for Welded Connections

Operation	Cost of fabricating	
	Per ton	Per cwt
Punching and framing beams and channels with end connections:		
6 to 18 in.	$52.00–$80.00	$2.60–$4.00
18 in. and larger	35.00– 56.00	1.75– 2.80
Framing beams with angles:		
6 to 18 in.	52.00– 80.00	2.60– 4.00
18 in. and larger	35.00– 56.00	1.75– 2.80
Fabricating plate and angle columns:		
Weight up to 40 lb per lin ft	120.00–155.00	6.00– 7.75
Weight 40 to 80 lb per lin ft	78.00–115.00	3.90– 5.75
Weight over 80 lb per lin ft	61.00– 87.00	3.05– 4.35
Fabricating built-up plate girders:		
Weight up to 200 lb per lin ft	120.00–150.00	6.00– 7.50
Weight 200 to 300 lb per lin ft	105.00–130.00	5.25– 6.50
Weight 300 to 400 lb per lin ft	95.00–120.00	4.75– 6.00
Weight 400 to 600 lb per lin ft	88.00–105.00	4.40– 5.25
Fabricating H columns, including bases, splices, beam connections, and end milling:		
Up to 8 in.	87.00–105.00	4.35– 5.25
Larger than 8 in.	70.00– 96.00	3.50– 4.80
Fabricating roof trusses:		
Weight up to 1,200 lb	160.00–200.00	8.00–10.00
Weight 1,200 to 2,400 lb	145.00–180.00	7.25– 9.00
Weight 2,400 to 3,600 lb	120.00–148.00	6.00– 7.40

TABLE 19-6 Cost of Applying a Shop Coat of Paint to Structural Steel

Member or structure	Sq ft per ton	Cost per ton			Cost per cwt
		Paint	Labor	Total	
Beams	200–250	$7.00	$5.55	$12.55	$0.63
Girders	125–200	3.90	4.80	8.70	0.44
Columns	200–250	7.00	5.55	12.55	0.63
Roof trusses	275–300	7.80	7.20	15.00	0.75
Bridge trusses	200–250	7.00	5.55	12.55	0.63

Table 19-6 gives the approximate cost for applying a coat of paint to structural steel for various types of members and structures. A painter, using a spray gun, should paint 1 to 2 tons per hr, depending on the size of the sections.

The cost of shop overhead and profit   If the prices given previously are for the actual costs of steel, shop drawings, and fabricating and painting steel, it will be necessary to add to these costs the cost of overhead and profit in order to determine the probable cost of the fabricated steel at the shop. The combined cost of overhead and profit will vary from 15 to 25 percent, depending on the size of the order.

The cost of transporting structural steel   The cost of transporting structural steel from the fabricating shop to the job will vary with the quantity of steel, the method of transporting it, and the distance from the shop to the job. If it is possible to haul the steel the entire distance by trucks instead of by a combination of railroad and trucks, the cost of an intermediate handling will be eliminated. A truck should haul approximately 10 tons per load.

The estimator should determine the freight or truck cost per ton or per cwt for the particular project in order that he may include the correct amount in his estimate.

If the steel is delivered by railroad to a siding near the project, unloaded onto trucks, and hauled to the project, the estimator must include the cost of freight, unloading from the cars, and hauling to the project.

The cost of fabricated structural steel delivered to a project   The following example illustrates a method of estimating the cost of structural steel delivered to a project.

Example   Estimate the total cost and the cost per cwt of structural steel for the columns, beams, and details for a framed steel building. All members are to be fabricated at a shop for bolted connections using high tensile-strength bolts for connections.

The list of members and details is given in the accompanying table:

No. of pieces	Description	Length each	Weight, lb per lin ft	Total weight, lb
18	Columns, W10 × 89	24 ft 9 in.	89	39,649
12	Columns, W10 × 112	24 ft 9 in.	112	33,264
18	Columns, W10 × 54	21 ft 6 in.	54	20,898
12	Columns, W10 × 72	21 ft 6 in.	72	18,576
Total weight of columns				112,387
30	16- × 16- × 1½-in. base plates			3,267
120	6- × 18- × ¾-in. splice plates			2,763
120	∟ 3 × 3 × ⅜	0 ft 8 in.	7.2	575
420	∟ 2½ × 2 × ³⁄₁₆	0 ft 8 in.	2.75	577
Total weight of column details				7,182
68	Beams, W14 × 48	17 ft 6 in.	48	57,120
54	Beams, W14 × 34	19 ft 9 in.	34	36,261
24	Beams, W12 × 32	16 ft 6 in.	32	12,672
18	Beams, W12 × 28	22 ft 6 in.	28	11,340
Total weight of beams				117,393
492	∟ 3 × 3 × ⅜	1 ft 0 in.	7.2	3,542
164	∟ 3 × 3 × ¼	0 ft 10 in.	4.9	670
Total weight of angles				4,212
Total weight of fabricated steel, lb				241,174
Total weight, tons				120.59

The total cost for the job will be

> Base cost for the fabricating shop:
>> Columns, 1,123.87 cwt @ $13.00 = $14,610.30
>> Beams, 1,173.93 cwt @ $13.00 = 15,261.00
>> Plates, 60.30 cwt @ $13.00 = 783.90
>> L 3 × 3, 47.87 cwt @ $13.00 = 622.31
>> L 2½ × 2, 5.77 cwt @ $13.00 = 75.01
> Size extra charges:
>> Columns, 1,123.87 cwt @ $0.50 = 561.99
>> Beams, 571.20 cwt @ $0.50 = 285.60
>> Beams, 362.61 cwt @ $0.55 = 199.46
>> Beams, 240.12 cwt @ $0.60 = 144.07
>> ∟ 3 × 3, 47.87 cwt @ $0.80 = 38.30
>> ∟ 2½ × 2, 5.77 cwt @ $0.90 = 5.19
> Quantity extra charges:
>> ∟ 3 × 3 × ¼, 6.70 cwt @ $2.25 = 15.08
>> ∟ 2½ × 2, 5.77 cwt @ $2.25 = 12.98
> Working drawings:
>> Total weight, 120.59 tons @ $35.00 = 4,220.65

Fabricating cost:

Columns, 56.20 tons @ $75.00		=	4,215.00
Beams, 58.70 tons @ $62.00		=	3,639.40

Shop painting:

Total weight, 120.59 tons @ $11.10		=	1,338.55
Total cost of materials		=	$46,028.79
Shop overhead, 10% of $46,028.79		=	4,602.88
Profit, 10% of $46,028.79		=	4,602.88
Total cost f.o.b. the shop		=	$55,234.55
Trucking cost to the job, 120.59 tons @ $4.20		=	506.48
Total cost delivered to the job		=	$55,741.03
Cost per ton, $55,741.03 ÷ 120.59 tons		=	462.40
Cost per cwt, $462.40 ÷ 20		=	23.12

Erecting structural steel When erecting a structural-steel building, the columns are erected first usually on previously prepared concrete foundations, with the anchor bolts already in place. The bases for the columns are supported above the concrete foundations by suitable means, such as with steel shims or steel wedges, to permit the bases to be set at the required elevations and grouted to these elevations. After the columns are erected, the beams are installed for the first tier of floors, usually two floors. The connections between the columns and the beams are temporarily bolted through rivet or other holes, using two or more bolts per connection. After the tier of columns and beams is in place, it is necessary to plumb the structure before installing the permanent bolts. This operation is repeated for subsequent tiers until the erection of the structure is completed.

Equipment for erecting structural steel The equipment used for erecting steel structures depends on the type of structure, the size of the structure and its component parts, and the location.

Roof trusses are usually delivered to the job partly or completely assembled and hoisted directly from the delivery trucks into place using power cranes. The mobility of a crane makes it very useful for such operations.

Multistoried framed steel buildings may be erected with cranes if the height is not excessive, usually about four stories.

If a building is so tall that a crane can not be used, the steel members may be placed with one or more guy derricks, or a tower crane may be used.

Labor erecting structural steel The cost of labor erecting structural steel will vary with the type of structure, the kind of equipment used, the sizes of the members, the kind of connections, the climatic conditions, and the prevailing wage rates.

A crew of steel erectors, classified as ironworkers, may vary from five to eight or more persons, excluding the individuals who bolt, rivet, or weld the connections.

If a power crane is used to erect a steel structure, with relatively light-weight members, the crew might include the indicated persons, at the specified wage rates.

Classification	Wage rate per hr	Total wages
1 foreman	$8.00	$ 8.00
1 crane operator	6.75	6.75
1 crane oiler	5.50	5.50
3 ironworkers*	7.25	21.75
Total cost per hr		$42.00

* One of these men might be classified as a helper at a reduced wage rate in some locations.

One of the ironworkers will hook the lifting line to the members to be lifted, while the other two will make the temporary or permanent bolt connections.

If a tower crane is used to handle the steel members, one or two additional crew members may be required. This crew will set the members in place,

TABLE 19-7   Approximate Rates of Erecting Structural Steel (Does Not Include Final Bolting or Welding)

Type of structure	Type of equipment	Size of crew	Crew-hour per ton
Roof trusses:			
Up to 1,200 lb	Crane	5	1.6
1,200–2,400 lb	Crane	5	1.3
2,400–3,600 lb	Crane	5	1.0
3,600–4,800 lb	Crane	6	0.8
4,800–6,000 lb	Crane	6	0.7
Purlins and braces	Crane	5	1.2
Framed steel structures:			
Up to 4 stories high	Crane	7	0.5
Up to 8 stories high	Tower crane	7	0.5
8 to 18 stories high	Tower crane	8	0.5
Mill buildings, factories, etc., columns and beams	Crane	6	0.8
Churches, theaters, etc.	Crane	5	1.2
Plate girders:			
5–10 tons	Crane	6	0.4
10–20 tons	2 cranes	8	0.3

temporarily bolt the connections, and plumb the structure for the final bolting or welding of the connections.

Table 19-7 gives the approximate rates of erecting steel for various types of structures.

Labor bolting structural steel When roof trusses are erected, one worker will be required at each end of the truss to install at least enough bolts to secure the truss to the supporting structure. One or two other workers may be required on the floor of the building to assist in moving the ends of the trusses into position for bolting. As the trusses are erected some or all of the purlins should be attached to the trusses to assure adequate rigidity for the trusses.

After the trusses are erected and plumbed, two ironworkers should be able to complete the permanent bolting. One helper might be required on the floor.

When a tower crane or a guy derrick is used to hoist the members and place them in positions, it is common practice to install enough bolts in the connections to permit the structure to be plumbed and braced as it is erected. If this procedure is followed, the rest of the permanent bolts will be installed later. Calibrated torque wrenches, selected to produce the desired tension in the bolts, may be used in tightening the nuts on the bolts. The size crew required for installing the bolts will vary with the number of bolts needed and the ease or difficulty in getting at the bolts.

Fig. 19-1 Two power cranes erecting steel roof trusses. (*Lima-Hamilton Corporation.*)

Fig. 19-2 Tower crane erecting structural steel. (*Mosher Steel Company.*)

WELDED STRUCTURES

General information With the improvements in the techniques of welding which have been developed in recent years, it is becoming increasingly common practice to use arc welding in fabricating and erecting steel structures. Arc welding has passed beyond the experimental stage and is now accepted as a safe and satisfactory method of connecting steel members, both in the shop and in the field. Laboratory tests may be applied to determine whether or not welders are qualified and welded connections are satisfactorily made.

Advantages of welded connections The use of arc-welded connections is in many instances more desirable than that of riveted connections. The advantages begin with the design of the structure and continue through the erection.

If a structure is designed for welded connections, it is possible to reduce the total weight of steel by as much as 15 to 25 percent. It is not necessary to punch rivet holes in members at points of critical stress for connection purposes. Where a few holes are required for temporarily bolting members in place, the holes may be punched at noncritical points, thus permitting the use of the full strength of a member at joints. As the welded joints can be made as strong as the full sections of the members, it is possible to design beams as continuous members over several supports, thus reducing the critical bending-moment stresses. As a result of these conditions, lighter beams may be used in a structure, which will reduce the total weight of steel required. Smaller and less-expensive foundation costs should result from reduction in dead weight of the structure.

The use of welded connections will simplify the preparation of plans for a steel structure, as less work is required in preparing the details for joint connections. It is not necessary to determine the number and spacing of rivets or to design gusset plates, as they are not used.

It is frequently less expensive to fabricate steel members for welded connections. The accurate location of rivet holes for shop and field assembling is eliminated. A somewhat greater freedom in dimensioning members is permissible as rivet holes for connecting members, which are punched separately, do not have to be matched exactly in assembling members. Members which are to be field-welded require fewer shop operations, thus reducing the costs and the fabricating time.

An experienced crew should be able to erect a welded-steel structure at no greater cost than for a riveted structure. This is especially true if the members are designed and fabricated in a manner which will facilitate welding, such as providing for down-hand welding of all joints where possible.

Erection equipment required for welded-steel structures In erecting a welded-steel structure one or more electric generators and welding equipment are substituted for the air compressor and riveting equipment. Otherwise the erection equipment is essentially the same as for a riveted structure.

Arc welding may be satisfactorily accomplished with either ac or dc electricity. In order to obtain good welds, it is necessary to provide electrical energy with the desirable characteristics, amperes, and volts for the particular job. The generator selected should be capable of meeting the demands that will be made on it. If this precaution is not observed, it will be difficult to obtain welds of the desired quality.

If alternating current is used for welding, the current may be obtained from commercial sources with a transformer. If direct current is used, the current may be provided with a generator. An electric-motor-driven generator is usually more satisfactory than a gasoline- or diesel-engine-driven generator.

Erecting steel structures with welded connections In erecting a steel structure with welded connections, the same general procedure is followed as for a structure with riveted connections. A few holes are punched in the members, at noncritical points, in order that bolts may be installed for temporarily connecting the members. After a unit of the structure is plumbed or brought to the correct position, welding the connections is started. In order to eliminate undesirable distortion of a structure, resulting from unequal heating at the connections, a definite pattern for welding should be established and rigidly followed. If beams are to be welded to opposite flanges of a column, the two welds should be performed simultaneously in order to eliminate unequal expansion of the two sides of the column, which would result from welding the beams separately.

In welding girders to columns or beams to columns and girders, the top flanges should be welded first and allowed to cool to the atmospheric temperature prior to welding the bottom flanges. As the welds for the bottom flanges

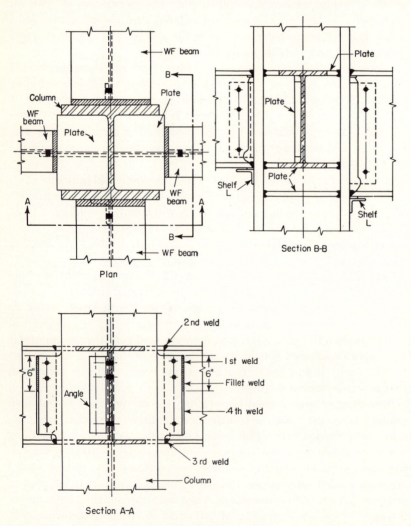

Fig. 19-3 Details of beam-to-column welds. (*Courtesy Boyd S. Myers.*)

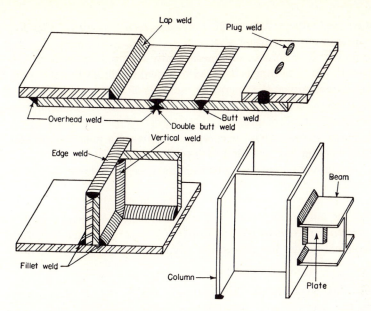

Fig. 19-4   Typical welds.

cool, the bottom flanges at the ends of the beams will be subjected to tension. Since this stress is opposite the stress to which the beams will be subjected under loaded conditions, the effect of welding is to reduce the ultimate bending stresses in the beams.

The correct procedure in welding beams to columns is illustrated in Fig. 19-3.

Types of welds   Names identifying the various types of welds have been standardized and should be used in specifying or referring to welds. Figure 19-4 illustrates various types of structural welds.

Arc-welding terminology   Below are given definitions of terms associated with arc welding.

Kilowatts   (Amperes × volts) ÷ 1,000.

Kilowatthours   Kilowatts × time in hours.

Efficiency of a transformer   The ratio of the output divided by the input for a transformer, expressed as a percent.

Electrodes   Electrodes are steel rods of various sizes and types, which are melted by the electric current to form the welding metal.

Melt-off rate of electrode   The rate at which the electric current melts an electrode, expressed in inches per minute.

Arc speed   The speed at which an electrode is moved along a weld, without interruptions, usually expressed in feet per hour.

Arc time   The time required to produce a given unit of welding, such as 100 ft, usually expressed in hours.

Interruption   A temporary stoppage in welding, as in changing electrodes, in moving from one weld to another, etc.

Operating factor   The percent of time actually devoted to welding. For example, if one-half the total time is devoted to actual welding while the other is consumed by interruptions, the operating factor is 50 percent.

Floor-to-floor speed   The net length of weld produced by a welder, in feet per hour, considering all interruptions.

Fit-up   The manner in which the component parts of a joint are brought together prior to welding, such as the width of the gap at a joint.

Passes   The number of times an electrode must be moved over a joint in order to complete a weld.

Deposition efficiency   The ratio of the amount of metal deposited in the weld divided by the amount of electrode consumed, expressed as a percent.

Methods of producing the most economical welds   In order to obtain the most economical welds, the procedures given below should be followed where possible:

1   Travel as fast as possible within the limits of good weld appearance by always keeping the electrode just ahead of the molten pool.

2   Use the largest electrode practical.

3   Use the highest current practical.

4   Use as short an arc as possible, dragging the electrode coating where practical.

5   Use a proper plate preparation and fit-up.

6   Keep the build-up to a minimum.

Electrodes   Metallic electrodes are manufactured in sizes varying from $\frac{1}{16}$ to $\frac{3}{8}$ in. in diameter or larger and in lengths of 14 and 18 in. For ordinary welding, a 2-in. stub end will be wasted for each electrode.

Electrodes are available for use with either direct or alternating current, and also with different coatings to satisfy the demands for the particular jobs for which they will be used. The proper electrode should be used in order to obtain the best results.

The cost of welding   The cost of welding steel structures includes the cost of electrodes, electricity, labor, and job overhead.

The effect on the cost of welding resulting from changes in the sizes of electrodes is illustrated in Table 19-8. The quantities and costs given are based on the following:

The deposition efficiency is 66-$\frac{2}{3}$ percent.

The operating factor is 50 percent.

The cost of electric energy is $0.02 kilowatthour.

A welder is paid $7.20 per hour.

The cost of job overhead is $6.00 per hour.

TABLE 19-8 The Effect of the Sizes of Electrodes on the Cost of Welding

Item	Size of electrode, in.					
	1/8	5/32	3/16	1/4	5/16	3/8
Cost of electrode per lb	$ 0.28	$ 0.28	$ 0.27	$ 0.27	$ 0.25	$ 0.25
Amperes required	110	130	150	250	320	425
Arc volts required	24	25	26	30	34	38
Kilowatts at arc	2.6	3.2	3.9	7.5	11.1	16.1
Efficiency of set, percent	47	50	51	55	59	59
Kilowatt input	5.6	6.5	7.6	13.6	18.8	27.3
Electrode consumed, lb per hr	1.3	1.7	2.0	3.8	5.4	8.1
Metal deposited, lb per hr	0.87	1.13	1.33	2.53	3.60	5.40
Kilowatt-hour per lb of metal deposited	3.20	2.90	2.85	2.70	2.60	2.55
Interruptions per lb of metal deposited	18	12	8	5	3	2
Cost per pound of metal deposited						
Labor	$ 8.330	$ 6.390	$ 5.400	$ 2.850	$ 2.000	$ 1.320
Overhead	6.940	5.320	4.500	2.270	1.670	1.110
Electricity	0.064	0.058	0.057	0.054	0.052	0.051
Electrode	0.420	0.420	0.396	0.396	0.367	0.367
Interruptions	0.263	0.173	0.115	0.073	0.042	0.026
Total cost per lb	$ 16.017	$ 12.361	$ 10.468	$ 5.743	$ 4.131	$ 2.874

TABLE 19-9   Cost of Fillet and Lap Welds for Horizontal and Flat Positions

Gauge size of fillet, in.	Electrode size, in.	Current, amp	Passes per weld	Electrode per ft of weld, lb	Cost per ft of joint, cents		
					Electrode	Electricity	Total
$5/32$	$3/16$	225	1	0.105	2.76	0.60	3.36
$3/16$	$1/4$	275	1	0.160	4.20	0.86	5.06
$1/4$	$1/4$	325	1	0.195	5.12	1.05	6.17
$5/16$	$1/4$	350	1	0.210	5.51	1.13	6.64
$3/8$	$5/16$	400	1	0.290	7.47	1.51	8.98
$7/16$	$5/16$	425	2	0.540	14.20	2.81	17.01
$1/2$	$5/16$	425	2	0.700	17.91	3.64	21.55
$5/8$	$5/16$	425	4	1.100	28.80	5.72	34.52
$7/8$	$5/16$	425	7	2.200	57.80	11.45	69.25

Labor cost

Welding speed, ft per hr	40	30	20	15	10	8	6	4
Labor cost, cents per ft of joint	18	24	26	48	72	90	120	175

TABLE 19-10   Cost of Fillet and Lap Welds, Vertical Position

Gauge size of fillet, in.	Electrode size, in.	Current, amp	Passes per weld	Electrode per ft of weld, lb	Cost per ft of joint, cents		
					Electrode	Electricity	Total
$5/32$	$5/32$	140	1	0.08	2.10	0.46	2.56
$3/16$	$3/16$	150	1	0.14	3.68	0.80	4.48
$1/4$	$3/16$	170	1	0.17	4.45	0.97	5.42
$5/16$	$3/16$	170	1	0.26	6.81	1.48	8.29
$3/8$	$3/16$	170	1	0.38	10.00	2.16	12.16
$7/16$	$3/16$	170	2	0.52	13.60	2.96	16.56
$1/2$	$3/16$	170	3	0.67	17.58	3.82	21.40
$5/8$	$3/16$	170	Varies	1.10	28.80	8.15	36.95
$7/8$	$3/16$	170	Varies	2.10	55.00	15.55	70.55

Labor cost

Welding speed, ft per hr	20	15	12	10	8	6	4	2
Labor cost, cents per ft of joint	36	48	60	72	90	120	180	360

TABLE 19-11   Cost of Butt Welds

Plate thickness, in.	Electrode size, in.	Current, amp	Passes per weld	Electrode per ft of weld, lb	Cost per ft of joint, cents		
					Electrode	Electricity	Total
¼	³⁄₁₆	220	2	0.25	6.00	1.43	7.43
⁵⁄₁₆	³⁄₁₆	220	2	0.27	6.45	1.54	7.99
⅜	¼	350	2	0.48	11.45	2.59	14.04
⁷⁄₁₆	¼	350	3	0.62	14.80	3.34	18.14
½	¼	350	3	0.82	19.50	4.42	23.92
⅝	¼	350	4	1.25	29.80	6.74	36.54
¾	¼	350	5	1.83	43.60	9.86	53.46
1	¼	350	7	3.13	74.50	16.88	91.38

Labor cost							
Welding speed, ft per hr			20	15	10	8	6	4	2
Labor cost, cents per ft of joint			36	48	72	90	120	180	300

Tables 19-9 and 19-10 give information on the cost of labor, electricity, and electrodes for fillet and lap welds. The costs are based on the following:

The cost of a welder is $7.20 per hour.

The cost of electricity is $0.02 per kilowatthour.

The cost of electrodes is $0.27 per pound.

Horizontal and flat welds are downhand.

Table 19-11 gives information on the cost of labor, electricity, and electrodes for butt welds. The joints will be made with a 60° groove, a ⅛-in. shoulder, and a ¹⁄₃₂-in. gap. The cost of a welder, electricity, and electrodes will be the same as for Table 19-8.

Quantity of details and welds for a structural-steel frame building
The relation between the quantities of structural-steel members, details, and welds is illustrated by the information given below. This structure was arc-welded in accordance with the method shown in Fig. 19-3.

Height of structure	16 stories
Average story height	12 ft 5½ in.
Floor area	253,000 sq ft
Volume of building	3,153,000 cu ft
Total weight of steel	2,274 tons
Weight of steel per sq ft	17.98 lb
Weight of steel per cu ft	1.44 lb

Weight of columns	745 tons
Weight of column details	101.5 tons
Weight of column base plates	47.9 tons
Weight of beams and girders	1,379.6 tons
Total weight of shop electrodes	14,844 lb
Total weight of field electrodes	9,210 lb
Ratios of weights of parts to total weight:	
Columns	32.76%
Column base plates	2.11%
Column details	4.46%
Columns, base plates, and details	39.33%
Beams and girders	60.67%
Job details	6.48%
Electrodes, shop	6.52 lb per ton
Electrodes, field	4.05 lb per ton

Analysis of the cost of welded connections compared with riveted connections In order to compare the cost of welded joints with riveted joints for a representative framed steel building, a bay with one column, girders, and beams was analyzed. The height of one story was 12 ft.

The total lengths of main members were as follows:

Member	Size and length	Weight, lb
Column	14WF211 × 12 ft 0 in.	2,532
Girder	24WF120 × 16 ft 4 in.	1,960
Girder	24WF110 × 13 ft 6 in.	1,485
Beam	18WF70 × 24 ft 6 in.	1,715
Beam	18WF55 × 39 ft 0 in.	2,145
Beam	18WF50 × 39 ft 0 in.	1,950
Total weight of main members	. . . .	11,787

List of details for riveted joints:

4 beams 18WF150 × 1 ft 2½ in.	725
4 beams 18WF140 × 1 ft 0 in.	560
4 ∟ 8 × 8 × ¾ × 0 ft 8½ in.	110
12 ∟ 4 × 3½ × ⅜ × 0 ft 11½ in.	105
Rivets	288
Total weight of details	1,788

List of details for welded joints:

12 ∟ 4 × 3½ × ⅜ × 0 ft 11½ in.	105
4 plates 8 × 1 × 1 ft 0½ in.	113
2 plates 8 × ⅞ × 1 ft 0½ in.	50
2 plates 12 × ½ × 1 ft 5 in.	58

1 plate 4 × ⅝ × 1 ft 7 in.	14
1 plate 4 × ½ × 1 ft 7 in.	12
Electrodes	34
	—
Total weight of details	386

For the riveted joints:

Weight of main members	11,787
Weight of details	1,788
	—
Total weight	13,575

Ratio weight of details to total weight, 1,788 ÷ 13,575 = 13.17%

For the welded joints:

Weight of main members	11,787
Weight of details	386
	—
Total weight	12,173

Ratio weight of details to total weight, 386 ÷ 12,173 = 3.17%

Reduction in total weight:

For riveted joints	13,575
For welded joints	12,173
	—
Reduction in weight	1,402

Ratio of reduction, 1,402 ÷ 13,575 = 10.33%

If a structure is designed for welded joints, it is frequently possible to reduce the sizes of main members. Thus the total reduction in the weight of steel will exceed the reduction due to joints only.

Painting structural steel   The cost of applying field coats of paint to structural steel will vary with the type of structure, the sizes of the members to be painted, and the ease or difficulty of gaining access to the steel members. The cost of painting a ton of steel for a roof truss will be considerably higher than for a framed steel building because of the greater area of steel per ton for the truss and also because of the difficulty of moving along the truss.

TABLE 19-12   Cost of Applying a Field Coat of Paint to Steel Structures Using a Spray Gun

Member or structure	Sq ft per ton	Cost per ton		
		Paint	Labor	Total
Beams	200–250	$5.90	$7.25	$13.15
Girders	125–200	3.55	6.28	9.83
Columns	200–250	5.90	7.25	13.15
Roof trusses	275–350	7.10	9.05	16.15
Bridge trusses	200–250	5.90	7.85	13.75

Paint is usually applied with a spray gun unless local restrictions ban the use of guns. Two field coats are usually applied.

Table 19-12 gives the approximate cost of applying a field coat of paint, using a spray gun, to structural steel for various types of members and structures. The costs are based on paying $9.50 per gallon for paint, and $7.50 per hour to the painter. A painter should paint $\frac{3}{4}$ to $1\frac{1}{2}$ tons per hour, depending on the structure.

20

Water-Distribution Systems

Cost of water-distribution systems The cost of a water-distribution system will include the materials, equipment, labor, and supervision to accomplish some or all of the following:

1 Clear the right of way for the trench
2 Remove and replace pavement
3 Excavate and backfill trenches
4 Relocate utility lines
5 Install pipe
6 Install fittings
7 Install valves and boxes
8 Install fire hydrants
9 Install service connections, meters, and meter boxes
10 Drill holes under roads and pavements and install casings for pipe line
11 Test and disinfect water pipe

Pipe lines The following types of pipes are used for water systems:

1 Cast iron
   a Bell and spigot
   b Mechanical joint

TABLE 20-1 Weights of Bell-and-spigot and Push-on Joint Cast-iron Pipe Centrifugally Cast in Metal Molds, AWWA C106-70 or ANSI A21.6-1970

Average weight, lb per lin ft

Pipe size, in.	Pressure 100* Length, ft		Pressure 150 Length, ft		Pressure 200 Length, ft		Pressure 250 Length, ft	
	18	20	18	20	18	20	18	20
3	12.0	11.9	12.0	11.9	12.0	11.9	12.0	11.9
4	16.1	16.0	16.1	16.0	16.1	16.0	16.1	16.0
6	25.6	25.6	25.6	25.6	25.6	25.6	25.6	25.6
8	36.9	36.8	36.9	36.8	36.9	36.8	36.9	36.8
10	49.0	48.7	49.0	48.7	49.0	48.7	49.0	48.7
12	63.4	63.1	63.4	63.1	63.4	63.1	68.3	67.9
14	78.2	77.8	78.2	77.8	83.8	83.4	89.5	89.0
16	94.5	94.0	94.5	94.0	100.9	100.4	109.0	108.4
18	113.9	113.3	113.9	113.3	122.8	122.2	131.8	131.1
20	134.9	134.2	134.9	134.2	144.9	144.2	154.8	154.0
24	177.3	176.4	189.4	188.4	203.6	202.6	203.6	202.6

* The numerals 100, 150, 200, and 250 designate an internal water pressure of 100 lb per sq in., etc.

    c   Push-on joint, or gasket-seal joint
    d   Threaded
    e   Cement lined
  2 Steel
    a   Threaded, black or galvanized
    b   Welded, plain or cement lined
  3 Reinforced concrete
    a   Prestressed
    b   Nonprestressed
  4 Asbestos cement
  5 Brass and copper
  6 Lead
  7 Plastic

Bell-and-spigot cast-iron pipe   Bell-and-spigot pipe is cast with a bell on one end and a spigot on the other. Joints are made by inserting the spigot end into the bell of an adjacent pipe, calking one or two encircling strands of yarning material into the bell, then filling the balance of the bell with a specified jointing material, such as lead, cement, or an approved substitute. Fittings and valves are installed in a pipe line in the same manner. A joint of pipe may be cut if necessary to permit a fitting or valve to be installed at a designated location.

    The inside surface of cast-iron pipe may be lined with portland cement to reduce tuberculation.

    Several types of cast-iron pipe are available. Table 20-1 gives the weights of bell-and-spigot and push-on joint pipe cast in metal molds, for water and other liquids.

Push-on joint cast-iron pipe   This pipe, which may be identified also as gasket-type joint, or gasket-seal joint, has a built-in rubber-type gasket in the bell or hub which produces a water-tight joint when the spigot end of the joining pipe is forced into the bell. This is now the most widely used type of cast-iron pipe in the water service.

    The American Water Works Association, designated as AWWA, and the American National Standards Institute, Inc., designated as ANSI, specify the same laying lengths, properties, and weights for this pipe as for bell-and-spigot cast-iron pipe for service under the same water pressure. Thus the information appearing in Table 20-1 applies to this pipe.

Fittings for bell-and-spigot cast-iron pipe   Fittings for bell-and-spigot cast-iron pipe include tees, crosses, bends, reducers, etc. The fittings may be purchased with any desired combination of bells and spigots. The joints are made with the same materials which are used in joining the pipe lines.

    Table 20-2 gives the approximate weight of lead and jute required per joint for cast-iron pipe and fittings.

TABLE 20-2 Approximate Weights of Lead and Jute
Required per Joint for Cast-iron Pipe and Fittings

Size pipe, in.	Weight of lead, lb			Weight of jute, lb		
		Per lin ft			Per lin ft	
		Length of pipe, ft			length of pipe, ft	
	Per joint	12	18	Per joint	12	18
3	6.50	0.54	0.36	0.18	0.015	0.010
4	8.00	0.67	0.45	0.21	0.018	0.012
6	11.25	0.94	0.63	0.31	0.026	0.017
8	14.50	1.21	0.81	0.44	0.037	0.025
10	17.50	1.46	0.97	0.53	0.044	0.030
12	20.50	1.71	1.14	0.61	0.051	0.034
14	24.00	2.00	1.33	0.81	0.068	0.045
16	33.00	2.75	1.84	0.94	0.078	0.052
18	36.90	3.07	2.05	1.00	0.083	0.056
20	40.50	3.37	2.25	1.25	0.104	0.070
24	52.50	4.38	2.92	1.50	0.125	0.084
30	64.75	5.40	3.60	2.06	0.172	0.115
36	77.25	6.45	4.30	3.00	0.250	0.167
42	104.25	8.70	5.80	3.50	0.292	0.195
48	119.00	9.93	6.62	4.00	0.334	0.222

Mechanical-joint cast-iron pipe   A joint for this pipe is made by inserting the plain end of one pipe into the socket of an adjoining pipe, then forcing a gasket ring into the socket by means of a cast-iron gland which is drawn to the socket by tightening bolts through the gland and socket. Fittings are installed in a similar manner. Joints may be made quickly with an unskilled crew, an ordinary ratchet wrench being the only tool required.

Table 20-3 gives the weights of mechanical-joint cast-iron pipe for working pressures of 100, 150, 200, and 250 psi for 18- and 20-ft lengths. In some locations 16-ft lengths are available.

Valves   Valves for cast-iron water pipes are usually cast-iron body, bronze-mounted, bell or hub type. Gate valves should be used. A cast-iron adjustable-length valve box should be installed over each wrench-operated valve to permit easy access when it is necessary to operate the valve.

Service lines   Service lines are installed from the water pipes to furnish water to the customers. These lines usually include a bronze corporation cock, which is tapped into the water pipe, copper pipe extending to the property line or meter, a bronze curb cock, and a meter set in a box. Customers install the service pipe from the meter to their house or business.

TABLE 20-3 Weights of Mechanical-joint Cast-iron Pipe Centrifugally Cast in Metal Molds, AWWA C106-70 or ANSI A21.6-1970

Average weight, lb per lin ft

Pipe size, in.	Pressure 100* Length, ft		Pressure 150 Length, ft		Pressure 200 Length, ft		Pressure 250 Length, ft	
	18	20	18	20	18	20	18	20
3	12.0	11.9	12.0	11.9	12.0	11.9	12.0	11.9
4	16.2	16.1	16.2	16.1	16.2	16.1	16.2	16.1
6	25.5	25.4	25.5	25.4	25.5	25.4	25.5	25.4
8	36.4	36.2	36.4	36.2	36.4	36.2	36.4	36.2
10	48.2	48.0	48.2	48.0	48.2	48.0	48.2	48.0
12	62.6	62.3	62.6	62.3	62.6	62.3	67.4	67.1
14	78.2	77.8	78.2	77.8	83.8	83.4	89.4	89.0
16	94.5	94.0	94.5	94.0	100.9	100.3	108.9	108.3
18	113.9	113.2	113.9	113.2	122.8	122.2	131.7	131.0
20	134.9	134.2	134.9	134.2	144.9	144.2	154.8	154.1
24	177.2	176.2	189.2	188.2	203.5	202.6	203.5	202.6

* The numerals 100, 150, 200, and 250 designate an internal water pressure of 100 lb per sq in. etc.

Fig. 20-1 Power-driven pipe cutter.

Fire hydrants Fire hydrants are specified by the type of construction, the size of the valve, the sizes and number of hose connections, the size of the hub for connection to the water pipe, and the depth of bury. It is good practice to install a gate valve between each hydrant and the main water pipe in order that the water may be shut off in the event repairs to the hydrant are necessary.

Tests of water pipes Specifications usually require the contractor to subject the water pipe to a hydrostatic test after it has been laid, prior to backfilling the trenches. If any joints show excessive leakage, they must be recalked. It is common practice to lay several blocks of pipe, install a valve temporarily, and subject the section to a test. If the test satisfies the specifications, the valve is removed and the trench is immediately backfilled. Additional lengths are laid and tested. This procedure is repeated until the system is completed.

Sterilization of water pipes   Prior to placing a water-distribution system in service, it should be thoroughly sterilized. Chlorine is most frequently used to sterilize water pipes. It should be fed continuously into the water which is used to flush the pipe lines. After the pipes are filled with chlorinated water, the water is permitted to remain in the pipes for the specified time; then it is drained out, and the pipes are flushed and placed in service.

Cost of cutting cast-iron pipe   Cast-iron pipe may be cut with chisels or with chain cutters. Chisel cutting is done by two or more men, using a steel chisel with a wood handle and a 6- to 8-lb hammer. Chain cutters may be operated by hand for pipes up to 12 in. in diameter, but for larger pipes a power-driven cutter should be used.

Figure 20-1 illustrates a power-driven cutter, which is operated by compressed air. This machine will cut pipe 10 to 60 in. in diameter. It requires 60 to 70 cu ft of air per minute, at a pressure of 85 psi. The saw is a portable milling machine on wheels, which travels around the pipe under two silent-type chains, which hold the machine to the pipe and act as a flexible ring gear for the feed sprockets. The machine moves, while the chains remain stationary.

TABLE 20-4   Approximate Time in Hours Required to Cut Cast-iron Pipe

| Size pipe, in. | Hand cutting | | | | Power cutting | |
| | With chisel | | With chain cutter | | | |
	Skilled	Laborer	Skilled	Laborer	Skilled	Laborer
4	0.30	0.30	0.25	0.25		
6	0.45	0.45	0.40	0.40		
8	0.60	0.60	0.55	0.55		
10	0.90	0.90	0.80	0.80	0.60	0.60
12	0.95	1.90	0.85	1.70	0.63	1.26
14	1.10	2.20	....	....	0.66	1.32
16	1.25	2.50	....	....	0.70	1.40
18	1.40	2.80	....	....	0.75	1.50
20	1.50	4.50	....	....	0.80	2.40
24	1.65	6.60	....	....	0.85	3.40
30*	1.25	3.75	....	....	0.95	2.85
36*	1.50	4.50	....	....	1.00	3.00
42*	2.00	6.00	....	....	1.10	3.30
48*	2.40	7.20	....	....	1.20	3.60
54*	2.80	8.40	....	....	1.30	3.90
60*	3.20	9.60	....	....	1.50	4.50

* For pipes larger than 24 in. in diameter, include the cost of a crane and an operator to handle the pipe.

A complete cut is made in one revolution around the pipe. It will require 1 min of cutting time for each inch of pipe diameter. Thus a 24-in. pipe is cut in 24 min of cutting time. An experienced crew of two individuals can install the machine on a pipe in approximately 15 min.

The total time allowed for cutting cast-iron pipe should include measuring, supporting on skids, if necessary, and cutting.

In cutting pipe larger than 24 to 30 in. it may be necessary to use a crane to handle the pipe.

Table 20-4 gives the approximate time in hours required to cut various sizes of cast-iron pipe. The time given includes measuring, setting up, and cutting for average conditions.

Labor required to lay cast-iron pipe   The installation of bell-and-spigot cast-iron pipe will include some or all of the following operations:

1   Cutting the pipe, if necessary
2   Lowering the pipe into the trench
3   Inserting the spigot into the bell
4   Yarning the bell
5   Attaching a runner and pouring the lead
6   Removing the runner and calking the lead

Table 20-4 gives representative rates for cutting cast-iron pipe.

Each joint of pipe is lowered into the trench by hand, or with a crane, tractor-mounted side boom, or other suitable equipment. After the spigot end is forced into the bell to full depth, two or more strands of yarning, which completely encircle the pipe, are calked into the bell to center the pipe and to prevent molten lead from flowing into the pipe. An asbestos runner is placed around the pipe against the bell, with an opening near the top to permit molten lead to be poured into the joint. The lead for a joint should be poured in one continuous operation, without interruption. After the lead cools to the temperature of the pipe, the runner is removed and the lead is calked by hand or with a pneumatic calking hammer.

The size of crew required to lay the pipe and the rate of laying will vary considerably with the following factors:

1   Class of soil
2   Extent of ground water present
3   Depth of the trench
4   Extent of shoring required
5   Extent of obstructions such as utilities, sidewalks, pavement, etc.
6   Size of pipe
7   Method of lowering the pipe into the trench
8   Extent of cutting required for fittings and valves

The crew required to dig the trenches, lay the pipe, and backfill the trenches for 12-in. pipe furnished in 18-ft lengths, using a tractor-mounted side boom to lower the pipe into the trench, in trenches 3 to 6 ft deep in firm earth, with no ground water and no shoring needed, might include the following:

1 trenching machine operator
2 laborers on bell holes
1 tractor operator
2 laborers on pipe
2 men centering pipe and installing yarning
1 man melting and supplying lead
1 man installing runners and pouring lead
2* men calking lead joints by hand
1 driver for utility truck
1 bulldozer operator backfilling trench
1 foreman

A crew should install four to six joints per hour, either pipe or fittings. The length of pipe laid will vary from 72 to 108 ft per hr, with 90 ft a fair average for the conditions specified.

Table 20-5 gives representative labor-hours required to lay cast-iron pipe in trenches 3 to 6 ft deep in firm soil with little or no shoring required, and no ground water, using calked-lead joints. For pipe laid under other conditions, the labor-hours should be altered to fit the conditions for the particular project. If a crane is used to lower the pipe, replace the tractor operator with a crane operator.

Labor required to lay cast-iron pipe with mechanical joints The operations required to lay cast-iron pipe with mechanical joints include lowering the pipe into the trench, installing the gasket and gland on the spigot, centering the pipe into the bell, pushing the gasket into the bell, and setting the gland by tightening the bolts. The nuts may be tightened with a hand or pneumatic wrench.

The crew required to dig the trench, lay 12-in. pipe furnished in 18-ft lengths, and backfill the trenches for trenches 3 to 6 ft deep in firm earth with no shoring needed, might include the following:

1 trenching machine operator
2 laborers on bell holes
1 tractor operator
2 laborers on pipe
2 men installing gland and gasket and centering the pipe

* If the joints are calked with a pneumatic hammer, use only one man. It will be necessary to include the cost of an air compressor, hose, and hammer.

TABLE 20-5 Labor-hours Required to Lay 100 Lin Ft of Cast-iron Pipe with Lead Joints

Size pipe, in.	Feet laid per hr	Labor-hours						
		Trenching machine operator	Tractor operators	Truck driver	Kettle man	Pipe layers*	Laborers†	Foreman
4	120	0.83	0.83	0.83	0.83	1.66	4.15	0.83
6	110	0.91	1.82	0.91	0.91	2.73	4.55	0.91
8	100	1.00	2.00	1.00	1.00	3.00	6.00	1.00
10	95	1.05	2.10	1.05	1.05	3.15	6.30	1.05
12	90	1.10	2.20	1.10	1.10	4.40	6.60	1.10
14	85	1.17	2.34	1.17	1.17	4.68	7.02	1.17
16	80	1.25	2.50	1.25	1.25	6.25	7.50	1.25
18	70	1.43	2.86	1.43	1.43	7.15	8.58	1.43
20	60	1.67	3.34	1.67	1.67	8.35	10.02	1.67
24	50	2.00	4.00	2.00	2.00	10.00	12.00	2.00

* Includes men pouring lead and calking joints by hand. If the joints are calked with a pneumatic hammer, reduce the time by one-third.

† Includes the men digging bell holes, handling and centering the pipe. Five men are used for the 4- and 6-in. pipe, and six men for the other sizes.

TABLE 20-6 Labor-hours Required to Lay 100 Lin Ft of Cast-iron Pipe with Mechanical Joints

Size pipe, in.	Feet laid per hr	Labor-hours					
		Trenching machine operator	Tractor operators	Truck driver	Pipe layers	Laborers	Foreman
4	140	0.72	1.44	0.72	1.44	3.60	0.72
6	130	0.77	1.54	0.77	1.54	3.85	0.77
8	120	0.83	1.66	0.83	1.66	4.15	0.83
10	110	0.91	1.82	0.91	1.82	4.55	0.91
12	100	1.00	2.00	1.00	2.00	5.00	1.00
14	90	1.11	2.22	1.11	2.22	5.55	1.11
16	85	1.18	2.36	1.18	2.36	5.90	1.18
18	80	1.25	2.50	1.25	2.50	6.25	1.25
20	75	1.33	2.66	1.33	2.66	6.65	1.33
24	65	1.54	3.08	1.54	3.08	7.70	1.54
30	55	1.82	3.64	1.82	3.64	9.10	1.82

1 man tightening the nuts
1 driver for utility truck
1 bulldozer operator backfilling trench
1 foreman

A crew should lay five to seven joints per hour, either pipe or fittings. The length of pipe laid will vary from 90 to 136 ft per hr, with 100 ft per hr a fair average.

Table 20-6 gives representative labor-hours required to lay cast-iron pipe with mechanical joints in trenches 3 to 6 ft deep in firm soil with little or no shoring required, and no ground water. For pipe laid under other conditions, the labor-hours should be altered to fit the conditions of the particular job. If a crane is used to lower the pipe, replace the tractor operator with a crane operator.

Labor required to lay push-on joint cast-iron pipe The operations required to lay cast-iron pipe and fittings with push-on joints include cutting the pipe to length, if necessary, lowering the pipe and fittings into the trench, using a tractor with a side boom, or a power crane, then forcing the spigot end of the pipe being laid into the bell end of the pipe previously laid. Bell holes should be dug in the trench for proper bedding of the pipe and joints.

The crew required to dig the trench, lay the pipe, and backfill the trench will vary with the requirements of the job, including the method of backfilling the earth around and over the top of the pipe.

Consider a project which requires the laying of 12-in. push-on joint cast-iron water pipe for a pressure of 150 psi, in 18-ft lengths, in a trench whose depth will vary from 4 to 6 ft in a soil that will not require shores or side support for the trenches. The pipe will be laid in a new subdivision of a city, under average conditions.

The trench, which will be 30 in. wide, will be dug with a gasoline-engine-operated wheel-type trenching machine. The backfill must be of select earth, free of rocks, hand-placed and compacted to a depth of 6 in. above the top of the pipe. The rest of the backfill may be placed with a bulldozer and compacted by running the bulldozer along the trench over the backfill, requiring an estimated three or four passes of the bulldozer.

The compaction around and 6 in. over the top of the pipe will be obtained using a hand-operated gasoline-engine-type self-contained tamper.

The pipe will be lowered into the trench with a tractor-mounted side boom. The pipe will be laid along one side of the trench prior to digging the trench.

The crew should lay four to seven joints per hour, for either pipe or fittings. For pipes furnished in 18-ft lengths, the total length laid in an hour should vary from 72 to 136 ft, with 100 ft per hr representing a reasonable average rate.

The crew might include the following:

1 trenching machine operator

2 laborers cleaning the trench and digging bell holes

1 tractor operator lowering pipe into trench

1 laborer assisting tractor operator

2 laborers in trench handling pipe

1 pipe layer with pipe and fittings

2 laborers backfilling trench

1 laborer operating the tamper

1 bulldozer operator backfilling trench

1 driver for the utility truck

1 foreman

Table 20-7 gives representative labor-hours required to lay push-on joint cast-iron pipe in trenches whose depths average around 6 ft, in firm soil with little or no shoring required, with no ground water, and no major obstacles to delay the progress.

In some locations it may be necessary to reclassify some of the individuals of the crew into semiskilled or skilled ratings.

The cost of a cast-iron water-distribution system In estimating the cost of installing a water-distribution system, the estimator must consider the many variables which will influence the cost of the project. No two projects will be alike. For one project there may be very favorable conditions such as a relatively level terrain, free of trees and vegetation, out in the open with no obstructions, no rocks, ground water, utility pipes, or pavement, and little rain to delay the project. The equipment may be in good physical condition. The construction gang may be well organized and experienced. The specifications may not require rigid exactness in construction methods, tests, and cleanup. For another project the conditions may be entirely different, with rough terrain, restricted working room, as in alleys, considerable rock, pavement, and unmarked utility pipes to contend with, and ground water and rain. The equipment may be in poor physical condition, and the construction gang may be poorly organized, with inexperienced workers. The specifications may be very rigid regarding construction methods, tests, and cleanup. As a result of the effect of these variable factors, an estimator should be very careful about using cost data from one project as the basis of estimating the probable cost of another project, especially if the conditions are appreciably different. The following example is intended to illustrate a method of estimating the cost of a water-distribution system but should not be used as the basis of preparing an estimate without appropriate modifications to fit the particular project.

TABLE 20-7 Representative Labor-hours Required to Lay 100 Lin Ft of Push-on Joint Cast-iron Pipe*

| Pipe size, in. | Feet laid per hr | Labor-hours | | | | | | |
		Trenching machine operator	Tractor operators	Truck driver	Pipe layer	Laborers	Foreman
4	140	0.72	1.44	0.72	0.72	5.70	0.72
6	130	0.77	1.54	0.77	0.77	6.15	0.77
8	120	0.83	1.66	0.83	0.83	6.65	0.83
10	110	0.91	1.82	0.91	0.91	7.25	0.91
12	100	1.00	2.00	1.00	1.00	8.00	1.00
14	90	1.11	2.22	1.11	1.11	8.88	1.11
16	85	1.18	2.36	1.18	1.18	9.44	1.18
18	80	1.25	2.50	1.25	1.25	10.00	1.25
20	75	1.33	2.67	1.33	1.33	10.64	1.33
24	65	1.54	3.08	1.54	1.54	12.32	1.54

* If for a given job the estimated number of feet of pipe laid per hour differs from the values given in this table, the information in the table can be used. If the estimated rate of laying 8-in. pipe is 100 ft per hr, the labor-hours for a rate of 100 ft per hr should be used.

Example Estimate the cost per unit in place for a cast-iron pipe water-distribution system consisting of the indicated quantities for each item.

The cast-iron pipe will be AWWA push-on joint for 150-psi water pressure, in 18-ft lengths. The fittings will be push-on joint AWWA class D.

The quantities will be

3,964 lin ft of 12-in. pipe

8 each 12 × 12 × 8 × 8 all bell crosses

6 each 12 × 12 × 6 × 6 all bell crosses

9 each 12 × 12 × 6 all bell tees

2 each cast-iron body bronze-mounted 12-in. gate valves

2 each cast-iron valve boxes for 3 ft 0 in. depth

84 each domestic water services including for each service:

> 1 corporation cock, ¾ in.
>
> 30 lin ft average length of ¾-in. type K copper pipe
>
> 2 copper-to-steel-pipe couplings, ¾ in.
>
> 1 brass curb cock, ¾ in.
>
> 1 water meter, ⅝-in. disk type
>
> 1 cast-iron meter box with cover

The trenches for the pipe will be dug to an average depth of 3 ft 6 in. and 30 in. wide, using a wheel-type trenching machine. The soil is firm sandy clay which will not require shores or trench supports.

In backfilling the trench around and over the pipe the earth will be placed by hand and compacted to a depth of 6 in. above the top of the pipe, using a self-contained gasoline-engine-operated tamper.

The pipe and fittings will be distributed along the trenches prior to digging the trenches. The cost of the pipe and fittings includes furnishing and delivering the materials along the trenches.

Assume that the job conditions will permit the pipe to be laid at a rate of 100 lin ft per hr.

Use Table 20-7 to estimate the cost of labor.

The cost per unit will be

12-in. cast-iron pipe, per lin ft	= $	7.840
Trenching machine, $6.90 per hr ÷ 100 ft per hr	=	0.069
Tractor lowering pipe into trench, $6.40 per hr	=	0.064
Bulldozer filling trench, $7.20 per hr	=	0.072
Utility truck and tools, $4.60 per hr	=	0.046
Earth tamper, $1.20 per hr	=	0.012
Trenching machine operator, $6.60 per hr	=	0.066
Tractor operator, $6.60 per hr	=	0.066
Bulldozer operator, $6.60 per hr	=	0.066
Truck driver, $4.80 per hr	=	0.048
Pipe layer, $5.90 per hr	=	0.059
Laborers, 8 men @ $4.80 per hr each	=	0.384
Foreman, $7.50 per hr	=	0.075
Testing and sterilizing pipe	=	0.050
Total direct cost per lin ft	= $	8.887

Cast-iron fittings, per ton:

Assume that the 12- × 12- × 6-in. tees are representative of all fittings.

Nominal weight of fitting, 458 lb

Fittings delivered to the job	= \$418.00

Trenching machine 4 lin ft × \$0.069 per lin ft = \$0.276 × $\dfrac{2,000}{458}$ =   1.21

Trenching machine operator, 4 lin ft × \$0.066 = \$0.264 × $\dfrac{2,000}{458}$ =   1.15

Labor digging 3 bell holes @ 8 holes per hr = ⅜ × \$4.80 = \$1.80 ×

$\dfrac{2,000}{458}$ =   7.88

Tractor lowering fittings into trench @ 1 ton per hr × \$6.40	= 6.40
Tractor operator, 1 hr per ton × \$6.60	= 6.60
Laborer helping lower fittings, 1 hr × \$4.80	= 4.80
Pipe layer installing fittings, 1 hr × \$5.90	= 5.90
Helpers for pipe layer, 2 men, 2 hr × \$4.80	= 9.60
Foreman, 1 hr × \$7.50	= 7.50
Testing and sterilizing fittings	= 1.75
Utility truck and tools, 1 hr × \$4.60	= 4.60
Total direct cost per ton	= \$475.39

12-in. gate valves, per valve:

Valve, delivered to the job	= \$218.50
Labor digging 2 bell holes @ 8 holes per hr = 2/8 × \$4.80	= 1.20
Pipe layer installing valve, 0.5 hr @ \$5.90	= 2.95
Laborers helping install valve, 3 men × 0.5 hr = 1.5 hr @ \$4.80	= 7.20
Foreman, 0.5 hr @ \$7.50	= 3.75
Utility truck and tools, 0.5 hr @ \$4.60	= 2.30
Total direct cost per valve	= \$235.90

Valve boxes, per box:

Valve box delivered to job	= \$ 19.85
Labor installing box, 0.5 hr @ \$4.80	= 2.40
Labor backfilling around box, 0.5 hr @ \$4.80	= 2.40
Foreman, 0.25 hr @ \$7.50	= 1.88
Utility truck and tools, 0.25 hr @ \$4.60	= 1.15
Total direct cost per box	= \$ 27.68

Domestic water services, per service:

Material for a service:

Corporation cock, ¾ in.	= \$ 4.05
Copper pipe, ¾ in., 30 lin ft @ \$1.04	= 31.20
Copper-to-steel-pipe couplings, 2 @ \$1.35	= 2.70
Curb cock, ¾ in.	= 4.20
Water meter, ⅝ in.	= 41.15
Meter box	= 9.25
Hauling materials to job, per service	= 2.85
Power-operated trencher, 0.5 hr @ \$2.20	= 1.10

Trencher operator, 0.5 hr @ $5.60   =   2.80
Labor tapping cast-iron pipe and installing corporation cock, 1 hr @
$4.80   =   4.80
Pipe layer installing copper pipe, meter, and meter box, 1.5 hr @ $5.90   =   8.85
Helper, 1 man, 1.5 hr @ $4.80   =   7.20
Labor backfilling trench, 1 hr @ $4.80   =   4.80
Foreman, 1 hr @ $7.50   =   7.50
Utility truck and tools, 0.5 hr @ $4.60   =   2.30
                                                                               ———————
Total direct cost per service   = $134.75

Reinforced-concrete pressure pipe   This pipe, which is available in sizes
from 12 to 144 in., or more, is used for water pipes for pressures from 0 to 150 ft
of head, or more. It is manufactured in lengths up to 32 ft, or more. Several
methods are used to manufacture the pipe, which may be prestressed or
nonprestressed.

One type is made with a cage of reinforcing steel, consisting of both
circumferential and longitudinal steel bars, embedded in a concrete wall.

Another type is made with a steel cylinder, lined on the inside with
concrete, reinforced on the outside with steel bars or wire, which may or may
not be prestressed, with the cylinder and reinforcing encased in a concrete
jacket.

Joints are made tight with one or two rubber gaskets which completely
encircle the spigot where it enters the bell of the adjacent pipe.

Fittings, which permit changes in direction and sizes, are available or may
be manufactured upon order.

After a pipe is lowered into the trench with a crane, and the spigot is
forced into the matching bell, the spaces between the ends of the pipe wall are
completely filled with portland cement mortar or grout, both inside and
outside the pipe.

21
Sewerage Systems

Items included in sewerage systems Sewerage systems include vitrified clay or concrete pipe, manholes, cleanout boots, booster pumping stations, and other appurtenances that are necessary to permit the system to collect waste water and sewage and transmit it to a suitable location where it may be disposed of in a satisfactory manner.

A sewerage system may be classified as one of the three types, depending on its primary function:

1 Sanitary sewer
2 Storm sewer
3 Combination sanitary and stormsewer

Sewer pipes Most sewer pipes are made of vitrified clay or concrete, in various sizes and lengths. Bell-and-spigot pipes are generally used. These pipes are available in different wall thicknesses, which are designated as single strength, double strength, etc. The higher strengths should be used when the loads acting on the pipe exceed the safe loads for single strength pipe. The pipes are laid in trenches dug to uniform slopes by hand or by trenching machines, backhoes, and draglines.

Several methods are used to join adjacent pipes into a continuous pipe line. One method is to insert the spigot end of a pipe into the bell of a pipe previously laid in the trench and then insert one or two strands of jute or oakum entirely around the pipe and force it into the annular space between the spigot and the bell. The remaining annular space is then filled with a mortar made of portland cement, sand and water, or with a heated asphaltic compound.

In recent years a modified type of vitrous-clay pipe has been developed, which is proving very satisfactory and economical. The pipe has a rubberlike gasket installed in the bell of the pipe, which engages the spigot of the inserted pipe to produce a flexible but water-tight joint.

The American Society for Testing Materials, ASTM, designates this pipe as compression-joint vitrified-clay bell-and-spigot pipe in ASTM publication Designation C425-71 [1].

Table 21-1 gives the sizes, lengths, and weights of pipes, and the quantities

TABLE 21-1   Sizes, Lengths, Weights, and Quantities of Materials Required for Joints for Vitrified-clay Sewer Pipe

| Size, in. | Length, ft | Weight per ft, lb | Quantity of material per joint | | | |
			Jute, lb	Asphalt,* lb	Oakum, lb	Mortar, cu in.
Single strength						
4	2	9	0.06	0.54	0.06	16
6	2½	15	0.17	0.80	0.09	34
8	3	24	0.36	1.34	0.17	50
10	3	33	0.52	1.60	0.31	68
12	3	45	0.62	1.94	0.36	87
15	3	65	0.78	3.61	0.45	122
18	3	85	0.93	4.28	0.87	182
21	3	120	1.29	5.94	1.12	259
24	3	150	1.48	6.80	1.53	345
30	3	260	3.30	11.30	2.93	646
36	3	360	6.10	16.80	5.54	1,003
Double strength						
15	3	75	0.78	3.61	0.45	122
18	3	105	0.93	4.28	0.87	182
21	3	145	1.29	5.94	1.12	259
24	3	185	1.48	6.80	1.53	345
30	3	300	3.30	11.30	2.93	646
36	3	385	6.10	16.80	5.54	1,003

* Include 10 percent for waste.

of materials required for joints for vitrified-clay sewer pipes, for joints made with asphaltic compounds or with cement and sand mortar.

Fittings for sewer pipe   Fittings for vitrified-clay and concrete sewer pipe and for compression-joint vitrified-clay pipe include single- and double-wye branches, single- and double-tee branches, $\frac{1}{8}$ curves, $\frac{1}{4}$ curves, increasers, and reducers.

Constructing a sewer system   In constructing a sewer system, it is common practice to start at the outlet end of the pipe line and proceed to the other end. Trenches are dug usually with a ladder-type trenching machine. Laborers, using round-point shovels, dig a round trench along the bottom of the machine-dug trench into which the lower one-fourth of the barrel of the sewer pipe can be placed for better bearing. A bell hole should be dug for each joint. At the same time a gauge line should be stretched tightly over batter boards located along and across the trench to assist in establishing a uniform grade for the pipe.

Installing sewer pipe   The sewer pipes are lowered into the trench, by hand or power equipment, depending on the sizes and weights. The bell ends should be placed in the direction of construction. Thus the spigot end of a pipe being installed will be fitted into the bell of the pipe previously installed. A strand of jute of the proper size and long enough to encircle completely the barrel of the pipe is then calked into the annular space between the spigot and the bell. The jute assists in centering the spigot and also prevents the joining material from flowing into the pipe, where it might partially obstruct the flow of sewage. The balance of the annular space is then filled with a heated asphaltic compound, a prefabricated asphaltic joining strip, or cement mortar. If a heated asphaltic compound is used, it is necessary first to place an asbestos runner entirely around the pipe to close the annular space in order to prevent the jointing material from flowing out while it is a liquid. A small opening is left at the top of the runner, through which the compound is poured. The entire joint should be poured without interruption. The runner can be removed within 5 to 10 min after the joint is poured. Two or more runners may be used if desirable.

If a mortar joint is used, the jute is installed in the annular space, after which the remaining space is calked full of a mortar, consisting of portland cement and sand in the proportions 1:1 or 1:2, by volume, with enough water to produce a plastic mix. The joints should be covered with wet earth to prevent the mortar from cracking while it is hardening.

When vitrified-clay pipes and fittings with compression joints are installed for a sewer pipe, the pipes and fittings are joined together by forcing the spigot of one into the bell of the other. No other jointing material is necessary, except a lubricant that should be spread on the cleaned spigot of the pipe and in the bell of the matching pipe to facilitate the jointing.

Service wye branches   It is good practice to install wye branches as a sanitary sewer pipe is laid. These branches are later used for service connections to the sewer pipe. The branch inlet is temporarily closed with a clay stopper, which is removed when the service is installed. Service branches should be installed for every potential user of the sewer system, and the exact location should be permanently recorded.

Manholes   It is necessary to install manholes at intervals along a sewer line to permit access to the lines for inspection and cleaning. Manholes should be installed at the juncture of laterals with the trunk line, at changes in grade or direction, and at changes in the size of sewer pipe. In general, spacing of manholes should not exceed 300 to 400 ft. The bottom of a manhole is usually concrete, with a split pipe or a split-wye branch to carry the sewage through. The wall of the manhole is constructed of bricks, either standard size, common, or special sewer manhole bricks, with mortar joints. The wall is circular, with an inside diameter of 3 to 4 ft usually, except near the top, where the diameter is gradually reduced to approximately 2 ft. The wall is usually about 8 in. thick. Cast-iron steps should be installed in the wall as it is constructed. A cast-iron ring and cover, set flush with the surface of the ground, is installed on top of each manhole. It is frequently specified that the outside of the wall shall be covered with a layer of cement mortar approximately ½ in. thick to

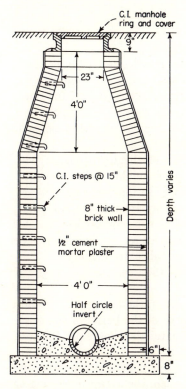

Fig. 21-1   Section through a typical sewer manhole.

TABLE 21-2 Approximate Labor-hours Required to Shape the Bottom of 100 Ft of Trench

Size pipe, in.	Length of pipe, ft	Volume of earth, cu yd	Labor-hr per 100 ft
6	2½	0.4	2
8	3	0.6	3
10	3	1.0	4
12	3	1.4	6
15	3	2.1	8
18	3	3.0	12
21	3	3.9	16
24	3	5.0	20
30	3	8.5	34
36	3	11.0	44

prevent or reduce the infiltration of ground water. Figure 21-1 illustrates a section through a typical manhole.

Cleanout boots Cleanout boots are sometimes installed at the upper ends of sewer lines to permit easy access to the lines for flushing and cleaning. The boots are cast iron, with a base, throat, and cover. The last joint of the sewer line is a wye branch, with a 6- or 8-in. branch opening. Sewer pipe of the proper diameter is installed from the branch opening to the throat of the cleanout boot.

Digging trenches for sewer pipe For rates of digging trenches with trenching machines, see Table 4-3. The estimator should modify the rates given in Table 4-3 to fit the conditions which apply to the particular project. If it is necessary to interrupt the trenching because of paved streets, sidewalks, driveways, or utility lines, the rate of digging will be considerably less than for straightaway digging. When the trenching machine is used to dig holes for the manholes, a proper allowance should be made for this extra digging.

If it is necessary to shore the trenches, the method described on page 107 should be followed in estimating the cost.

Labor shaping the bottom of the trench When the bottom of the main trench is shaped to receive the bottom one-fourth of the barrel of the sewer pipe, with a bell hole at each joint, the approximate labor-hours per 100 ft of trench should be as given in Table 21-2.

Labor laying sewer pipes The labor laying sewer pipes will vary with the size and weight of the pipes, the depth of the trench, the extent of shoring or sheeting required, and the type of joint. If the pipes are distributed along the trench, less labor will be required than if the pipes must be hauled from stockpiles along the trench. If ground water is present in the bottom of the trench, the rate of laying pipes will be reduced.

An estimator should determine the size of crew, by classification, required to lower the pipes into the trench, supply the jointing material, and make the joints. If the number of feet of pipe laid per hour is known, the cost per foot or per 100 ft can be determined. For example, in laying 10-in. vitrified-clay pipes which have been distributed along the trench, two laborers handling and lowering the pipes, one kettle man heating jointing compound, one ladle man, and one joint man will be required. Under average conditions, in a dry trench, 15 to 20 joints of 3-ft-long pipes should be laid per hour by an experienced gang.

If mortar joints are used, one man mixing mortar, one man supplying mortar and jute to the joint man, one joint man, and two men handling pipes will be required. The rate of laying pipes will be about the same as for joints using asphaltic compound.

Table 21-3 gives representative labor-hours required to lay vitrified-clay or concrete pipes when the joints are made with asphaltic compounds or with cement and sand mortar.

When compression-joint clay pipe is laid in a properly graded and prepared trench, the only operations required are digging the bell holes, lowering the pipe into the trench, wiping the bell and spigot of the pipe clean, applying

TABLE 21-3   Labor-hours Required to Lay 100 Ft of Vitrified-clay or Concrete Sewer Pipe

Size pipe, in.	Joints per hr	Lin ft per hr*	Labor-hr per 100 lin ft		
			Joint man, kettle man, ladle man, each	Laborers	Crane operator
6	24	60	1.7	1.7	
8	20	60	1.7	1.7	
10	18	54	1.9	3.8	
12	15	45	2.2	4.4	
15	13	39	2.6	7.8	
18	11	33	3.0	9.0	
21	10	30	3.4	10.6	
24	8	24	4.2	8.6	4.2
30	5	15	6.7	13.4	6.7
36	3	9	11.0	22.0	11.0

* The number of linear feet per hour is based on using 2.5-ft lengths for 6-in. pipes and 3-ft lengths for all other sizes.

TABLE 21-4   Representative Labor-hours Required to Lay 100 Lin Ft of Compression-joint Vitrified-clay Sewer Pipe

Pipe size, in.	Length, ft	Weight, lb per lin ft	Joints made per hr	Lin ft per hr	Labor-hr per 100 lin ft			
					Pipe layer	Laborers	Foreman	Tractor operator
6	4	15	20	80	1.25	5.00	1.25	
8	5	23	18	90	1.11	5.55	1.11	
10	5	34	16	80	1.25	7.50	1.25	
12	5	50	14	70	1.43	8.58	1.43	
15	5	75	12	60	1.67	8.35	1.67	1.67
18	5	115	10	50	2.00	10.00	2.00	2.00
21	5	138	9	45	2.23	11.15	2.23	2.23
24	5	180	8	40	2.50	15.00	2.50	2.50
27	5	225	6	30	3.33	20.00	3.33	3.33
30	5	270	5	25	4.00	24.00	4.00	4.00

a lubricant to the bell and spigot, and then forcing the spigot into the bell of the previously laid joint of pipe or fitting.

Two laborers should be able to lower joints of pipes 5 ft long and up to 12 in. in diameter into a trench using a rope. When pipes larger than 12 in. in diameter are laid, a small wheel-type tractor with a side boom, or a power crane, may be used to lower the pipe. A pipe layer and one to three helpers will be needed in the trench to handle the pipe and to assist the pipe layer in making the joints.

A representative crew to lay 12-in. pipe in a trench 4 ft to about 12 ft deep should be as follows, if the pipe is already distributed along the trench.

1 laborer preparing the trench and digging bell holes
2 laborers lowering the pipe
1 pipe layer
2 laborers in the trench helping the pipe layer
1 foreman

This crew should lay 11 to 16 joints per hour, with 14 joints a reasonable number for average conditions.

Table 21-4 gives representative labor-hours required to lay compression-joint vitrified-clay sewer pipe under average conditions.

Backfilling and tamping earth around sewer pipes Specifications governing the installation of sewer pipe usually require that the backfill

---

TABLE 21-5 Labor-hours Required to Backfill and Tamp Earth by Hand to 12 In. Above the Top of the Sewer Pipe per 100 Lin Ft*

Pipe size, in.	Volume of earth, cu yd	Labor-hours
6	9	9
8	10	10
10	11	11
12	12	12
15	14	14
18	16	16
21	18	18
24	20	20

* If the backfill is placed by a machine, but tamped by hand, the number of labor-hours may be reduced by 50 percent.

TABLE 21-6 Machine-hours and Labor-hours Required to Excavate the Earth for Manholes

Depth of manholes, ft	Hr per manhole		
	Machine	Operator	Labor
Up to 6	0.5	0.5	4
6–12	0.75	0.75	5
12–18	1.00	1.00	7

material be free of rocks or solid objects that might damage the pipe, to a depth of 12 to 24 in. above the top of the pipe. It may be necessary to compact this earth using hand tools or a small self-contained power-operated tamper. After this earth has been placed and tamped, the rest of the earth may be placed in the trench with a bulldozer or other suitable equipment.

Table 21-5 gives representative labor-hours required to backfill and hand-tamp earth to a depth of 12 in. above the top of the pipe. If the backfill to be tamped can be placed in a satisfactory manner with mechanical equipment, the hours given in Table 21-5 can be reduced by about 50 percent.

Excavating for sewer manholes The trenching machine can be used to excavate most of the earth from the manhole. Some of the sides and the base will have to be shaped by hand. If the inside diameter of the finished manhole is 4 ft and walls are 8 in. thick and if 6 in. of clear space is provided outside the wall for plastering, the total diameter of the hole will be approximately 6.5 ft. The total volume of earth to be excavated by hand should not exceed ½ cu yd per manhole, varying somewhat with the depth.

Table 21-6 gives the approximate machine-hours and labor-hours required to excavate the earth from manholes. The rates are based on ordinary earth, with no shoring required for the manhole proper.

Building sewer manholes Brick manholes are frequently built with a plain concrete base about 8 in. thick and 6.5 ft in diameter. After the brick walls are partly completed and the sewer pipe is laid through the manhole, additional concrete is placed inside the manhole equal in depth to one-half the diameter of the sewer pipe, with the surface sloping upward toward the wall of the manhole, as illustrated in Fig. 21-1.

If standard bricks are used for the wall, the thickness of the mortar joint for bedding should be about ⅜ in. The wedge-shaped vertical joints should be completely filled with mortar. A ½-in.-thick mortar plaster is frequently applied to the outside of the wall to reduce the infiltration of water.

As the brick wall is built, cast-iron steps are installed, about 15 in. apart, from the bottom to the top of the manhole.

A cast-iron ring and cover is set in a mortar bed on the top course of brick for each manhole.

Quantity of concrete for manholes   The quantity of concrete required for a manhole will include the base and the inside fill. The quantity required for the base will remain constant, while the quantity for the inside fill will vary with the size of the pipe.

Table 21-7 gives the total quantity of concrete required for a manhole having a base thickness of 8 in. and a diameter of 6.5 ft, for varying sizes of sewer pipe.

Quantity of brick for manholes   In calculating the number of bricks required for a manhole, the estimator should determine the number required per foot of depth for the constant diameter. The number required for the neck will remain constant so long as the dimensions of the neck are constant.

For an inside diameter of 4 ft, 40 standard-size bricks per course will be required, with no allowance for a vertical mortar joint at the inner surface. If a ⅜-in.-thick bedding joint is used, the net thickness of the brick and mortar joint will be 2⅝ in. The number of bricks per foot of wall depth will be 183, with no allowance for breakage.

The number of bricks required for the neck should be determined by assuming a uniform reduction in the inside diameter of the neck. If the number of courses for the depth of the neck and the number of bricks per course are determined, the total number of bricks for the neck can be calculated. For the neck of Fig. 21-1, the total number of bricks will be 483 when a ⅜-in.-thick mortar joint for bedding is used.

Table 21-8 gives the number of standard-size bricks required to build manholes for various thicknesses of mortar joints for bedding the bricks. All

TABLE 21-7   Quantity of Concrete Required for Bases for Sewer Manholes

Size pipe, in.	Quantity of concrete,* cu yd
6	0.94
8	0.97
10	1.00
12	1.02
15	1.04
18	1.06
21	1.08
24	1.08

* No allowance is included for waste.

TABLE 21-8 Number of Standard-size Bricks Required for Manholes for Various Thicknesses of Mortar Joints for Bedding

| Thickness of mortar joints, in. | Number of bricks* | |
	Per ft of wall depth	For the neck
¼	192	508
⅜	183	483
½	175	462
⅝	167	442
¾	159	423

Inside diameter of manhole, 4 ft
Inside diameter of neck at top, 23 in.
Depth of neck, 4 ft
Size of bricks, 2¼ × 3¾ × 8 in.
* No allowance is included for breakage.

manholes are 4 ft inside diameter, with the neck varying uniformly from 4 ft to 23 in. in a depth of 4 ft.

Quantity of mortar for manholes In determining the quantity of mortar, the estimator should divide the manhole into two parts, the wall with a constant diameter and the neck. The quantity of mortar for the joints and for plastering the outside surface should be calculated for each part separately.

The quantity of mortar required for joints only for 1 ft of depth of wall having a constant diameter will be equal to the gross volume of the wall less the net volume of the bricks. For a wall 8 in. thick with an inside diameter of 4 ft, the gross volume per foot of depth is 9.75 cu ft. If the bedding mortar joint is ⅜ in. thick, 183 bricks will be used for each foot of height. The volume of the bricks is 7.14 cu ft, leaving 2.61 cu ft to be filled with mortar.

If the outside of the wall is covered with a mortar plaster ½ in. thick, the quantity of mortar for a wall depth of 1 ft will be 0.7 cu ft.

The gross volume of the neck is 28.7 cu ft. If the bedding mortar joint is ⅜ in. thick, 483 bricks will be used. The volume of the bricks is 18.8 cu ft, leaving 9.9 cu ft to be filled with mortar.

The quantity of mortar required to cover the outside wall of the neck with a ½-in.-thick plaster will be 2.2 cu ft.

Table 21-9 gives the quantities of mortar required for manholes with an inside diameter of 4 ft, a neck diameter of 23 in. at the top, and a neck depth of 4 ft, using standard bricks. The quantities are given for various thicknesses of bedding mortar joints and are separated to show the amounts required for joints only and for covering the outside wall with a ½-in. mortar plaster. Add 15 to 20 percent to the quantities given for waste.

Labor building manholes An experienced bricklayer and a helper should lay about 125 bricks per hour. The bricks, sand, cement, and lime should be delivered by truck to the site of each manhole. The helper will mix the mortar and supply the mortar and bricks to the bricklayer. The bricklayer will lay the brick, and plaster the outside of the manhole as it is constructed.

In order to assure a good and permanent bond between the bricks and the mortar, the bricks should be wet before they are laid. The plaster will cure more satisfactorily if the space around the manhole is backfilled with moist earth soon after the manhole is completed.

The mortar which falls inside the manhole should be removed in order to preserve a smooth surface in the pipe and on the concrete bottom.

Example An extension to a sewer system requires the laying of compression-joint vitrified-clay sewer pipe and the construction of manholes of the indicated quantities.

Sewer pipe

Length, lin ft	Pipe size, in.	Maximum depth, ft	Minimum depth, ft	Average depth, ft
8,645	12	12.6	6.4	8.2

Brick manholes

Number	Size of pipe, in.	Average depth, ft
24	12	8.2

The soil is dense sandy clay, with no ground water. The trenches will be dug with a ladder-type trenching machine. After the pipe is laid, the backfill will be hand-placed and tamped to a depth of 12 in. above the top of the pipe. The rest of the backfill will be placed with a bulldozer.

The manholes will be constructed as illustrated in Fig. 21-1, using standard-size common bricks, with ⅜-in. bed joints. A 5-ft-long split pipe will be laid across the bottom of each manhole.

Service wye branches, with 4-in. inlets, will be spaced at 30-ft intervals along the pipe line.

Shoring will be required for the full lengths of all trenches. Each shore will consist of two 2-in. by 12-in. by 12-ft 0-in. pieces of lumber. The shores will be spaced at intervals of 6 ft, with trench jacks spaced not more than 3 ft 0 in. apart vertically between the two pieces of lumber. The maximum length of trench requiring shores will be limited to 300 lin ft. Thus, the maximum number of shores will be 300 ÷ 6 = 50. This will require 2 × 50 = 100 pieces of lumber at any given time.

TABLE 21-9   Quantities of Mortar Required for Man-holes Using Standard-size Bricks and Various Thicknesses of Bedding Joints

| Thickness of mortar joint, in. | Cu ft of mortar* | | | |
| | Per ft of wall | | For the neck | |
	Joints	Plaster	Joints	Plaster
¼	2.3	0.7	8.9	2.2
⅜	2.6	0.7	9.9	2.2
½	2.9	0.7	10.7	2.2
⅝	3.2	0.7	11.5	2.2
¾	3.6	0.7	12.3	2.2

* Based on using a ½-in.-thick plaster. Add 15 to 20 percent for waste.

The quantity of lumber required will be

100 pc, 2 in. × 12 in. × 12 ft 0 in.	= 2,400 fbm
Add for replacement lumber	= 1,200 fbm
Total quantity of lumber required	= 3,600 fbm

Estimate the total direct cost of the project.

The specified costs of materials includes delivering them to the project.

The quantities of pipe and fittings required are determined as follows:

Total length of pipe line	= 8,645 lin ft
Deduct the length of the split pipe laid through the manholes, 24 × 5 ft	= 120 lin ft
Subtotal length of pipe	= 8,525 lin ft
No. of wye branches, 8,645 ÷ 30 = 288	
Deduct the length of wye branches, 288 × 5 ft	= 1,440 lin ft
Length of standard pipe	= 7,085 lin ft
Add for breakage, etc., 2% of 7,085	= 145 lin ft
Gross length of standard pipe	= 7,230 lin ft
Gross length of wye branches, including 2% for breakage	= 1,470 lin ft

The direct costs will be

Sewer pipe:

Sewer pipe, 7,230 lin ft @ $2.22	= $10,050.60
Wye branches, 288 plus 6 = 294 @ $14.34	= 4,215.96
Split pipe, 120 plus 5 = 125 lin ft @ $1.56	= 195.00
Vitrified-clay stoppers, 294 @ $0.71	= 208.74
Trenching machine, 8,645 lin ft ÷ 60 lin ft per hr = 144 hr @ $13.20	= 1,900.80
Machine operator, 144 hr @ $5.60	= 806.40

Helpers, 2 men, 288 hr @ $4.80	=	1,382.40
Lumber for shoring, 3,600 fbm @ $260.00 per M fbm	=	936.00
Trench jacks, 8,645 lin ft ÷ 6 ft = 1,440 shores × 3 jacks = 4,320 uses @ $0.06	=	259.20
Labor installing and removing shores, 1,440 × 0.5 hr per shore = 720 hr @ $4.80	=	3,456.00
Pipe layer, 8,645 lin ft × 1.43 hr per 100 lin ft = 124 hr @ $5.90	=	731.60
Laborers, (see Table 21-4), 8,645 lin ft × 8.58 hr per 100 lin ft = 742 hr @ $4.80	=	3,561.60
Laborers backfilling and tamping trench, 8,645 lin ft × 12 hr per 100 100 lin ft = 1,038 hr @ $4.80	=	4,982.40
Bulldozer backfilling trench, 8,645 lin ft ÷ 70 lin ft per hr = 124 hr @ $6.75	=	837.00
Bulldozer operator, 124 hr @ $5.60	=	694.40
Foreman, 144 hr @ $7.50	=	1,080.00
Total direct cost	=	$41,298.10

Manholes:

Trenching machine excavating 24 manholes × 0.5 hr each = 12 hr @ $13.20	= $	158.40
Machine operator, 12 hr @ $5.60	=	67.20
Laborers excavating, 24 × 5.8 hr each = 139.2 hr @ $4.80	=	668.16
Concrete, 24 × 1.02 cu yd = 24.5 cu yd		
Cement, 24.5 cu yd × 5 sk + 5% waste = 129 sk @ $1.60	=	206.40
Sand, 24.5 cu yd × 1,400 lb per cu yd + 20% waste = 20.5 tons @ $3.85	=	78.93
Gravel, 24.5 cu yd × 1,900 lb per cu yd + 15% waste = 26.7 tons @ $4.30	=	114.81
Concrete mixer, 24 manholes × 0.5 hr per manhole = 12 hr @ $1.96	=	23.52
Laborers mixing and placing concrete, 24.5 cu yd × 2.5 hr per cu yd = 61 hr @ $4.80	=	292.80
Pipe layer, 16 hr @ $5.90	=	94.40

Bricks:

Total depth of manholes, 24 × 8.2	= 196.8 ft
Deduct 24 rings and covers × 9 in.	= 18.0 ft
Total depth of brick	= 178.8 ft
Deduct 24 necks × 4 ft	= 96.0 ft
Total depth of 4-ft-diameter manhole =	82.8 ft

Bricks for 4-ft-diameter wall, 82.8 ft × 183 per ft = 15.15 M @ $52.00	=	787.80
Bricks for necks, 24 × 432 = 10.38 M @ $52.00	=	539.76
Add for waste brick, 2% = 0.47 % @ $52.00	=	24.44

Mortar, including ½-in.-thick plaster on walls:

For walls, 82.8 ft × 3.3 cu ft per ft + 10% waste = 290 cu ft
For necks, 24 × 12.1 + 10% waste	= 300 cu ft
Total quantity of mortar	= 590 cu ft

Mortar ratio, hydrated lime, 1 part, portland cement, 5 parts, sand, 15 parts by volume

Sand, 590 cu ft ÷ 27 = 22 cu yd @ $3.80   =   83.60

Cement, $\dfrac{5}{15}$ × 590 = 197 sk @ $1.60   =   315.20

Lime, $\dfrac{1}{15}$ × 590 = 40 sk @ $1.25   =   50.00

Cast-iron steps, 179 ft × $\dfrac{12}{15}$ in. = 143 @ $1.05   =   150.15

Cast-iron rings and covers, 24 @ $45.30   =   1,087.20
Bricklayers, 26,000 bricks ÷ 110 per hr = 235 hr @ $8.23   =   1,934.05
Bricklayers' helpers, 235 hr @ $4.80   =   1,128.00
Backfilling around manholes is included under laying pipe, no charge
    here
Foreman, based on 2 bricklayers, 117.5 hr @ $7.50   =   881.25

    Total direct cost of manholes   = $ 8,686.07
Moving onto the job and back to the storage yard   = $ 575.00

The total direct cost of the project will be

Sewer pipe   = $41,298.10
Manholes   =   8,686.07
Move to and away from the job   =   575.00

    Total cost   = $50,559.17

REFERENCES

1   Standard Specifications for Compression Joints for Vitrified Clay Bell-and-spigot Pipe, ASTM Designation: C 425-71, American Society for Testing and Materials, 1916 Race Street, Philadelphia, Pa. 19103.

22
Total Cost of
Engineering Projects

The previous chapters of this book have discussed methods of estimating the costs of constructing engineering projects. However, the cost of construction is not the only cost which the owner of the project must pay. The total cost to the owner may include, but is not necessarily limited to, the following items.

1   Land, purchase of, right of way, easements
2   Legal expense
3   Bond expense, or cost of obtaining money to finance the project
4   Cost of construction
5   Engineering and/or architects expense
6   Interest during construction
7   Contingencies

Each of these costs will be described.

Cost of land, right of way, and easements   If it is necessary to purchase land or obtain rights to use land in constructing an engineering project, the owner of the project to be constructed must provide the money to finance these acquisitions.

The land on which a project is to be constructed may be purchased by the owner. In general, the cost of acquiring the land should be included in the total cost of the project.

If the project includes the construction of pipe lines, power lines, telephone lines, or other extended items, the owner of the project may prefer to obtain a continuing right to construct and maintain a facility through the property without actually purchasing it. This right may be defined as an easement, for which the owner of the project may pay the owner of the land.

Legal expenses The construction of a project frequently involves actions and services which require the employment or use of an attorney. The actions requiring an attorney may be the acquisition of land and easements, holding a bond election for a government agency, printing and obtaining approval of bonds to be sold for providing the money to finance the project, or assistance to a private corporation. The legal fees paid for these services should be included in the cost of the project.

Bond expense Before a project can be constructed by a governmental agency, it is usually necessary to hold a bond election for the purpose of permitting the qualified voters to approve or reject the project. In the event the voters approve the project, it is necessary to print, register, and sell the bonds, usually through a qualified underwriting broker, who charges a fee for his services. Private corporations frequently sell bonds to finance new construction. In any event, it is correct to charge to the project the costs for these services.

Cost of construction The cost of constructing a project is usually an estimate only, made in advance of receiving bids from contractors, prepared by an engineer or an architect. It is the amount that he believes the owner will have to pay for the construction of the project.

The estimate may be a lump-sum cost, such as for a building, or it may be a unit-price estimate for a project that requires the construction of various items whose exact quantities are not exactly known in advance of construction. For example, the contract for constructing a highway may specify a given payment to the contractor for each ton of asphaltic material placed in the pavement.

For a project that involves the construction of units of work, the bid form will provide for the bidders to state the amounts which they will charge for constructing each specified unit.

Engineering expense When an owner desires to have a project constructed, he engages an engineer or an architect to make the necessary surveys and studies, prepare the plans and specifications, and assist in securing bids for the construction and supervising the construction of the project. The cost for this service is usually based on an agreed percentage of the cost of construction. The fee usually varies from 5 to 10 percent of the cost, depending on several factors.

Interest during construction Most construction contracts provide that the owner will pay to the contractor at the end of each month during the period of construction a specified percent of the value of the work completed during the month, frequently about 90 percent. In addition to the amounts that the owner must pay to the contractor, he likely will have to pay other costs prior to completion of the project. Thus the owner will have considerable funds invested in the project during the time it is under construction. As the owner must pay interest on the money required to finance these costs, there will be an interest cost during construction, which should be included in the total cost of the project.

The amount of interest chargeable to the project during construction is usually estimated at the time the total cost of the project is estimated. A method that is sometimes used is to assume that one-half the cost of the project will require the payment of interest during the full period of construction. However, if the full cost of the project is secured in advance of the beginning of construction, and if interest is paid on the full amount during this period, the total cost of interest should be included in the cost of the project.

Contingencies If the exact cost of the project is not known in advance of raising funds to finance the cost, as is frequently the case, it is good practice to provide additional funds to cover any additional costs that may occur during construction. An estimator should rely on his judgment to determine the amount of contingencies desirable.

Example illustrating an estimate for the total cost of an engineering project An example illustrating an estimate for the total cost of a project involves drilling three water wells and furnishing all materials, labor, and equipment required to provide additional sources of water for a city. The project will require acquiring land and easements, drilling wells, and installing pumps and cast-iron water pipes, fittings, and valves to bring the water to the city.

A bond election will be held to provide the money to finance the total cost of the project. The estimated total cost is determined as shown in the table on the following page.

This is the minimum amount that should be included in a bond election for the project.

Item		Estimated cost
1	Land and easements	= $ 8,650
2	Legal expense	= 1,200
3	Bond expense	= 1,875
4	Cost of construction:	
	a Water wells with pumps, 3 @ $42,650 = $127,950	
	b Pump houses, 3 @ $720 = 2,160	
	c Extend electric power lines to wells, 9,420 lin ft @ $3.80 = 35,796	
	d 12-in. Class 150 cast-iron pipe, 8,780 lin ft @ $8.95 per ft = 78,581	
	e Cast-iron fittings, 8.64 tons @ $574.40 = 4,963	
	f 12-in. gate valves, 4 @ $386.00 = 1,544	
	Total cost of construction = $250,994	= 250,994
5	Engineering expense, 7% of $250,994	= 17,569
6	Interest during construction, estimated to be 10 months @ 7% per year =	

$$\frac{\$250,994}{2} \times 0.07 \times 10/12 \qquad\qquad = \qquad 7,321$$

7	Contingencies	= 12,391
8	Estimated total cost	= $300,000

Appendix A
Compensation Insurance Base Rates for Construction Workers

TABLE A-1   Base Rates on Compensation Insurance for Construction Rate is per $100 payroll, compiled by Herbert L. Jamison Company, Insurance Advisors and Auditors. Effective July 1, 1973

Classification of work	Ala.	Alaska	Ariz.	Ark.	Calif.	Colo.	Del.	D.C.	Fla.	Ga.	Hawaii	Idaho	Ill.
Carpentry, 1-2 family residences	2.01	2.72	6.52	4.06	4.88	3.24	2.45	3.26	5.25	2.78	5.36	4.81	2.74
3 stories or less	2.19	3.89	6.46	4.60	8.41	4.01	2.45	3.26	5.25	2.78	3.44	4.81	2.82
Interior trim, cabinet work	1.31	2.13	6.40	3.40	8.41	1.74	2.45	3.07	4.97	1.84	5.46	3.02	1.92
General	2.95	6.25	8.71	5.87	8.41	3.46	2.45	4.30	7.69	3.80	16.44	6.42	4.00
Chimney construction; brick, concrete	7.88	12.41	9.48	20.76	13.15	9.69	a	31.30	29.41	6.15	19.02	23.96	10.63
Concrete work, bridges, culverts	4.06	5.32	13.66	7.86	9.52	4.70	1.65	11.17	12.89	4.32	6.45	6.03	5.08
Dwelling, 1-2 family	1.43	5.70	9.00	3.28	2.98	3.56	1.65	7.26	6.19	1.06	6.06	3.91	1.61
N.O.C.	2.59	3.27	9.94	5.50	6.98	3.73	1.65	6.44	7.88	2.72	5.49	3.48	4.86
Concrete or cement walks, floors	1.60	3.85	5.01	2.18	2.98	2.10	.77	3.75	4.69	1.68	2.82	1.91	2.12
Electrical wiring	1.27	2.15	3.73	1.79	2.54	1.15	4.75	3.12	3.34	1.72	2.93	2.10	1.50
Excavation, earth	2.59	4.69	5.30	4.85	3.89	3.18	4.75	5.38	6.07	3.18	10.50	8.31	2.63
Rock	4.40	4.03	7.78	4.99	8.21	3.71	a	21.93	12.88	5.99	10.67	4.31	13.82
Glazing	3.03	4.71	5.26	5.58	5.98	3.62	2.45	4.52	6.47	3.61	6.15	2.94	2.85
Insulating	1.89	5.22	6.49	2.63	5.09	2.49	1.75	4.30	4.88	1.84	3.20	5.19	3.24
Lathing	1.47	3.28	3.54	3.43	2.96	1.69	2.50	4.02	2.58	1.68	4.53	3.68	1.39
Masonry	1.43	5.78	8.82	3.28	6.05	3.56	2.30	6.38	4.92	1.49	6.06	3.91	3.02
Painting and decorating	2.16	4.98	4.25	3.22	5.38	2.10	6.00	3.70	5.17	2.48	4.07	2.79	2.80
Pile driving	10.36	13.17	19.30	17.67	13.96	10.61	1.75	19.82	15.23	4.86	12.51	19.43	7.92
Plastering	1.89	2.98	7.18	2.63	7.16	2.49	1.40	4.30	4.52	1.84	3.20	5.19	1.79
Plumbing	1.77	2.80	4.35	2.35	2.99	2.01	4.25	3.20	4.16	1.74	2.74	2.45	2.03
Roofing	3.38	7.21	10.32	10.01	12.66	5.96	4.25	10.68	12.37	4.56	16.28	13.35	7.93
Sheet metal work	2.05	1.95	3.96	2.98	3.67	2.55	7.60	4.04	4.29	2.71	4.93	3.42	2.40
Steel erection, doors & sash	2.03	6.19	2.77	3.87	5.22	2.78	7.60	5.13	4.71	2.10	3.63	5.24	2.63
Interior ornament	2.03	6.19	2.77	3.87	5.22	2.78	7.60	5.13	4.71	2.10	3.63	5.24	2.63
Structural	5.29	20.75	20.37	9.28	17.20	15.08	7.60	29.76	13.23	6.30	18.04	4.91	12.48
Dwelling, 2 stories	3.80	8.71	14.60	6.95	14.45	5.01	7.60	13.93	10.76	5.71	12.48	7.08	7.30
N.O.C.	3.72	13.48	12.42	9.01	17.09	6.20	1.10	37.83	9.64	9.51	21.33	5.11	8.54
Tile work, interior	1.06	.96	2.51	2.35	3.45	1.55	incl	5.82	4.32	1.06	2.82	2.28	1.37
Timekeepers and watchmen	2.20	5.67	2.53	4.01	4.15	3.16	2.30	6.40	4.97	2.61	5.07	5.68	2.59
Waterproofing, interior (brush)	2.16	4.98	4.25	3.22	5.38	2.10	1.75	3.70	5.17	2.48	4.07	2.79	2.80
Trowel (interior)	1.89	2.98	7.18	2.63	7.16	2.49	2.50	4.30	4.52	1.84	3.20	5.19	1.79
Trowel (exterior)	1.43	5.78	8.82	3.28	6.05	3.56	1.65	6.38	4.92	1.49	6.06	3.91	3.02
Pressure gun	2.59	3.27	9.94	5.50	6.98	3.73		6.44	7.88	2.72	5.49	3.48	4.86
Wrecking	10.59	14.72	25.95	27.16	31.65	16.08	7.15	9.68	30.41	9.84	34.17	37.04	15.12

Classification of work	Ind.	Iowa	Kans.	Ky.	La.	Maine	Md.	Mass.	Mich.	Minn.	Miss.	Mo.
Carpentry, 1-2 family residences	2.02	2.47	2.82	5.96	6.38	2.52	3.07	5.08	4.99	3.79	3.94	3.85
3 stories or less	1.81	2.36	2.82	5.75	5.64	2.52	3.07	5.08	5.76	3.79	3.94	3.85
Interior trim, cabinet work	.98	1.03	1.32	2.96	3.32	1.36	2.65	2.82	2.93	3.79	2.22	2.84
General	2.03	2.89	3.69	6.01	10.85	4.76	3.57	9.65	4.96	6.95	5.55	4.73
Chimney construction; brick, concrete	8.19	7.74	11.52	17.18	29.77	12.12	12.92	11.29	15.70	16.36	15.70	17.71
Concrete work, bridges, culverts	3.21	4.32	4.95	11.71	8.23	6.65	5.30	4.44	7.43	5.08	10.21	6.25
Dwelling, 1-2 family	1.67	1.90	2.23	3.78	3.45	2.65	4.03	3.77	5.87	2.20	1.93	2.40
N.O.C.	2.08	2.49	3.74	5.04	6.65	2.99	4.46	7.72	7.79	6.03	5.19	6.27
Concrete or cement walks, floors	1.01	1.66	1.75	4.48	5.00	1.70	2.65	3.79	4.17	2.50	2.11	2.87
Electrical wiring	.94	1.15	1.94	2.08	3.24	1.05	1.84	2.25	1.91	2.20	2.40	3.15
Excavation, earth	1.62	2.41	2.68	5.55	6.75	3.26	2.58	3.49	4.04	5.03	5.50	4.54
Rock	4.27	4.43	4.16	5.41	21.20	5.80	6.22	4.16	7.64	5.48	10.66	4.80
Glazing	2.28	1.99	2.54	5.49	9.43	4.35	7.47	4.08	3.83	5.90	5.32	4.27
Insulating	1.64	1.49	2.88	3.74	6.11	3.20	2.51	3.75	3.80	3.47	2.43	3.31
Lathing	.91	1.38	1.86	3.96	4.94	2.01	2.34	3.20	2.04	1.99	2.86	2.21
Masonry	1.54	1.90	2.23	3.78	3.45	2.65	2.51	3.20	5.25	3.08	1.93	3.96
Painting and decorating	2.09	1.80	2.20	5.93	4.82	3.23	3.09	a	a	3.14	2.72	4.02
Pile driving	4.19	5.07	7.49	15.72	22.88	5.49	5.54	10.11	11.02	10.10	14.98	9.85
Plastering	1.64	1.49	2.88	3.74	4.66	3.20	2.51	3.82	3.80	3.01	2.43	2.97
Plumbing	1.39	1.54	2.08	2.82	5.91	1.30	2.54	2.23	2.37	2.94	2.25	2.90
Roofing	3.20	5.54	8.17	10.59	10.03	9.06	10.13	a	12.94	8.65	9.18	9.14
Sheet metal work	1.68	1.18	2.20	3.24	6.01	2.04	3.12	3.61	3.11	2.73	4.90	3.19
Steel erection, doors & sash	1.81	1.73	2.00	5.44	6.37	2.82	4.10	4.51	4.78	3.69	3.92	3.39
Interior ornament		1.75	2.00	5.44	6.37	2.82	4.10	4.57	4.78	3.69	3.92	3.39
Structural	6.36	7.84	8.60	19.21	13.65	9.63	12.04	21.26	12.15	16.86	10.56	10.57
Dwelling, 2 stories	3.35	4.82	5.80	11.23	14.04	5.54	7.78	17.68	8.15	8.92	7.39	9.94
N.O.C.	5.67	5.30	6.68	13.03	15.18	9.75	3.05	17.68	8.93	10.53	8.09	12.56
Tile work, interior	1.29	.98	1.25	2.08	2.80	1.30	1.92	3.25	2.53	3.30	2.35	2.01
Timekeepers and watchmen	2.15	2.29	3.08	5.69	7.59	2.93	3.09	6.26	3.61	4.34	3.52	5.57
Waterproofing, interior (brush)	2.09	1.80	2.20	5.93	4.82	3.23	3.09	a	a	3.14	2.72	4.02
Trowel (interior)	1.64	1.49	2.88	3.74	4.66	3.20	2.51	3.82	3.80	3.01	2.43	2.97
Trowel (exterior)	1.54	1.90	2.23	3.78	3.45	2.65	2.51	4.39	5.25	3.08	1.93	3.96
Pressure gun	2.08	2.49	3.74	5.04	6.65	2.99	4.46	7.72	7.79	6.03	5.19	6.27
Wrecking	9.73	19.93	15.39	20.82	45.82	21.06	20.39	14.19	a	14.89	12.88	21.10

TABLE A-1—Continued

Classification of work	Mont.	Nebr.	Nev.	N.H.	N.J.	N.M.	N.Y.	N.C.	N.D.	Ohio	Okla.	Oreg.	Pa.
Carpentry, 1-2 family residences	5.16	2.82	7.07	3.11	5.34	4.35	3.40	2.39	3.66	2.12	5.56	6.94	2.00
3 stories or less	5.16	3.45	7.07	3.11	3.49	3.11	4.00	2.39	3.66	2.07	6.06	6.52	2.00
Interior trim, cabinet work	2.24	1.12	3.07	1.22	3.49	1.82	No rate	2.06	3.66	1.13	3.79	6.05	2.00
General	6.28	2.93	8.60	3.58	5.63	5.63	4.10	3.59	6.72	2.15	8.00	9.95	2.00
Chimney construction; brick, concrete	19.71	10.00	27.00	10.55	12.12	8.68	18.50	9.14	15.82	9.33	26.44	20.77	1.65
Concrete work; bridges, culverts	9.90	3.18	13.56	4.03	6.55	7.69	5.30	2.93	4.91	3.67	16.66	12.39	2.50
Dwelling, 1-2 family	5.07	1.67	6.95	1.73	3.96	1.76	No rate	1.29	2.12	1.72	3.19	5.47	2.50
N.O.C.	6.98	2.58	9.56	2.92	6.25	4.46	5.30	2.66	5.83	2.52	9.00	7.21	2.50
Concrete or cement walks, floors	2.77	1.78	3.79	1.63	3.85	2.31	2.30	1.29	2.42	.94	4.70	4.02	2.50
Electrical wiring	1.78	1.04	2.44	1.51	2.81	1.91	1.60	1.75	2.12	1.33	3.43	2.34	1.05
Excavation, earth	7.44	2.69	10.19	3.47	5.19	2.82	3.80	2.15	4.86	2.10	7.95	6.96	2.75
Rock	8.76	7.33	12.00	5.17	8.44	6.46	3.80	5.01	5.30	4.80	6.17	7.40	2.75
Glazing	6.69	3.21	9.17	3.84	5.97	4.36	4.30	2.49	5.70	2.35	6.54	4.28	1.40
Insulating	5.90	2.14	8.08	1.78	4.29	4.18	2.50	2.28	3.35	1.50	4.01	8.39	2.00
Lathing	2.96	1.50	1.57	1.78	3.36	1.55	2.80	1.49	1.92	1.03	2.72	7.96	1.50
Masonry	5.07	1.67	1.74	1.73	5.10	3.06	4.10	1.29	2.98	1.97	5.13	7.24	1.65
Painting and decorating	6.77	1.83	2.07	2.62	7.23	3.06	3.60	1.90	3.03	1.93	4.89	6.41	3.40
Pile driving	17.75	6.45	6.98	5.09	7.25	10.76	5.30	5.07	9.76	4.27	15.52	20.32	6.15
Plastering	5.90	2.14	2.04	1.78	3.36	4.18	4.30	2.28	2.91	1.50	4.23	5.79	1.50
Plumbing	3.31	1.16	1.28	1.68	3.74	2.51	2.70	1.36	2.84	1.25	4.47	4.24	1.25
Roofing	12.17	7.77	8.30	8.53	16.92	7.75	No rate	4.91	8.36	3.75	18.87	19.80	3.60
Sheet metal work	4.25	2.34	2.22	1.76	3.51	2.56	3.40	1.91	2.64	1.63	4.91	4.13	3.60
Steel erection, doors & sash	4.23	2.22	2.37	2.50	5.02	2.44	2.30	1.90	3.57	2.32	3.80	7.93	7.25
Interior ornament	4.23	2.22	2.37	2.50	5.02	2.44	2.30	1.90	3.57	2.32	3.80	7.93	7.25
Structural	22.30	9.55	9.03	9.32	13.14	5.19	10.10	9.92	16.30	6.91	30.99	13.27	7.25
Dwelling, 2 stories	9.35	4.87	5.11	5.85	9.38	8.61	4.10	4.72	8.62	3.62	17.06	9.63	7.25
N.O.C.	14.67	3.75	4.73	7.41	8.53	8.52	7.60	6.62	10.18	4.97	17.38	15.54	7.25
Tile work, interior	1.84	1.21	1.29	1.19	3.78	2.17	2.60	1.29	3.19	1.45	3.36	3.68	1.40
Timekeepers and watchmen	4.22	2.32	2.43	2.64	4.74	3.63	2.90	2.30	4.19	2.32	5.34	11.39	incl.
Waterproofing, interior (brush)	6.77	1.83	2.43	2.67	7.23	3.06	3.60	1.90	3.03	1.93	4.89	6.41	1.65
Trowel (interior)	5.90	2.14	2.07	1.78	3.36	4.18	4.30	2.28	2.91	1.50	4.23	5.79	1.65
Trowel (exterior)	5.07	1.67	2.04	1.73	5.10	3.06	4.10	1.29	2.98	1.97	5.13	7.24	1.65
Pressure gun	6.98	2.58	1.74	2.92	6.25	4.46	5.30	2.66	5.83	2.52	9.00	7.21	1.65
Wrecking	28.21	10.95	12.67	13.78	24.49	14.38	No rate	8.34	14.39	8.73	42.93	65.44	1.10

Classification of work	R.I.	S.C.	S.D.	Tenn.	Tex.	Utah	Vt.	Va.	Wash.	W.Va.	Wis.	Wyo.
Carpentry, 1-2 family residences	2.93	3.00	3.00	4.01	5.02	A	1.78	2.24	3.87	5.96	2.83	a
3 stories or less	3.73	3.00	3.00	4.01	5.88	A	1.78	1.85	3.98	5.75	1.79	a
Interior trim, cabinet work	3.07	1.59	1.40	1.59	3.07	A	1.20	1.48	1.27	2.96	2.83	a
General	6.25	7.10	2.68	4.40	6.87	3.19	2.89	2.50	5.11	6.01	3.54	2.80
Chimney construction; brick, concrete	21.23	13.75	8.70	11.87	21.87	10.48	10.24	8.52	a	17.18	13.19	9.22
Concrete work; bridges, culverts	4.95	5.54	5.77	5.88	10.05	3.51	4.25	5.54	5.78	11.71	7.98	3.08
Dwelling, 1-2 family	2.84	1.54	1.81	2.26	3.61	2.71	1.74	1.18	3.66	3.78	2.27	2.38
N.O.C.	6.05	4.25	3.72	3.06	6.90	3.04	3.12	3.05	4.20	5.04	3.63	2.67
Concrete or cement walks, floors	3.33	1.94	2.07	2.09	3.61	2.09	1.49	1.13	1.94	4.48	2.25	1.84
Electrical wiring	1.48	1.63	1.26	2.13	2.79	1.58	1.20	1.43	1.95	2.08	1.44	1.39
Excavation, earth	3.81	3.57	3.09	4.45	4.97	2.68	3.41	2.35	4.40	5.55	3.60	2.35
Rock	10.16	7.70	4.98	3.90	7.20	3.84	3.41	2.80	9.32	5.41	9.60	3.38
Glazing	5.16	4.05	3.52	3.53	3.51	1.58	3.61	2.58	4.35	5.49	3.67	1.39
Insulating	2.81	2.39	2.10	2.29	5.10	2.74	2.03	1.48	3.40	3.74	2.45	2.41
Lathing	3.25	2.34	1.66	2.45	2.87	1.91	1.64	1.75	3.30	3.96	1.79	1.68
Masonry	2.84	1.54	1.81	2.26	4.38	2.59	1.75	1.53	3.77	3.78	2.14	2.28
Painting and decorating	3.35	3.53	2.33	2.45	3.70	2.64	2.93	2.32	4.19	5.93	2.97	2.32
Pile driving	10.23	7.83	9.40	10.98	10.49	5.01	4.78	5.01	9.79	15.72	8.19	4.40
Plastering	2.81	2.39	2.10	2.29	3.88	2.74	2.03	1.48	3.40	3.74	2.05	2.41
Plumbing	3.22	1.56	1.34	1.70	3.70	1.60	1.09	1.69	1.72	2.82	1.85	1.40
Roofing	7.16	9.04	7.40	7.19	10.96	7.00	7.81	5.07	a	10.59	8.67	6.16
Sheet metal work	2.26	3.58	2.21	2.57	4.73	2.68	2.93	2.33	3.75	3.24	1.80	2.35
Steel erection, doors & sash	3.45	1.83	2.45	2.37	4.65	1.81	2.32	2.20	a	5.44	2.53	1.59
Interior ornament	3.45	1.83	2.45	2.37	4.65	1.81	2.32	2.20	a	5.44	2.53	1.59
Structural	21.38	11.25	11.57	7.57	15.39	a	12.96	7.60	15.86	19.21	7.04	a
Dwelling, 2 stories	10.80	6.26	4.75	6.08	15.39	5.97	6.42	4.73	7.51	11.23	5.39	5.25
N.O.C.	30.29	7.12	4.80	7.23	9.92	6.18	8.39	10.13	10.56	13.03	11.06	5.43
Tile work, interior	2.10	1.34	1.22	1.70	2.49	1.50	1.14	1.11	1.38	2.08	1.75	1.32
Timekeepers and watchmen	4.80	2.92	2.82	4.04	1.66	2.79	2.44	2.47	4.18	5.69	3.91	2.45
Waterproofing, interior (brush)	3.35	3.53	2.33	2.45	3.70	2.64	2.93	2.32	4.19	5.93	2.97	2.32
Trowel (interior)	2.81	2.39	2.10	2.29	3.88	2.74	2.03	1.48	3.40	3.74	2.05	2.41
Trowel (exterior)	2.84	1.54	1.81	2.26	4.38	2.59	1.75	1.53	3.77	3.78	2.14	2.28
Pressure gun	6.05	4.25	3.72	3.06	6.90	3.04	3.12	3.05	4.20	5.04	3.63	2.67
Wrecking	17.63	16.74	14.96	21.66	28.46	18.13	14.61	7.49	21.52	20.82	8.68	15.95

SOURCE: *Engineering News-Record*, September 20, 1973.

a Specially rated—refer to company.
b Rates indicated are published by National Council on Compensation and Insurance. Subscription to State Fund Mandatory; Industrial Commission, Carson City, Nev.; Workmen's Compensation Bureau, Bismarck, N.D.; The Industrial Commission of Ohio, Columbus; Department of Labor and Industry, Olympia, Wash.; State Compensation Commission, Charleston, W.Va.; Workmen's Compensation, State Treasurer's Office, Cheyenne, Wyo.

Appendix B

Cost of Owning and Operating Construction Equipment

The costs given in this appendix are intended to indicate the approximate costs per hour for ownership, fuel, and other expenses and the total ownership and operating costs per working hour for construction equipment.

The cost per hour includes depreciation, major overhauling and repairs, minor repairs, fuel, lubrication, interest, insurance, taxes, and storage.

The annual cost of depreciation is based on the original cost of the equipment and the estimated economical life of the equipment, using the straight-line method of determining the cost. Thus the average annual cost of depreciation for equipment whose estimated useful is 4 years, with no salvage value, is 25 percent of the original cost.

The average cost of repairs is based on experience obtained from the operation of the equipment and will vary with the equipment and the conditions under which it is operated. The cost of minor repairs, which are usually made in the field, are included under fuel and other expenses related to fuel.

The annual cost of interest, insurance, taxes, and storage is based on the average value of equipment as given in Table 2-3. The rates used are: interest, 9 percent; insurance, taxes, and storage, 5 percent. The total is 14 percent of the average value of the equipment. The following table gives the factors that may be used to determine the average annual investment cost based on the original total cost and the estimated useful life of the equipment.

Estimated life, yr	Calculations	Average annual investment cost as a percent of the original cost
2	0.7500 × 14	10.5
3	0.6667 × 14	9.3
4	0.6250 × 14	8.8
5	0.6000 × 14	8.4
6	0.5833 × 14	8.2
7	0.5714 × 14	8.0
8	0.5625 × 14	7.9
9	0.5555 × 14	7.8
10	0.5500 × 14	7.7

The values given in column 3 of the appendix table are rounded out to the nearest whole number, as indicated. If an estimator wishes to avoid the rounding out, he may use the appropriate value appearing in the above table.

The costs appearing in column 7 of the appendix-table do not include any provision for possible salvage value of the equipment at the end of its estimated useful life.

The indicated hours used per year are for average conditions and will vary between jobs, with different owners, and with geographical locations. Estimators should modify the estimated number of hours that equipment is operated, if necessary, to represent more nearly the actual conditions under which the equipment will be used. If equipment is used more than the number of hours given in the appendix, the ownership cost should be reduced; whereas, if it is used fewer hours than given in the appendix, the cost per hour should be increased. The ownership cost per hour appearing in column 7 is obtained by multiplying the cost of the equipment to the owner by the total percent of cost in column 4, and then dividing the product by the hours used per year.

The cost per hour for fuel and other expenses includes fuel, lubricating oil, grease, filters, air cleaners, and minor repairs for equipment powered by internal combustion engines. For equipment powered by electric motors, the cost in column 8 includes the cost of electric energy and minor repairs.

The costs given in column 8 for equipment powered by diesel engines are obtained using the following information:

Cost of diesel fuel, $0.40 per gal
Fuel consumed per hp-hr at full load, 0.04 gal
Assume the engine will operate at an average of 66-⅔ percent of the rated hp
Fuel used per rated hp-hr, ⅔ × 0.04 = 0.027 gal
Fuel cost per rated hp-hr, 0.027 × $0.40   = $0.011
Cost per rated hp-hr for oil, grease, filters, minor repairs, etc.   = 0.006
                                                                           ─────────
Total cost per rated hp-hr   = $0.017

The costs given in column 8 for equipment powered by gasoline engines are determined by using the following information:

Cost of gasoline, $0.40 per gal
Fuel consumed per rated hp-hr at full load, 0.06 gal
Assume the engine will operate at an average of 66-⅔ percent of the rated hp
Fuel consumed per rated hp-hr, ⅔ × 0.06 = 0.04 gal
Fuel cost per rated hp-hr, 0.04 × $0.40 = $0.016
Cost per rated hp-hr for oil, grease, filters,
    minor repairs, etc.,   = 0.008
                                            ─────────
Total cost per rated hp-hr   = $0.024

The costs given in column 8 for equipment which has no direct power unit, such as a bulldozer blade or a rock ripper, include greasing and minor repairs.

The total operating cost per working hour, as given in column 9, is the sum of the costs given in columns 7 and 8. The costs do not include the wages paid to the operators of the equipment, nor do they include any costs for transporting the equipment to a job and back to the storage yard. These costs must be included in a total estimate for a job, as separate items.

The costs given for construction equipment are based on 1974 prices. The estimator should modify the information given in the appendix to allow for changes in prices.

Estimated Hourly Ownership and Operating Cost for Construction Equipment

Equipment	Average annual expense, percent of cost				Cost to owner	Hr used per year	Cost per working hr		
	(1) Depreciation	(2) Major repairs	(3) Interest, taxes, insurance	(4) Total percent of cost	(5)	(6)	(7) Ownership	(8) Fuel and other expenses	(9) Total operating
Air compressors, portable, free air at 100 psi:									
Gasoline engine:									
75 cfm	25	15	9	49	$ 4,370	1,200	$ 1.78	$1.00	$ 2.78
105 cfm	25	15	9	49	6,430	1,200	2.62	1.50	4.12
125 cfm	25	15	9	49	6,660	1,200	2.69	1.68	4.37
250 cfm	20	15	9	44	11,720	1,200	4.30	3.10	7.40
315 cfm	20	15	9	44	15,860	1,200	5.82	3.60	9.42
Diesel engine:									
105 cfm	25	15	9	49	8,750	1,200	3.56	0.74	4.30
125 cfm	25	15	9	49	9,310	1,200	3.80	0.90	4.70
185 cfm	25	15	9	49	11,670	1,200	4.75	1.26	6.01
250 cfm	20	15	9	44	14,850	1,200	5.45	1.50	6.95
365 cfm	20	15	9	44	21,820	1,200	8.02	2.16	10.18
600 cfm	20	15	9	44	29,425	1,200	10.80	3.30	14.10
Air tools, no hose or steel:									
Drifters:									
Light	25	15	9	49	1,525	1,200	0.63	.....	0.63
Medium	25	15	9	49	1,710	1,200	0.70	.....	0.70
Heavy	25	15	9	49	2,070	1,200	0.85	.....	0.85

Jackhammers:								
Light	33	15	9	57	765	1,200	0.37	.....	0.37
Medium	33	15	9	57	905	1,200	0.43	.....	0.43
Heavy	33	15	9	57	975	1,200	0.46	.....	0.46
Paving breakers	33	15	9	57	765	1,200	0.37	.....	0.37
Wagon drills:								
Light	15	15	8	38	3,650	1,200	1.16	.....	1.16
Medium	15	15	8	38	4,260	1,200	1.36	.....	1.36
Heavy	15	15	8	38	5,875	1,200	1.87	.....	1.87
Air hose, 50 ft, with couplings:								
½ in., 2 braid	50	15	10	75	54	1,200	0.04	.....	0.04
¾ in., 2 braid	50	15	10	75	66	1,200	0.05	.....	0.05
1 in., 2 braid	50	15	10	75	85	1,200	0.06	.....	0.06
1¼ in., 2 braid	50	15	10	75	112	1,200	0.07	.....	0.07
1½ in., 2 braid	50	15	10	75	145	1,200	0.09	.....	0.09
2 in., 3 braid	50	15	10	75	225	1,200	0.14	.....	0.14
Batching equipment for concrete aggregates:								
Batcher, only, weigher:								
2 materials, 1 cu yd	25	17	9	51	3,915	1,200	1.66	.....	1.66
2 materials, 2 cu yd	25	17	9	51	4,665	1,200	1.98	.....	1.98
3 materials, 1 cu yd	25	17	9	51	4,500	1,200	1.92	.....	1.92
3 materials, 2 cu yd	25	17	9	51	5,200	1,200	2.22	.....	2.22
Bins, steel, portable:								
1 compartment, 15 tons	20	15	9	44	2,100	1,200	0.77	.....	0.77
1 compartment, 27 tons	20	15	9	44	2,560	1,200	0.95	.....	0.95
2 compartments, 37 tons	20	15	9	44	3,175	1,200	1.16	.....	1.16
2 compartments, 50 tons	20	15	9	44	4,350	1,200	1.59	.....	1.59
3 compartments, 60 tons	20	15	9	44	5,100	1,200	1.87	.....	1.87
3 compartments, 100 tons	20	15	9	44	6,375	1,200	2.34	.....	2.34
Bituminous equipment:								
Distributor, gas engine:								
600 gal with truck	20	17	9	46	11,750	1,600	3.38	3.60	6.98

Estimated Hourly Ownership and Operating Cost for Construction Equipment—(Continued)

Equipment	Average annual expense, percent of cost				(5) Cost to owner	(6) Hr used per year	Cost per working hr		
	(1) Depreciation	(2) Major repairs	(3) Interest, taxes, insurance	(4) Total per cent of cost			(7) Owner-ship	(8) Fuel and other expenses	(9) Total operating
1,000 gal with truck	20	17	9	46	15,225	1,600	4.38	3.85	8.23
1,500 gal with truck	20	17	9	46	16,950	1,600	4.88	4.65	9.53
Paver, complete, 100 tph:									
Crawler, gasoline	25	15	9	49	38,600	1,600	11.80	1.80	13.60
Crawler, diesel	20	15	9	44	39,550	1,600	10.90	1.65	12.55
Spreader, gasoline engine:									
5–12 ft crawler	25	15	9	49	12,850	1,600	3.94	1.56	5.50
5–12 ft tires	25	17	9	51	12,000	1,600	3.82	1.56	5.38
7–13 ft crawler	25	15	9	49	13,550	1,600	4.15	1.70	5.85
7–15 ft tires	25	17	9	51	12,950	1,600	4.13	1.70	5.83
Buggies, concrete, power driven									
9 cu ft	33	25	9	67	1,280	1,400	0.61	0.30	0.91
11 cu ft	33	25	9	67	1,485	1,400	0.71	0.36	1.07
13 cu ft	33	25	9	67	1,985	1,400	0.95	0.40	1.35
Concrete buckets; bottom dump:									
¾ cu yd	20	15	9	44	660	1,200	0.24	.....	0.24
1 cu yd	20	15	9	44	765	1,200	0.28	.....	0.28
1½ cu yd	20	15	9	44	1,090	1,200	0.40	.....	0.40
2 cu yd	20	15	9	44	1,465	1,200	0.54	.....	0.54
3 cu yd	20	15	9	44	2,140	1,200	0.78	.....	0.78

Concrete mixers, construction:									
6S	40	12	10	62	3,980	1,600	1.54	0.42	1.96
11S	40	12	10	62	4,610	1,600	1.79	0.76	2.55
16S	33	12	9	54	6,160	1,600	2.09	0.90	2.99
28S on skids	25	12	9	46	11,400	1,600	3.28	1.64	4.92
Concrete mixers, paving:									
Diesel engine:									
27E single	25	12	9	46	40,250	1,400	13.25	1.95	15.20
27E double	25	12	9	46	53,400	1,400	17.35	2.15	19.70
34E single	20	12	9	41	48,150	1,400	14.10	2.15	16.25
34E double	20	12	9	41	65,500	1,400	19.22	3.70	22.92
Concrete mixers, truck mounted:									
3 cu yd	25	15	9	49	18,550	2,000	4.55	4.25	8.80
4 cu yd	25	15	9	49	23,400	2,000	5.73	5.05	10.78
5 cu yd	25	15	9	49	29,300	2,000	7.18	5.95	13.13
7 cu yd	25	15	9	49	40,950	2,000	10.05	7.80	17.85
Concrete pumps, complete:									
Single, 15–20 cu yd per hr, with 800 ft of pipe	25	15	9	49	26,900	1,400	9.42	1.24	10.66
Single, 25–33 cu yd per hr, with 1,000 ft of pipe	20	15	9	44	42,350	1,400	13.35	2.46	15.81
Double, 50–65 cu yd per hr, with 1,000 ft of pipe	20	15	9	44	57,650	1,400	18.15	2.85	21.00
Concrete screeds and finishers, gasoline engine:									
10–15 ft adjustable	25	15	9	49	12,050	1,600	3.69	0.72	4.41
20–25 ft adjustable	25	15	9	49	12,610	1,600	3.87	0.72	4.59
Concrete screeds, vibrating, adjustable:									
6 ft, gasoline engine	25	15	9	49	965	1,600	0.30	0.08	0.38
8 ft, gasoline engine	25	15	9	49	980	1,600	0.30	0.08	0.38
10 ft, gasoline engine	25	15	9	49	1,055	1,600	0.32	0.08	0.40

Estimated Hourly Ownership and Operating Cost for Construction Equipment—(Continued)

Equipment	Average annual expense, percent of cost				(5) Cost to owner	(6) Hr used per year	Cost per working hr		
	(1) Depreciation	(2) Major repairs	(3) Interest, taxes, insurance	(4) Total per cent of cost			(7) Ownership	(8) Fuel and other expenses	(9) Total operating
12 ft, gasoline engine	25	15	9	49	1,125	1,600	0.35	0.09	0.44
16 ft, electric motor	25	12	9	46	1,320	1,600	0.41	0.11	0.52
Concrete vibrators, internal:									
Electric motor:									
2 hp, 7-ft shaft	25	10	9	44	595	1,600	0.17	0.10	0.27
2½ hp, 14-ft shaft	25	10	9	44	732	1,600	0.20	0.11	0.31
2½ hp, 21-ft shaft	25	10	9	44	840	1,600	0.23	0.11	0.34
2½ hp, 28-ft shaft	25	10	9	44	916	1,600	0.25	0.11	0.36
Gasoline engine:									
2 hp, 7–14 ft shaft	33	12	9	54	440	1,600	0.15	0.08	0.23
4 hp, 21-ft shaft	33	12	9	54	890	1,600	0.30	0.15	0.45
Cranes, complete:									
Crawler, gasoline engine:									
4 tons, 10-ft radius	25	10	9	44	26,250	1,600	7.20	1.26	8.46
8 tons, 10-ft radius	25	10	9	44	37,600	1,600	10.35	1.57	11.92
16 tons, 10-ft radius	20	10	9	39	44,250	1,600	10.80	2.82	13.62
8 tons, 12-ft radius	20	10	9	39	36,550	1,600	8.93	1.95	10.88
12 tons, 12-ft radius	20	10	9	39	43,750	1,600	10.60	2.47	13.07
20 tons, 12-ft radius	20	10	9	39	54,150	1,600	13.25	2.82	16.07

Crawler, diesel engine:									
6 tons, 12-ft radius	20	10	9	39	34,920	1,600	8.53	0.78	9.31
12 tons, 12-ft radius	20	10	9	39	47,500	1,600	11.60	1.08	12.68
20 tons, 12-ft radius	17	10	9	36	59,075	1,600	13.22	1.25	14.57
35 tons, 12-ft radius	17	10	9	36	85,260	1,400	22.00	1.71	23.71
Truck-mounted, gas engine:									
4 tons, 10-ft radius	25	15	9	49	28,700	1,600	8.90	1.92	10.82
8 tons, 10-ft radius	20	15	9	44	39,325	1,600	11.85	2.40	14.25
12 tons, 10-ft radius	20	15	9	44	54,150	1,400	17.30	3.38	20.38
16 tons, 10-ft radius	20	15	9	44	77,000	1,400	24.20	4.28	28.48
Truck-mounted, diesel engine:									
8 tons, 10-ft radius	20	15	9	44	43,175	1,400	13.60	1.13	14.73
35 tons, 12-ft radius	17	10	9	36	121,800	1,400	31.20	2.63	33.83
Draglines, including rope and bucket:									
Crawler-mounted, gas engine:									
½ cu yd	25	15	9	49	35,300	2,000	8.65	2.32	10.97
¾ cu yd	25	15	9	49	44,850	2,000	11.00	2.70	13.70
1 cu yd	20	15	9	44	55,950	2,000	12.30	3.05	15.35
1½ cu yd	17	15	9	41	74,800	1,600	19.25	3.98	23.23
2 cu yd	17	15	9	41	104,750	1,600	26.90	5.65	32.25
Crawler-mounted, diesel engine:									
½ cu yd	25	15	9	49	36,000	2,000	8.85	1.10	9.95
¾ cu yd	25	15	9	49	47,500	2,000	11.70	1.25	12.95
1 cu yd	20	15	9	44	60,500	2,000	13.35	1.38	14.73
1½ cu yd	17	15	9	41	86,950	1,600	22.30	1.75	24.05
2 cu yd	17	15	9	41	121,960	1,600	31.20	2.55	33.75
Hoisting units, including power units:									
Gasoline engine:									
Single drum, 10 hp	20	15	9	44	2,690	1,600	0.74	0.25	0.99
Single drum, 15 hp	20	15	9	44	3,430	1,600	0.95	0.35	1.30
Double drum, 20 hp	17	10	9	36	5,020	1,600	1.13	0.47	1.60

Estimated Hourly Ownership and Operating Cost for Construction Equipment—(Continued)

Equipment	Average annual expense, percent of cost				(5) Cost to owner	(6) Hr used per year	Cost per working hr		
	(1) Depreciation	(2) Major repairs	(3) Interest, taxes, insurance	(4) Total per cent of cost			(7) Ownership	(8) Fuel and other expenses	(9) Total operating
Three drum, 40 hp	17	10	9	36	9,450	1,600	2.13	0.95	3.08
Three drum, 60 hp	17	10	9	36	11,580	1,600	2.61	1.42	4.03
Electric motor:									
Single drum, 10 hp	17	10	9	36	2,515	1,600	0.57	0.35	0.92
Double drum, 10 hp	17	10	9	36	4,105	1,600	0.93	0.35	1.28
Double drum, 20 hp	13	10	8	31	7,325	1,600	1.42	0.68	2.10
Double drum, 40 hp	13	10	8	31	10,000	1,600	1.95	1.35	3.30
Three drum, 40 hp	13	10	8	31	13,200	1,600	2.56	1.35	3.91
Three drum, 60 hp	13	10	8	31	16,525	1,600	3.20	1.98	5.18
Motor graders, diesel engine:									
12-ft blade, 80 hp	25	15	9	49	24,150	2,000	5.92	1.35	7.27
12-ft blade, 115 hp	25	15	9	49	27,750	2,000	6.80	1.85	8.65
14-ft blade, 150 hp	25	15	9	49	36,550	2,000	8.95	2.60	11.55
16-ft blade, 225 hp	25	15	9	49	71,500	2,000	17.40	3.80	21.20
Pile driving hammers:									
Single-acting steam:									
3,600 ft lb	20	12	9	41	7,500	1,400	2.20	.....	2.20
7,500 ft lb	20	12	9	41	10,350	1,400	3.04	.....	3.04
15,000 ft lb	17	12	9	38	12,950	1,400	3.51	.....	3.51
30,000 ft lb	17	12	9	38	20,150	1,200	6.38	.....	6.38

Double-acting steam:									
3,600 tf lb	20	12	9	41	8,060	1,400	2.37	.....	2.37
7,500 ft lb	20	12	9	41	11,350	1,400	3.32	.....	3.32
15,000 ft lb	17	12	9	38	14,400	1,400	3.90	.....	3.90
25,000 ft lb	17	12	9	38	21,850	1,400	5.93	.....	5.93
36,000 ft lb	17	12	9	38	45,250	1,200	14.35	.....	14.35
Diesel:									
4,300 ft lb	20	20	9	49	14,650	1,400	5.15	1.20	6.35
9,000 ft lb	20	20	9	49	22,400	1,400	7.84	2.10	9.94
16,000 ft lb	20	20	9	49	32,200	1,400	11.30	2.40	13.70
25,000 ft lb	20	20	9	49	46,350	1,400	16.20	4.10	20.30
Power plants, portable, gasoline engine:									
1½ kw, mounted	15	15	9	39	1,045	1,600	0.26	0.30	0.56
2½ kw, mounted	15	15	9	39	1,395	1,600	0.34	0.45	0.79
5 kw, mounted	15	10	9	34	2,795	1,600	0.60	0.75	1.35
7½ kw, mounted	15	10	9	34	3,515	1,600	0.75	0.98	1.73
5 kw, on skids	15	10	9	34	2,550	1,600	0.55	0.75	1.30
Pumps, centrifugal:									
Electric, portable:									
1½ in., 4,000 gph	17	15	9	41	560	1,200	0.20	0.10	0.30
2 in., 7,000 gph, low head	17	15	9	41	695	1,200	0.24	0.20	0.44
2 in., 7,000 gph, high head	17	15	9	41	885	1,200	0.31	0.25	0.56
Pumps, centrifugal, electric:									
3 in., 12,000 gph, low head	17	15	9	41	810	1,200	0.28	0.35	0.53
3 in., 20,000 gph, high head	17	15	9	41	1,160	1,200	0.40	0.55	0.95
4 in., 30,000 gph, low head	17	15	9	41	1,615	1,200	0.55	1.20	1.75
4 in., 40,000 gph, high head	17	15	9	41	2,090	1,200	0.72	1.60	2.32
Gasoline, portable:									
2 in., 10,000 gph	20	20	9	49	815	1,200	0.34	0.30	0.64
3 in., 20,000 gph	20	20	9	49	1,215	1,200	0.50	0.46	0.96
4 in., 40,000 gph	20	20	9	49	2,660	1,200	1.10	1.25	2.35
6 in., 90,000 gph	20	20	9	49	4,190	1,200	1.72	1.60	3.32

Estimated Hourly Ownership and Operating Cost for Construction Equipment—(Continued)

Equipment	Average annual expense, percent of cost					(6) Hr used per year	Cost per working hr		
	(1) Depreciation	(2) Major repairs	(3) Interest, taxes, insurance	(4) Total percent of cost	(5) Cost to owner		(7) Ownership	(8) Fuel and other expenses	(9) Total operating
Rollers, earth:									
Pneumatic, self-propelled:									
10 tons	33	30	10	73	9,110	1,400	4.75	1.40	6.15
12 tons	33	30	10	73	9,720	1,400	5.10	1.55	6.65
25 tons	33	30	10	73	17,450	1,400	9.10	2.25	11.35
35 tons	33	30	10	73	31,600	1,400	16.50	3.35	19.85
Pneumatic, towed type:									
8 tons	33	30	10	73	1,980	1,400	1.03	0.15	1.18
10 tons	33	30	10	73	2,500	1,400	1.31	0.20	1.51
50 tons	33	30	10	73	11,300	1,400	5.90	0.35	6.25
100 tons	33	30	10	73	30,150	1,400	15.70	0.70	16.40
Sheep's-foot, tamping:									
Self-propelled, diesel:									
4 × 4 dual drum	25	15	9	49	28,100	2,000	6.90	2.62	9.52
5 × 5 dual drum	25	15	9	49	37,100	2,000	9.10	3.38	12.48
Towed type									
4 × 4 dual drum	17	15	9	41	3,540	2,000	0.78	......	0.78
5 × 5 dual drum	17	15	9	41	5,975	2,000	1.23	......	1.23
5 × 6 dual drum	17	15	9	41	8,250	2,000	1.70	......	1.70

Tandem wheel:								
3–5 tons, gasoline	15	12	9	36	6,900	2,000	1.24	0.90	2.14
5–8 tons, gasoline	15	12	9	36	11,470	2,000	2.06	1.67	3.73
8–12 tons, diesel	15	12	9	36	16,220	2,000	2.92	1.35	4.27
12–14 tons, diesel	15	12	9	36	17,050	2,000	3.07	1.39	4.46
Three-wheel, diesel:								
6–8 tons	15	12	9	36	12,500	2,000	2.25	0.90	3.15
8–10 tons	15	12	9	36	13,815	2,000	2.49	1.05	3.54
10–12 tons	15	12	9	36	15,720	2,000	2.84	1.25	4.09
12–14 tons	15	12	9	36	16,910	2,000	3.05	1.30	4.35
Vibrating, drum type, self-propelled:								
5 × 6 single drum	25	15	9	49	20,250	1,600	6.20	1.80	8.00
Towed type:								
5 × 6 single drum, 5 tons	20	15	9	44	11,950	1,600	3.29	1.65	4.94
5½ × 6½ single drum, 12 tons	20	15	9	44	22,150	1,600	6.10	2.90	9.00
Saws:								
Chain, gasoline:								
18-in. cut	33	15	10	58	350	1,200	0.17	0.12	0.29
24-in. cut	33	15	10	58	475	1,200	0.23	0.16	0.39
36-in. cut	33	15	10	58	490	1,200	0.24	0.17	0.41
48-in. cut	33	15	10	58	620	1,200	0.30	0.18	0.48
Hand, electric, no blade:								
4-in. blade	33	15	10	58	55	1,400	0.03	0.03	0.06
6-in. blade	33	15	10	58	65	1,400	0.03	0.03	0.06
8-in. blade	33	15	10	58	120	1,400	0.05	0.03	0.08
Masonry, for 18-in. blade, gasoline or electric	25	25	9	59	995	1,200	0.50	0.06	0.56
Tilting table, electric:								
8-in. blade	25	10	9	44	380	1,600	0.11	0.04	0.15
10-in. blade	25	10	9	44	565	1,600	0.16	0.06	0.22

Estimated Hourly Ownership and Operating Cost for Construction Equipment—(Continued)

Equipment	Average annual expense, percent of cost				(5) Cost to owner	(6) Hr used per year	Cost per working hr		
	(1) Depreciation	(2) Major repairs	(3) Interest, taxes, insurance	(4) Total percent of cost			(7) Ownership	(8) Fuel and other expenses	(9) Total operating
12-in. blade	25	10	9	44	915	1,600	0.25	0.08	0.33
14-in. blade	25	10	9	44	1,405	1,600	0.39	0.10	0.49
Scrapers, tractor-pulled: Struck capacity:									
7 cu yd, 125 hp	20	15	9	44	34,725	2,000	7.64	2.25	9.89
12 cu yd, 200 hp	20	15	9	44	51,720	2,000	11.42	3.60	15.02
14 cu yd, 300 hp	20	15	9	44	74,450	2,000	16.40	4.22	20.62
21 cu yd, 400 hp	20	15	9	44	109,900	2,000	24.20	5.64	29.84
28 cu yd, 500 hp	20	15	9	44	127,600	2,000	28.20	7.13	35.33
32 cu yd, 500 hp	20	15	9	44	144,050	2,000	31.75	7.66	39.41
Shovels, power, complete: Crawler, gasoline engine:									
½ cu yd	25	15	9	49	33,500	2,000	8.20	2.61	10.81
¾ cu yd	25	15	9	49	43,450	2,000	10.62	2.96	13.58
1 cu yd	20	15	9	44	49,825	2,000	11.00	3.34	14.34
1½ cu yd	20	15	9	44	97,520	1,600	26.80	4.34	31.14
Crawler, diesel engine:									
½ cu yd	25	15	9	49	36,625	2,000	9.05	1.36	10.41
¾ cu yd	25	15	9	49	46,310	2,000	11.40	1.66	12.06
1 cu yd	20	15	9	44	53,650	2,000	11.85	1.81	13.66

1½ cu yd	20	15	9	44	88,500	2,000	19.50	2.26	21.76
2 cu yd	17	15	9	41	100,850	1,600	25.95	3.30	29.25
Towers, elevator, tube steel:									
Light, single, 25 ft	20	10	9	39	1,200	1,600	0.29	.....	0.29
Light, single, 50 ft	20	10	9	39	1,665	1,600	0.41	.....	0.41
Light, double, 25 ft	20	10	9	39	2,050	1,600	0.50	.....	0.50
Light, double, 50 ft	20	10	9	39	2,910	1,600	0.71	.....	0.71
Heavy, single, 50 ft	20	10	9	39	2,140	1,600	0.52	.....	0.52
Heavy, single, 100 ft	20	10	9	39	3,310	1,600	0.81	.....	0.81
Heavy, double, 50 ft	20	10	9	39	3,410	1,600	0.83	.....	0.83
Heavy, double, 100 ft	20	10	9	39	5,355	1,600	1.31	.....	1.31
Tractors, diesel, crawler:									
70 dbhp	25	15	9	49	19,500	2,000	4.78	1.27	6.05
90 dbhp	20	15	9	44	29,950	2,000	6.60	1.62	8.22
120 dbhp, direct drive	20	15	9	44	35,350	2,000	7.78	2.35	10.13
180 fwhp, direct drive	20	15	9	44	50,000	2,000	11.02	3.60	14.62
270 fwhp, direct drive	20	15	9	44	68,500	2,000	15.15	5.25	20.40
385 fwhp, power shift	20	15	9	44	106,000	2,000	23.40	7.50	30.90
Four wheel:									
100 fwhp	20	15	9	44	28,520	2,000	6.49	2.25	8.74
200 fwhp	20	15	9	44	37,780	2,000	8.32	4.05	12.37
250 fwhp	20	15	9	44	43,950	2,000	9.67	5.40	15.09
Tractor attachments:									
Angle dozers, cable:									
13-ft blade	20	15	9	44	4,475	1,600	1.23	.....	1.23
15-ft blade	20	15	9	44	6,580	1,600	1.81	.....	1.81
Bulldozers, cable:									
8-ft blade	20	15	9	44	2,880	2,000	0.64	.....	0.64
10-ft blade	20	15	9	44	3,420	2,000	0.75	.....	0.75
12-ft blade	20	15	9	44	4,270	2,000	0.94	.....	0.94
Bulldozers, hydraulic:									
8-ft blade	20	15	9	44	4,050	2,000	0.89	.....	0.89
10-ft blade	20	15	9	44	4,800	2,000	1.06	.....	1.06

Estimated Hourly Ownership and Operating Cost for Construction Equipment—(Continued)

Equipment	Average annual expense, percent of cost				(5) Cost to owner	(6) Hr used per year	Cost per working hr		
	(1) Depreciation	(2) Major repairs	(3) Interest, taxes, insurance	(4) Total percent of cost			(7) Ownership	(8) Fuel and other expenses	(9) Total operating
12-ft blade	20	15	9	44	5,975	2,000	1.32	.....	1.32
14-ft blade	20	15	9	44	8,550	2,000	1.90	.....	1.90
Rippers, hydraulic, for indicated size tractors, points and shanks not included:									
120 fwhp	20	15	9	44	6,620	1,400	2.04	.....	2.04
180 fwhp	20	15	9	44	8,035	1,400	2.53	.....	2.53
270 fwhp	20	15	9	44	9,125	1,400	2.88	.....	2.88
385 fwhp	20	15	9	44	10,150	1,400	3.19	.....	3.19
Tractors and bottom-dump wagons:									
20 tons, 200 fwhp, diesel	20	15	9	44	60,500	2,000	13.35	3.60	16.95
38 tons, 240 fwhp, diesel	20	15	9	44	81,200	2,000	17.95	4.35	22.30
40 tons, 300 fwhp, diesel	20	15	9	44	88,400	2,000	19.50	5.40	24.90
Trenching machines: Ladder type, diesel engine:									
6-ft depth	25	20	9	54	32,300	2,000	8.75	1.10	9.85
9-ft depth	25	20	9	54	34,200	2,000	9.25	1.35	10.60
12-ft depth	25	20	9	54	42,150	2,000	11.45	1.75	13.20

15-ft depth	20	9	49	48,850	1,600	15.00	2.05	17.05
17-ft depth	20	9	49	53,100	1,600	16.35	2.35	18.70
Wheel type:								
4-ft depth	20	9	49	16,200	2,000	3.97	1.30	5.27
6-ft depth	20	9	49	20,850	2,000	5.10	1.75	6.85
Trowel, concrete floor, gasoline engine:								
36-in. diameter	20	10	63	745	1,400	0.34	0.10	0.44
46-in. diameter	20	10	63	745	1,400	0.40	0.21	0.61
Trucks:								
Dump, gasoline engine:								
4 cu yd	15	9	44	12,850	2,000	2.82	2.88	5.70
6 cu yd	15	9	44	15,900	2,000	3.50	3.62	7.12
10 cu yd, heavy duty	12	9	41	30,580	1,600	7.83	6.11	13.94
Dump, diesel engine:								
7 cu yd, 10 tons	15	9	44	24,800	1,800	6.07	2.35	8.42
10 cu yd, 15 tons	12	9	41	38,200	1,800	8.70	3.80	12.50
15 cu yd, 20 tons	12	9	41	55,600	1,600	14.30	6.20	20.50
20 cu yd, 35 tons	12	9	41	84,950	1,600	21.80	8.05	29.85
Stake body, gasoline engine:								
½ ton	15	10	58	3,150	2,000	0.92	1.54	2.44
1 ton	15	10	58	4,340	2,000	1.26	1.86	3.12
2 tons	12	9	46	10,350	2,000	2.38	2.20	4.58
4 tons	12	9	46	12,450	2,000	2.87	2.42	5.29
5 tons	12	9	41	14,480	2,000	2.98	2.76	5.74
Welding equipment, with cables, holders, etc.:								
Electric drive, portable:								
200 amp dc	12	9	41	543	2,000	0.11	0.36	0.47
300 amp dc	12	9	41	1,050	2,000	0.22	0.45	0.67
400 amp dc	12	9	41	1,185	2,000	0.25	0.53	0.78
600 amp dc	12	9	41	1,660	2,000	0.34	0.78	1.12

Estimated Hourly Ownership and Operating Cost for Construction Equipment—(Continued)

Equipment	Average annual expense, percent of cost				(5) Cost to owner	(6) Hr used per year	Cost per working hr		
	(1) Depreciation	(2) Major repairs	(3) Interest, taxes, insurance	(4) Total percent of cost			(7) Ownership	(8) Fuel and other expenses	(9) Total operating
Diesel engine drive, portable:									
300 amp dc	20	12	9	41	4,260	2,000	0.88	0.70	1.58
400 amp dc	20	12	9	41	4,870	2,000	1.00	0.82	1.82
600 amp dc	20	12	9	41	5,650	2,000	1.16	1.20	2.36
Gasoline-engine drive, portable:									
200 amp dc	20	15	9	44	1,840	2,000	0.38	0.76	1.14
300 amp dc	20	15	9	44	2,710	2,000	0.56	1.64	2.20
400 amp dc	20	15	9	44	3,065	2,000	0.62	2.10	2.72
600 amp dc	20	15	9	44	3,490	2,000	0.72	2.95	3.67

Index